I0064448

Statistical Distributions

Statistical Distributions

Edited by **Calanthia Wright**

Willford Press
New York

Published by Willford Press,
118-35 Queens Blvd., Suite 400,
Forest Hills, NY 11375, USA
www.willfordpress.com

Statistical Distributions
Edited by Calanthia Wright

© 2016 Willford Press

International Standard Book Number: 978-1-68285-164-7 (Hardback)

This book contains information obtained from authentic and highly regarded sources. Copyright for all individual chapters remain with the respective authors as indicated. All chapters are published with permission under the Creative Commons Attribution License or equivalent. A wide variety of references are listed. Permission and sources are indicated; for detailed attributions, please refer to the permissions page and list of contributors. Reasonable efforts have been made to publish reliable data and information, but the authors, editors and publisher cannot assume any responsibility for the validity of all materials or the consequences of their use.

The publisher's policy is to use permanent paper from mills that operate a sustainable forestry policy. Furthermore, the publisher ensures that the text paper and cover boards used have met acceptable environmental accreditation standards.

Trademark Notice: Registered trademark of products or corporate names are used only for explanation and identification without intent to infringe.

Printed in the United States of America.

Contents

	Preface	**VII**
Chapter 1	**Extinction in a branching process: why some of the fittest strategies cannot guarantee survival** Sterling Sawaya and Steffen Klaere	**1**
Chapter 2	**Bayesian reference analysis for exponential power regression models** Marco AR Ferreira and Esther Salazar	**14**
Chapter 3	**A closed-form universal trivariate pair-copula** Werner Hürlimann	**34**
Chapter 4	**The Kumaraswamy-geometric distribution** Alfred Akinsete, Felix Famoye and Carl Lee	**59**
Chapter 5	**Joint distribution of rank statistics considering the location and scale parameters and its power study** Wan-Chen Lee	**80**
Chapter 6	**Geometric disintegration and star-shaped distributions** Wolf-Dieter Richter	**96**
Chapter 7	**Bounds on the mean residual lifetime of progressive type II right censored order statistics** Mohammad Z Raqab	**120**
Chapter 8	***T*-normal family of distributions: a new approach to generalize the normal distribution** Ayman Alzaatreh, Carl Lee and Felix Famoye	**132**
Chapter 9	**On the distribution theory of over-dispersion** Evdokia Xekalaki	**150**
Chapter 10	**The Marshall-Olkin extended Weibull family of distributions** Manoel Santos-Neto, Marcelo Bourguignon, Luz M Zea, Abraão DC Nascimento and Gauss M Cordeiro	**172**
Chapter 11	**Chi-*p* distribution: characterization of the goodness of the fitting using L^p norms** George Livadiotis	**196**

Chapter 12 **Univariate and multivariate Pareto models** **210**
Barry C Arnold

Permissions

List of Contributors

Preface

Statistical distribution aims to measure different subsets of a possible outcome by assigning a probability and is beneficial in random surveys. It is further divided into various sub-categories and has wide ranging applications in diverse scientific and engineering fields. The book focuses upon discrete, continuous and mixed probability distributions with detailed examples. It provides various graphs and methodologies to measure the sample space of random variables. Applied probability is extensively discussed in this book through different researches which makes it a very helpful reference source for students, researchers and academicians.

Various studies have approached the subject by analyzing it with a single perspective, but the present book provides diverse methodologies and techniques to address this field. This book contains theories and applications needed for understanding the subject from different perspectives. The aim is to keep the readers informed about the progress in the field; therefore, the contributions were carefully examined to compile novel researches by specialists from across the globe.

Indeed, the job of the editor is the most crucial and challenging in compiling all chapters into a single book. In the end, I would extend my sincere thanks to the chapter authors for their profound work. I am also thankful for the support provided by my family and colleagues during the compilation of this book.

Editor

1

Extinction in a branching process: why some of the fittest strategies cannot guarantee survival

Sterling Sawaya[1,3*] and Steffen Klaere[2]

*Correspondence: sterlingsawaya@gmail.com
[1] Institute for Behavioral Genetics, University of Colorado, Boulder, Colorado, USA
[3] Formerly: Department of Anatomy, and Allan Wilson Centre for Molecular Ecology and Evolution, University of Otago, Dunedin, New Zealand
Full list of author information is available at the end of the article

Abstract

Biological fitness is typically measured by the expected rate of reproduction, but strategies with high fitness can also have high probabilities of extinction. Likewise, gambling strategies with a high expected payoff can also have a high risk of ruin. We take inspiration from the gambler's ruin problem to examine how extinction is related to population growth. Using moment theory we demonstrate how higher moments can impact the probability of extinction and how the first few moments can be used to find bounds on the extinction probability, focusing on *s*-convex ordering of random variables. This approach generates "best case" and "worst case" scenarios to provide upper and lower bounds on the probability of extinction.

MSC Codes: 92D15, 60J80, 60E15

Keywords: Extinction; Branching process; S-convex

1 Introduction

Reproduction is necessary for the survival of populations. However, a population can have a high expected reproductive rate but nevertheless go extinct with near certainty (Lewontin and Cohen 1969). For example, populations with large variation in reproductive success can sometimes have a high probability of extinction, even if they have a high expected growth (Tuljapurkar and Orzack 1980).

Similarly, investors and gamblers can avoid Gambler's Ruin through growth of capital. However, a gambler should not simply apply the strategy with the highest expected growth rate as it may also have a high risk of ruin. For example, investors can use the Kelly ratio (Kelly 1956) to maximize expected geometric growth of their capital but strict adherence to this ratio can be risky, and playing a more conservative strategy is often recommended (MacLean et al. 2010).

To estimate the probability of Gambler's Ruin, one can use approximations based on moments (Ethier and Khoshnevisan 2002; Canjar 2007; Hürlimann 2005). Here we apply these approaches to estimate the probability of extinction in a branching process. The mathematics of Gambler's Ruin are very similar to that of extinction in a branching process (Courtois et al. 2006). Both statistical models involve a random variable (payoff/offspring number), resulting in a random walk (change in capital/change in population size), and an absorbing state (ruin/extinction). Moreover, both processes are assumed to be Markovian, and finding the probability of ruin/extinction involves solving for the root of a convex function.

Here we examine the random variable representing the number of offspring, and investigate how the moments of this random variable are related to the probability of extinction. We demonstrate an important relationship between these moments and extinction: odd moments favor survival and even moments favor extinction. The first moment of the offspring distribution, its mean, has the biggest influence on extinction. However, the first moment alone is not usually informative about extinction probabilities. In fact, strategies with arbitrarily large first moments can nevertheless go extinct with near certainty. Some of the "fittest" strategies can be highly unlikely to survive.

Using the first few moments of the offspring distribution, one can obtain bounds on the ultimate probability of extinction (Courtois et al. 2006; Daley and Narayan 1980). These bounds provide "best case" and "worst case" distributions. We present these bounds, termed s-convex extremal random variables, adapted from actuarial science and research on the gambler's ruin problem (Denuit and Lefevre 1997; Hürlimann 2005; Courtois et al. 2006). The extremal distributions for discrete processes have been developed previously, using up to four moments (Hürlimann 2005). Here we find the conditions under which these extremal distributions provide non-trivial bounds. Using some simple examples, we demonstrate how these methods can be used to rank distributions using only their moments. We then discuss how these bounds can be used to better understand the evolutionary process.

2 Methods

2.1 Extinction in the Galton-Watson branching process

To investigate biological extinction, we use a Galton-Watson branching process in which, at each discrete time interval, every individual generates i discrete offspring with probability p_i, and zero offspring with p_0. Without loss of generality we assume that an individual produces its offspring and then dies, so that each individual in a population is restricted to a single generation. The offspring number is a random variable, which we denote by X. Let n be the maximum value of X so that X takes values in the state space $\mathcal{D}_n = \{0, 1, 2, \ldots, n\}$

At any given time t, the size of a population (Z_t) is the number of individuals in the branching process. We set $Z_0 \equiv 1$ unless otherwise specified. The probability of extinction of a branching process is $q \equiv \lim_{t\to\infty} P(Z_t = 0|Z_0 = 1)$. If the starting size of the population is greater than one, then the overall probability of extinction can be defined as

$$\lim_{t\to\infty} P(Z_t = 0|Z_0 = N) = q^N.$$

So we can solve for extinction in the case of $Z_0 = 1$ and extend the results to larger starting populations if necessary.

The recursive formula for finding q can be found through a first step analysis (Kimmel and Axelrod 2002). The probability that the lineage of a single individual eventually goes extinct is the probability that it dies without offspring (p_0) plus the probability that it produces a single offspring whose lineage dies out ($p_1 q$) plus the probability that it produces two offspring whose joint lineages die out ($p_2 q^2$), and so on.

This leads to the formal definition of the probability generating function:

$$f(q) = \mathbb{E}[q^X] = p_0 + p_1 q + p_2 q^2 + p_3 q^3 \ldots p_n q^n = \sum_{k=0}^{n} p_k q^k. \tag{1}$$

The probability of extinction of a branching process starting with a single individual is the smallest root of the equation $f(q) = q$ for $q \in [0, 1]$. The solution $q = 1$ is always a root of (1) and is not necessarily the smallest positive root. In some cases, the probability of extinction is trivially obvious. For instance, if $p_0 = 0$ individuals always produces at least one offspring, therefore $q = 0$. Furthermore, cases where $\mathbb{E}[X] \leq 1$ always yield $q = 1$ (Kimmel and Axelrod 2002).

Inferring the probability of extinction analytically for branching processes with $p_0 > 0$ and $\mathbb{E}[X] > 1$ can be difficult because (1) has n complex-valued roots according to the fundamental law of algebra. In the following we illustrate how (1) can be seen in terms of moments of the offspring distribution, and discuss how this approach can be used to estimate q.

2.2 Moments of the branching process

Let $m_k \equiv \mathbb{E}[X^k]$ denote the kth moment of the branching process generator X. The first moment, m_1, is equivalent to the average offspring number. Higher moments can be used to obtain other summary statistics of the distribution, such as the variance $\sigma^2 = m_2 - m_1^2$.

The Laplace transform of (1) can be used to (recursively) express extinction in terms of the moments of the branching process

$$\begin{aligned} f(q) &= \mathbb{E}\left[q^X\right] = \mathbb{E}\left[e^{X \log q}\right] \\ &= 1 + m_1 \log q + m_2 \frac{(\log q)^2}{2} + m_3 \frac{(\log q)^3}{6} + \ldots \\ &= \sum_{k=0}^{\infty} m_k \frac{(\log q)^k}{k!} \end{aligned}$$

where $m_0 = 1$. Note that $m_k > 0$ for all $k \geq 0$. Furthermore, with $q \in (0, 1)$ we have $\log q < 0$. Therefore, even moments increase the probability of extinction while odd moments decrease it. Additionally, if $q \in (e^{-1}, 1)$ then $\log q \in (-1, 0)$ and the series converges with $\log q$. Thus, approximations, $f^*(q)$, which take the form

$$f^*(q) = \sum_{k=0}^{s-1} m_k \frac{(\log q)^k}{k!} + o\left((\log q)^s\right)$$

for $s \geq 3$ are only accurate when q is large and the moments are small. As $q \downarrow 0$, the series requires more and more terms to provide accurate approximation. Therefore, when q is small the first few moments are not necessarily informative about the probability of extinction.

2.3 *s*-Convex orderings of random variables

Here we demonstrate how the first few moments of the offspring distribution can be used to find bounds on the probability of extinction. The random variable X is bound by zero and its maximum, n, conveniently allowing for s-convex ordering (Denuit and Lefevre 1997; Hürlimann 2005; Courtois et al. 2006). Following (Hürlimann 2005) denote by Δ

the forward difference operator for $g : \mathcal{D}_n \to \mathbb{R}$ by $\Delta g(i) = g(i+1) - g(i)$ for all $i \in \mathcal{D}_{n-1}$. Analogously for $k \in \mathcal{D}_n$ the k-th order forward difference operator is defined recursively by $\Delta^0 g = g$ and for $k \geq 1$ by $\Delta^k g(i) = \Delta^{k-1} g(i-1) - \Delta^{k-1} g(i)$ for all $i \in \mathcal{D}_{n-k}$. Then, for two random variables X and Y valued in $\mathcal{D}_n$ we say X precedes Y in the s-convex order, written $X \leq_{s-cx}^{\mathcal{D}_n} Y$ if $\mathbb{E}[g(X)] \leq \mathbb{E}[g(y)]$ for all s-convex real functions g on $\mathcal{D}_n$. A convenient consequence is that if $X \leq_{s-cx}^{\mathcal{D}_n} Y$ then

$$\mathbb{E}\left(X^k\right) = \mathbb{E}\left(Y^k\right) \quad \text{for } k = 1, 2, \ldots, s-1$$
$$\mathbb{E}\left(X^k\right) \leq \mathbb{E}\left(Y^k\right) \quad \text{for } k \geq s.$$

Define the *moment space* for all random variables with state set $\mathcal{D}_n$ and fixed first $s-1$ moments $m_1, \ldots, m_{s-1}$ by

$$\mathfrak{B}_{s,n}^{\vec{m}} \equiv \mathfrak{B}\left(\mathcal{D}_n, m_1, m_2, \ldots, m_{s-1}\right).$$

Since the random variable X is strictly positive, its moment space only contains positive elements. Further, we are only interested in cases where the mean is greater than 1 so that extinction is not certain. This provides a moment space with well behaved properties. The study of the moment problem (see e.g., (Karlin and McGregor 1957; Prékopa 1990)) yields an important relationship between consecutive moments on $\mathfrak{B}_{s,n}^{\vec{m}}$ conditional on $m_1 \geq 1$

$$(m_i)^{\frac{i+1}{i}} \leq m_{i+1} \leq n m_i. \tag{2}$$

Minimum and maximum extrema distributions on $\mathfrak{B}_{s,n}^{\vec{m}}$ can be found for any distribution on $\mathcal{D}_n$, with fixed first s moments $m_1, m_2, \ldots, m_s$ (Denuit and Lefevre 1997). The random variables for these distributions are denoted $X_{\min}^{(s)}$ and $X_{\max}^{(s)}$ such that

$$X_{\min}^{(s)} \leq_{s-cx}^{\mathcal{D}_n} X \leq_{s-cx}^{\mathcal{D}_n} X_{\max}^{(s)} \quad \text{for all } X \in \mathcal{D}_n.$$

Extrema have been derived for $s = 2, 3, 4, 5$ (Denuit and Lefevre 1997; Denuit et al. 1999; Hürlimann 2005). Here, we reiterate these results providing the inferred distributions and their utility when obtaining bounds on the probability of extinction. We begin on $\mathfrak{B}_{2,n}^{\vec{m}}$ with the maximal random variable, $X_{\max}^{(2)}$, defined as:

$$X_{\max}^{(2)} = \begin{cases} 0 & \text{with } p_0 = 1 - \dfrac{m_1}{n} \\ n & \text{with } p_n = \dfrac{m_1}{n}. \end{cases}$$

For $X_{\max}^{(2)}$ we observe $m_{i+1} = n m_i$, so by (2) this can clearly be seen as the maximum extrema. Intuitively, this is the "long shot" distribution on $\mathcal{D}_n$, a worst case scenario. Because the values and respective probabilities of $X_{\max}^{(2)}$ are known, q can be solved explicitly by finding the least positive root of the generating function:

$$f(q) = p_0 + p_n q^n.$$

This provides an upper limit on extinction because this generating function will be greater than or equal to the generating function for all other random variables with the same m_1 and n, on $q \in [0, 1]$.

$\mathfrak{B}_{2,n}^{\vec{m}}$ is a very general moment space and the first moment does not often provide much information about an unknown distribution. Therefore, $X_{\max}^{(2)}$ is not likely to be a tight upper bound when n is large or unknown. However, if m_1 is near n, then the distribution can be fairly well approximated by $X_{\max}^{(2)}$.

Unlike $X^{(2)}_{\max}$, $X^{(2)}_{\min}$ does not provide a useful bound on the probability of extinction. $X^{(2)}_{\min}$ is defined as:

$$X^{(2)}_{\min} = \begin{cases} \alpha & \text{with } p_\alpha = \alpha + 1 - m_1 \\ \alpha + 1 & \text{with } p_{\alpha+1} = m_1 - \alpha \end{cases} \tag{3}$$

where α is the integer on $\mathcal{D}_n$ such that

$$\alpha < m_1 \leq \alpha + 1.$$

This extremal random variable represents a best case scenario. However, since $m_1 > 1$, α must be larger than zero and this branching process has no chance of death (i.e. $p_0 = 0$) and consequently no chance of extinction ($q = 0$). Therefore $X^{(2)}_{\min}$ does not provide a useful bound on the probability of extinction as the bound $q \geq 0$ is obvious.

This bound and all other bounds examined here can be found using discrete Chebyshev systems (Denuit and Lefevre 1997). However, extremal bounds are perhaps more intuitive for continuous random variables, to which the discrete cases can be seen as similar (Shaked and Shanthikumar 2007; Hürlimann 2005; Denuit et al. 1999). For example, $X^{(2)}_{\min}$ in the continuous case has only one possible value, m_1 with $p_{m_1} = 1$. By (2) this is clearly an extrema because $(m_i)^{(i+1)/i} = m_{i+1} = (m_1)^{i+1}$. In comparison, the discrete case (3) has similar properties.

The following notation helps extending these calculations to higher order systems (Denuit et al. 1999). Let $w, x, y, z \in \mathcal{D}_n$, and set $m_0 = 1$. Then:

$$\begin{aligned} m_{j,z} &:= z \cdot m_{j-1} - m_j, \quad j = 1, 2, \ldots; \\ m_{j,z,y} &:= y \cdot m_{j-1,z} - m_{j,z}, \quad j = 2, 3, \ldots; \\ m_{j,z,y,x} &:= x \cdot m_{j-1,z,y} - m_{j,z,y}, \quad j = 3, 4, \ldots; \\ m_{j,z,y,x,w} &:= w \cdot m_{j-1,z,y,x} - m_{j,z,y,x}, \quad j = 4, 5, \ldots. \end{aligned}$$

The reader should recognize this notation as it is simply the iterative forward difference operator Δ^k for moments.

If the first two moments are known, then a tighter upper bound can be found. On $\mathfrak{B}^{\vec{m}}_{3,n}$ the minimal distribution in the 3-convex sense is given by:

$$X^{(3)}_{\min} = \begin{cases} 0 & \text{with } p_0 = 1 - p_\alpha - p_{\alpha+1} \\ \alpha & \text{with } p_\alpha = \dfrac{m_{2,\alpha+1}}{\alpha} \\ \alpha + 1 & \text{with } p_{\alpha+1} = \dfrac{-m_{2,\alpha}}{\alpha+1} \end{cases}$$

where

$$\alpha < \frac{m_2}{m_1} \leq \alpha + 1.$$

This bound is already known in the branching process literature (Daley and Narayan 1980). Similar to $X^{(2)}_{\max}$, the extremal random variable $X^{(3)}_{\min}$ represents a worst case scenario, this time using two moments. The root of the equation

$$f(q) = q = p_0 + p_\alpha q^\alpha + p_{\alpha+1} q^{\alpha+1} \tag{4}$$

provides an upper bound to the probability of extinction, so that (4) has greater values at any $q \in [0, 1)$ than the probability generating functions of any other random variable in $\mathfrak{B}^{\vec{m}}_{3,n}$.

In contrast to $X^{(2)}_{\max}$, the minimum extrema on $\mathfrak{B}^{\vec{m}}_{3,n}$ yields the upper limit for the probability of extinction. The alternation between minimum and maximum for the worst case scenarios is due to the convexity of (1). Again, this extrema is perhaps more intuitive in the continuous sense, in which

$$X^{(3)}_{\text{min, cont.}} = \begin{cases} 0 & \text{with } p_0 = 1 - p_{m_2/m_1} \\ \dfrac{m_2}{m_1} & \text{with } p_{m_2/m_1} = \dfrac{(m_1)^2}{m_2}. \end{cases}$$

In this case, successive moments simply grow by m_2/m_1, so that $m_{i+1} = m_i(m_2/m_1)$, providing a minimum on $\mathfrak{B}^{\vec{m}}_{3,n}$. And, as was the case for the minimum on $\mathfrak{B}^{\vec{m}}_{2,n}$, the discrete minimum extrema on $\mathfrak{B}^{\vec{m}}_{3,n}$ has similar properties to the continuous minimum extrema.

For both $\mathfrak{B}^{\vec{m}}_{2,n}$ and $\mathfrak{B}^{\vec{m}}_{3,n}$ the discrete cases are simply discretization of the continuous case. However, this is not necessarily the case for higher moment spaces (Courtois et al. 2006). While the continuous cases provide more intuitive extrema, derivation of the discrete case for higher moments is not as simple as deriving the continuous case and discretizing.

Next, we examine the maximum extrema on $\mathfrak{B}^{\vec{m}}_{3,n}$:

$$X^{(3)}_{\max} = \begin{cases} \alpha & \text{with } p_\alpha = \dfrac{m_{2,n,\alpha+1}}{n-\alpha} \\ \alpha+1 & \text{with } p_{\alpha+1} = \dfrac{-m_{2,n,\alpha}}{n-\alpha-1} \\ n & \text{with } p_n = 1 - p_\alpha - p_{\alpha+1} \end{cases}$$

where

$$\alpha < \frac{nm_1 - m_2}{n - m_1} \leq \alpha + 1.$$

Since $X^{(3)}_{\max}$ can only provide non-trivial information about q if $p_0 > 0$, this extremal distribution is only informative about extinction when $\alpha = 0$ and $p_\alpha > 0$, which is the case whenever $nm_1 - m_2 < n - m_1$. Although this requirement may appear restrictive, some classes of distributions have simple rules under which $X^{(3)}_{\max}$ is informative. For example, for binomial distributions, $B_{n,p}$, $X^{(3)}_{\max}$ will provide a non-zero lower bound if $1/n < p \leq 1/(n-1)$.

We move on to $\mathfrak{B}^{\vec{m}}_{4,n}$. The use of three moments can improve bounds on the probability of extinction, but as with all of the maximal random variables, $X^{(4)}_{\max}$ requires the knowledge of the maximum, n. $X^{(4)}_{\max}$ is defined as:

$$X^{(4)}_{\max} = \begin{cases} 0 & \text{with } p_0 = 1 - p_\alpha - p_{\alpha+1} - p_n \\ \alpha & \text{with } p_\alpha = \dfrac{m_{3,n,\alpha+1}}{\alpha(n-\alpha)} \\ \alpha+1 & \text{with } p_{\alpha+1} = \dfrac{-m_{3,n,\alpha}}{(\alpha+1)(n-\alpha-1)} \\ n & \text{with } p_n = \dfrac{m_{3,\alpha,\alpha+1}}{n(n-\alpha)(n-\alpha-1)} \end{cases}$$

where

$$\alpha < \frac{m_2 n - m_3}{m_1 n - m_2} \leq \alpha + 1.$$

While this is a potential improvement to the lower bound given by $X^{(3)}_{\min}$, the improvement is sometimes negligible. As $n \to \infty$, the difference between $X^{(4)}_{\max}$ and $X^{(3)}_{\min}$ vanishes because

$$\lim_{n\to\infty} \frac{m_2 n - m_3}{m_1 n - m_2} = \frac{m_2}{m_1}$$

and furthermore, if $n \to \infty$ then $p_n \to 0$. Therefore, the resulting generating function for $X^{(4)}_{\max}$ is identical to (4) if the maximal value is unknown. So, like the first moment, the third moment is uninformative about extinction when n is unknown, unless assumptions are made about the distribution (see e.g., (Daley and Narayan 1980; Ethier and Khoshnevisan 2002)).

The minimal extrema for $\mathfrak{B}^{\vec{m}}_{4,n}$, $X^{(4)}_{\min}$ is given by

$$X^{(4)}_{\min} = \begin{cases} \alpha, & \text{with } p_\alpha = \dfrac{m_{3,\beta,\beta+1,\alpha+1}}{(\beta-\alpha)(\beta+1-\alpha)} \\ \alpha+1, & \text{with } p_{\alpha+1} = \dfrac{-m_{3,\beta,\beta+1,\alpha}}{(\beta-\alpha)(\beta-1-\alpha)} \\ \beta, & \text{with } p_\beta = \dfrac{m_{3,\alpha,\alpha+1,\beta+1}}{(\beta-\alpha)(\beta-1-\alpha)} \\ \beta+1, & \text{with } p_{\beta+1} = \dfrac{-m_{3,\alpha,\alpha+1,\beta}}{(\beta-\alpha)(\beta+1-\alpha)} \end{cases}$$

where α and β are given by

$$\alpha < \frac{m_{3,\beta,\beta+1}}{m_{2,\beta,\beta+1}} \le \alpha+1, \quad \beta < \frac{m_{3,\alpha,\alpha+1}}{m_{2,\alpha,\alpha+1}} \le \beta+1.$$

Again, this bound is only useful if $p_0 > 0$. Unfortunately there is no short form equation to identify which spaces $\mathfrak{B}^{\vec{m}}_{4,n}$ fit this requirement. However, one can easily determine if a given $\mathfrak{B}^{\vec{m}}_{4,n}$ has a useful $X^{(4)}_{\min}$. Assuming $\alpha = 0$, $\widehat{\beta}$ is simply bound by

$$\widehat{\beta} < \frac{m_3 - m_2}{m_2 - m_1} < \widehat{\beta} + 1.$$

And if $m_{3,\widehat{\beta},\widehat{\beta}+1} < m_{2,\widehat{\beta},\widehat{\beta}+1}$, then the bound is useful because the resulting $X^{(4)}_{\min}$ has $p_0 > 0$. Alternatively, if $m_{3,\widehat{\beta},\widehat{\beta}+1} \ge m_{2,\widehat{\beta},\widehat{\beta}+1}$ the supports for $X^{(4)}_{\min}$ have $p_0 = 0$ and consequently $q = 0$.

If the first four moments are known, the extremal variable $X^{(5)}_{\min}$ can be obtained. Its distribution takes a simple form, but the equations used to find its values and relative probabilities are relatively large. From (Hürlimann 2005), $X^{(5)}_{\min}$ is defined as:

$$X^{(5)}_{\min} = \begin{cases} 0 & \text{with } p_0 = 1 - p_\alpha - p_{\alpha+1} - p_\beta - p_{\beta+1} \\ \alpha & \text{with } p_\alpha = \dfrac{m_{4,\beta,\beta+1,\alpha+1}}{\alpha(\beta-\alpha)(\beta+1-\alpha)} \\ \alpha+1 & \text{with } p_{\alpha+1} = \dfrac{-m_{4,\beta,\beta+1,\alpha}}{(\alpha+1)(\beta-\alpha)(\beta-1-\alpha)} \\ \beta & \text{with } p_\beta = \dfrac{m_{4,\alpha,\alpha+1,\beta+1}}{\beta(\beta-\alpha)(\beta-1-\alpha)} \\ \beta+1 & \text{with } p_{\beta+1} = \dfrac{-m_{4,\alpha,\alpha+1,\beta}}{(\beta+1)(\beta-\alpha)(\beta+1-\alpha)} \end{cases}$$

where

$$\alpha < \frac{m_{4,\beta,\beta+1}}{m_{3,\beta,\beta+1}} \le \alpha+1, \quad \beta < \frac{m_{4,\alpha,\alpha+1}}{m_{3,\alpha,\alpha+1}} \le \beta+1. \tag{5}$$

Courtois et al. (Courtois et al. 2006) explained that there is no analytic form to directly obtain α and β for $X_{\min}^{(5)}$. They showed this by disproving the intuitive idea that the discrete support encloses the continuous support. To find α and β, we iteratively search all possible supports on D_n until both inequalities are satisfied. This exhaustive method for finding the supports for this extrema is not ideal, especially if D_n is dense. Linear programing can be used to easily find the extremal supports and their probabilities (Prékopa 1990), but such approaches are not necessary when D_n is sparse (e.g., when n is relatively small).

Hürlimann (Hürlimann 2005) also presents a form for the upper extremal variable in $\mathfrak{B}_{5,n}^{\vec{m}}$. The process $X_{\max}^{(5)}$ is defined as:

$$X_{\max}^{(5)} = \begin{cases} \alpha, & \text{with } p_\alpha = \dfrac{m_{4,n,\beta,\beta+1,\alpha+1}}{(\beta-\alpha)(\beta+1-\alpha)(n-\alpha)} \\ \alpha+1, & \text{with } p_{\alpha+1} = \dfrac{-m_{4,n,\beta,\beta+1,\alpha}}{(\beta-\alpha)(\beta-\alpha-1)(n-\alpha-1)} \\ \beta, & \text{with } p_\beta = \dfrac{m_{4,n,\alpha,\alpha+1,\beta+1}}{(\beta-\alpha)(\beta-\alpha-1)(n-\beta)} \\ \beta+1, & \text{with } p_{\beta+1} = \dfrac{-m_{4,n,\alpha,\alpha+1,\beta}}{(\beta-\alpha)(\beta+1-\alpha)(n-\beta-1)} \\ n, & \text{with } p_n = 1 - p_\alpha - p_{\alpha+1} - p_\beta - p_{\beta+1} \end{cases}$$

where

$$\alpha < \frac{m_{4,n,\beta,\beta+1}}{m_{3,n,\beta,\beta+1}} \le \alpha+1, \quad \beta < \frac{m_{4,n,\alpha,\alpha+1}}{m_{3,n,\alpha,\alpha+1}} \le \beta+1.$$

As was the case for $X_{\min}^{(4)}$, one can determine if $X_{\max}^{(5)}$ has $p_0 > 0$ by assuming $\alpha = 0$ and solving for $\widehat{\beta}$ with

$$\widehat{\beta} < \frac{m_{4,n,0,1}}{m_{3,n,0,1}} \le \widehat{\beta}+1.$$

If the resulting $\widehat{\beta}$ in the inequality $m_{4,n,\widehat{\beta},\widehat{\beta}+1} < m_{4,n,\widehat{\beta},\widehat{\beta}+1}$ holds, the bound for $X_{\max}^{(5)}$ is informative.

All $X_{\max}^{(j)}$ extrema rely on the maximum offspring number, n. Similar to $X_{\max}^{(4)}$, when n is unknown or infinity $X_{\max}^{(5)}$ goes to the minimum on the lower moment space, here $X_{\min}^{(4)}$. Thus if n is unknown, $X_{\max}^{(j)}$ goes to $X_{\min}^{(j-1)}$, at least for the cases examined here.

The Chebychev approach can be used to extend this approach to higher moments (Hürlimann 2005). However, moments above the fourth are rarely used, and higher moments can be difficult to estimate from small samples. Further, the equations for the supports and probabilities for moments above the fourth become increasingly complex.

3 Results and discussion

Here we discuss some example distributions, graph their generating functions, and also graph generating functions for the extremal distributions. The plot of the probability generating function, $f(q)$, on $q \in (0, 1)$ is a useful way to visualize how the moments are related to extinction. The probability generating function takes the value p_0 at $q = 0$. At small q, $f(q)$ has a slope of approximately p_1. In this part of the function, when q is small, there can be a weak relationship between $f(q)$ and moments. In comparison, when q is close to 1, the moments are closely related to $f(q)$. For example $f'(1) = m_1$. Higher moments begin to influence the function as q moves away from 1.

The probability of extinction of a process is found when $f(q) = q$, i.e at the intersect between its probability generating function $f(q)$ and the diagonal q. Thus, processes with a high probability of extinction will cross the diagonal near $q = 1$, in the domain of q in which the probability generating function is often closely related to its first few moments.

Plotting the probability generating functions for the extremal distributions helps demonstrate why they act as bounds on extinction. In these examples (Figure 1), we compare two distributions with identical first moment and maximum ($m_1 = 2$, $n = 20$), i.e. both distributions are in $\mathfrak{B}^2_{2,20}$. In particular, we look at a binomial distribution and a truncated geometric distribution. For each of these plots we also plot the generating functions for some of the extremal distributions. The extremal distributions provide clear bounds: best case extrema are found below the plot of the generating function, worst case extrema are found above. For example, the extremal distribution based on one moment, $X^{(2)}_{\max}$, provides an upper bound on the probability of extinction, and can be seen as the upper line in both plots. Because they share an identical first moment and maximum, $X^{(2)}_{\max}$ is the same for both distributions. Clearly, one moment does not provide a good bound in these examples. As more moments are used, the bounds become tighter. The extrema using four moments provide relatively accurate upper and lower bounds for both examples. The lower bounds provide the best case extrema, which are useful in both cases only when three or four moments are known. The lower bound using two moments is

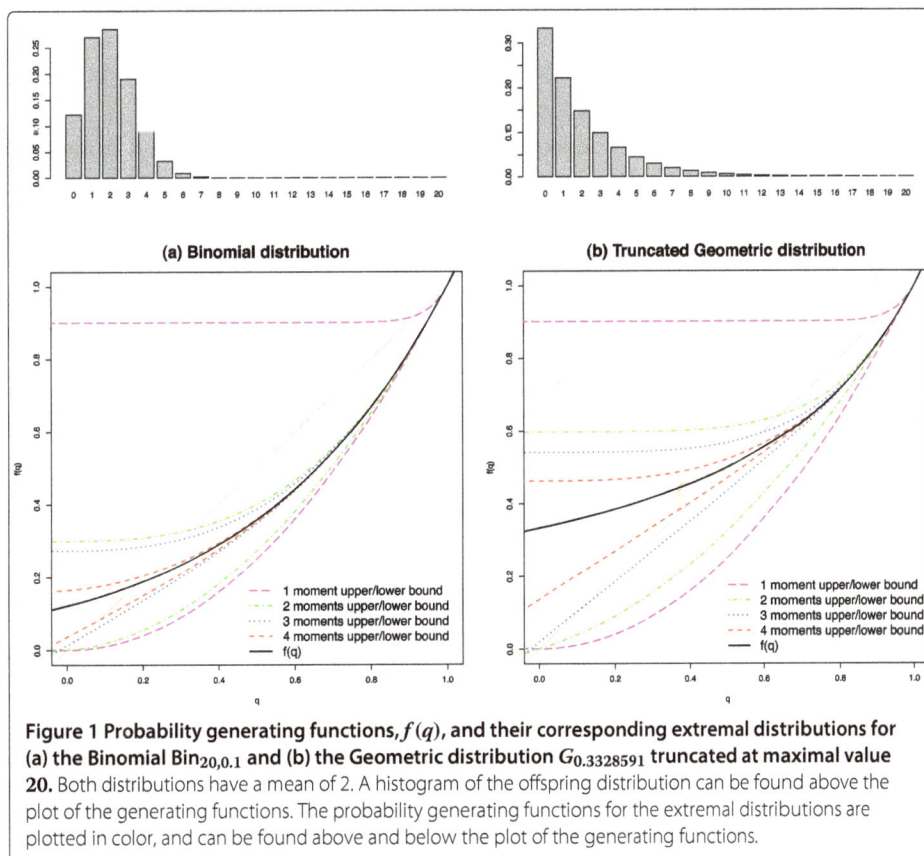

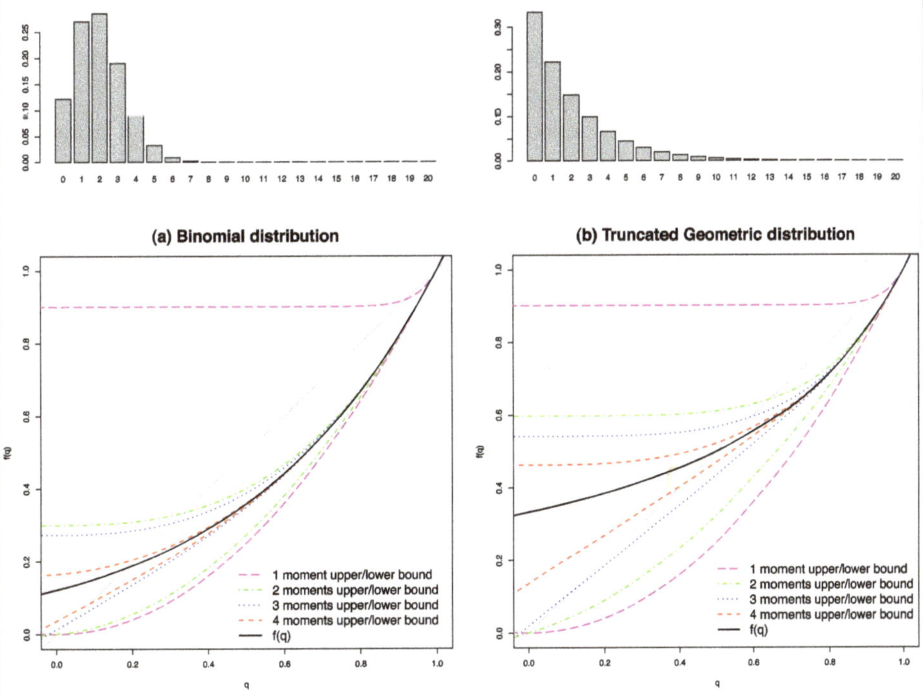

Figure 1 Probability generating functions, $f(q)$, and their corresponding extremal distributions for (a) the Binomial $\mathrm{Bin}_{20,0.1}$ and (b) the Geometric distribution $G_{0.3328591}$ truncated at maximal value 20. Both distributions have a mean of 2. A histogram of the offspring distribution can be found above the plot of the generating functions. The probability generating functions for the extremal distributions are plotted in color, and can be found above and below the plot of the generating functions.

not useful here in either case, as its probability generating function crosses the diagonal at zero so its probability of extinction is zero. The lower bound using only one moment was not included because its generating function is trivial and always uninformative about extinction.

Importantly, these examples demonstrate why higher moments are often necessary to compare strategies. These two distributions have identical first moments ($m_1 = 2$) so classically their fitness value would be equal. However, the binomial example is more likely to survive. If entire distributions are known, then extinction probabilities can be calculated explicitly using (1). This requires solving a polynomial of degree 20 for the examples shown here, which were solved in R (R Development Core Team 2011) with the package "rootSolve" (Soetaert and Herman 2009).

If instead the moments are known, the extremal distributions can be found and the roots of their generating functions can be solved to find the bounds on extinction. These roots can again be solved in R (R Development Core Team 2011) with the package "rootSolve" (Soetaert and Herman 2009). However, some of these extremal generating functions are relatively simple and can be solved by hand. For example, our binomial distribution ($\text{Bin}_{20,0.1}$) has $m_1 = 2$ and $m_2 = 5.8$. The resulting $X^{(3)}_{\text{min}}$ has supports at 0, 2 and 3 with respective probabilities of 0.3, 0.1 and 0.6, leading to its generating function

$$q = 0.3 + 0.1q^2 + 0.6q^3.$$

Using the knowledge that the generating function has a root at $q = 1$, this equation can be factorized as:

$$0 = (q - 1)(3q - 1)(2q + 3).$$

The probability of extinction is the smallest positive root of the above equation, $1/3$ (Table 1), providing an upper bound on extinction.

If four moments are known, one can conclude that the truncated geometric distribution has a higher probability of extinction. Compare the extremal distributions when four moments are known, paying attention to where they cross the diagonal. The value at the intersect is the probability of extinction for the extrema, which we display in Table 1 and Table 2, respectively for the binomial example and the truncated geometric example. Using four moments, the best case for the truncated geometric example (0.404, Table 2) is worse than the worst case for binomial example (0.207, Table 1). In fact, the worst case for the binomial example using two moments (0.333) is already better than the best case for the truncated geometric using four moments (0.404). These examples highlight how moment spaces can be used to rank branching processes by their extinction probabilities when only moments of their distributions are known.

Table 1 Extinction probabilities and supports for the extremal distributions of the Binomial example $B_{20,0.1}$

	1 moment	2 moments	3 moments	4 moments
Supports	{2}	{1,2,20}	{0,1,3,4}	{0,1,3,4,20}
Lower bound (best)	0.000	0.000	0.034	0.083
Supports	{0,20}	{0,2,3}	{0,2,3,20}	{0,1,2,4,5}
Upper bound (worst)	0.918	0.333	0.306	0.207

The actual probability of extinction for this process is 0.181.

Table 2 Extinction probabilities and supports for the extremal distributions of the truncated Geometric example

	1 moment	2 moments	3 moments	4 moments
Supports	{2}	{1,2,20}	{0,1,7,8}	{0,1,6,7,20}
Lower bound (best)	0.000	0.000	0.110	0.404
Supports	{0,20}	{0,4,5}	{0,4,5,20}	{0,3,4,11,12}
Upper bound (worst)	0.918	0.641	0.592	0.534

The actual probability of extinction for this process is 0.499.

And finally, these examples can be used to better understand how ranking distributions using their s-convex extrema can be useful in investing and gambling (Canjar 2007; Courtois et al. 2006; Denuit and Lefevre 1997; Ethier and Khoshnevisan 2002; Hürlimann 2005). If these distributions were returns on an investment or gamble, then by comparing their moments an investor could determine that the binomial distribution is a superior investment model. Both distributions would provide the same expected growth on capital, but the geometric distribution would have a higher probability of gambler's ruin. Being wary of gambler's ruin is especially important for an investor with limited initial funds for their investment.

4 Conclusion

The work here is intended to highlight the relationship between the moments of the offspring distribution and the probability of extinction. Extinction can be defined in terms of moments, but the first few moments are only informative about extinction under certain conditions. Nevertheless, for all offspring distributions there exists an interesting relationship with even and odd moments: high even moments favor extinction, high odd moments favor survival. This relationship between even and odd moments is also seen in the stochastic Price equation, where relative growth rates increase with increasing odd moments, and decrease with increasing even moments (Rice 2008).

The relationship between moments and extinction can provide insight into the evolutionary process. A high first moment can favor survival, but worst case extrema ("long shots") represent the strategies that are least likely to survive. Strategies with a relatively low second moment (low variance) will always have a lower probability of extinction than their corresponding "long shot" extrema. When two moments are known, the worst case distributions have the lowest third moment (strongest right skew). Therefore, strategies with identical first and second moments and relatively high third moments (strong left skew) will always have a better chance at survival than the extrema with the lowest third moment. Worst case extrema using three moments have the highest possible fourth moment (excessive kurtosis). The relative importance of higher moments depends on the distribution, and in some cases higher moments can have a big influence on extinction.

Strategies with a high probability of extinction are unlikely to be found in natural populations, even if their expected reproductive rate is high (Tuljapurkar and Orzack 1980). New alleles will often arrive in a population as a singlet, and extinction is permanent unless the same mutation occurs more than once. In such cases, survival is more important than the average rate of reproduction. Using moments of the offspring distribution one can find bounds on extinction using their s-convex extrema. If the best case extrema for a set of moments has a high probability of extinction, then strategies with these moments will be evolutionarily unlikely, regardless of how fit these strategy would be if they avoided extinction.

Gamblers can avoid strategies with a high risk of ruin by calculating their odds. In natural populations, such calculations are not required to prevent the occurrence of high risk strategies. Instead, risky strategies will be naturally unlikely, especially considering that many arrive as a single allele with one chance at survival. Similarly, gamblers and investors who begin with limited funds and chose risky strategies are likely to "go extinct" through gambler's ruin. Risk is not solely determined by mean growth, and strategies with a high mean can sometimes have high risk. Unfortunately, these high risk and high reward strategies are unlikely to return anything without sufficient investment, so natural avoidance of risk can result in missed opportunity for growth.

Competing interests

The authors declare that they have no competing interests.

Authors' contributions

SS and SK both conceived, designed and developed the methods and results, which were then interpreted by both authors. The manuscript was primarily written by SS, with help from SK. Both authors read and approved the final manuscript.

Acknowledgement

The authors like to thank Ninh Anh for a fruitful discussion on calculating the supports and probabilities for the extremal distributions, and the many constructive comments by the editor and two anonymous referees.

Author details

[1]Institute for Behavioral Genetics, University of Colorado, Boulder, Colorado, USA. [2]Department of Statistics and School of Biological Sciences, University of Auckland, Auckland, New Zealand. [3]Formerly: Department of Anatomy, and Allan Wilson Centre for Molecular Ecology and Evolution, University of Otago, Dunedin, New Zealand.

References

Canjar, RM: Gambler's Ruin revisited: The effects of skew and large jackpots. In: Ethier, S, Eadington, W (eds.) *Optimal Play: Mathematical Studies of Games and Gambling. Institute for the Study of Gambling and Commercial Gaming*, pp. 439–469. University of Nevada, Reno, (2007)

Courtois, C, Denuit, M, van Bellegem, S: Discrete *s*-convex extremal distributions: theory and applications. Appl. Math. Lett. **19**, 1367–1377 (2006)

Daley, DJ, Narayan, P: Series expansions of probability generating functions and bounds for the extinction probability of a branching process. J. Appl. Probab. **17**, 939–947 (1980)

Denuit, M, Lefevre, C: Some new classes of stochastic order relations among arithmetic random variables, with applications in actuarial sciences. Insur. Math. Econ. **20**, 197–213 (1997)

Denuit, M, de Vylder, E, Lefevre, C: Extremal generators and extremal distributions for the continuous *s*-convex stochastic orderings. Insur. Math. Econ. **24**, 201–217 (1999)

Denuit, M, Lefevre, C, Mesfioui, M: On *s*-convex stochastic extrema for arithmetic risks. Insur. Math. Econ. **25**, 143–155 (1999)

Ethier, SN, Khoshnevisan, D: Bounds on gambler's ruin probabilities in terms of moments. Methodol. Comput. Appl. Probab. **4**, 55–68 (2002). 10.1023/A:1015705430513

Hürlimann, W: Improved analytical bounds for gambler's ruin probabilities. Methodol. Comput. Appl. Probab. **7**, 79–95 (2005). 10.1007/s11009-005-6656-4

Karlin, S, McGregor, JL: The differential equations of birth-and-death-processes, and the stieltjes moment problem. T. Am. Math. Soc. **85**, 489–546 (1957)

Kelly, J: A new interpretation of information rate. Bell Sys. Tech. J. **35**, 917–926 (1956)

Kimmel, M, Axelrod, DE: Branching Processes in Biology. Springer, New York (2002)

Lewontin, RC, Cohen, D: On population growth in a randomly varying environment. P. Natl. Acad. Sci. USA. **62**, 1056–1060 (1969)

MacLean, LC, Thorp, EO, Ziemba, WT: Good and bad properties of the Kelly criterion. In: *The Kelly Capital Growth Investment Criterion: Theory and Practice*, pp. 563–574. World Scientific Publishing, Singapore, (2010)

Prékopa, A: The discrete moment problem and linear programming. Discrete Appl. Math. **27**, 235–254 (1990)

R Development Core Team: R: a language and environment for statistical computing. R Foundation for Statistical Computing, Vienna, Austria (2011). http://www.R-project.org

Rice, SH: A stochastic version of the Price equation reveals the interplay of deterministic and stochastic processes in evolution. BMC Evol. Biol. **8**, 262 (2008)
Shaked, M, Shanthikumar, JG: Stochastic Orders. Springer, New York (2007)
Soetaert, K, Herman, PMJ: A practical guide to ecological modelling. Using r as a simulation platform. Springer, New York (2009)
Tuljapurkar, S, Orzack, SH: Population dynamics in variable environments I. Long-run growth rates and extinction. Theor. Popul. Biol. **18**, 314–342 (1980)

2

Bayesian reference analysis for exponential power regression models

Marco AR Ferreira[1*] and Esther Salazar[2]

*Correspondence:
ferreiram@missouri.edu
[1] Department of Statistics, University of Missouri, Columbia, USA
Full list of author information is available at the end of the article

Abstract

We develop Bayesian reference analyses for linear regression models when the errors follow an exponential power distribution. Specifically, we obtain explicit expressions for reference priors for all the six possible orderings of the model parameters and show that, associated with these six parameters orderings, there are only two reference priors. Further, we show that both of these reference priors lead to proper posterior distributions. Furthermore, we show that the proposed reference Bayesian analyses compare favorably to an analysis based on a competing noninformative prior. Finally, we illustrate these Bayesian reference analyses for exponential power regression models with applications to two datasets. The first application analyzes per capita spending in public schools in the United States. The second application studies the relationship between sold home videos versus profits at the box office.

MSC: 62F15; 62F35; 62J05

Keywords: Bayesian inference; Exponential power errors; Frequentist properties; Reference prior; Robustness

1 Introduction

A flexible way to deal with outliers in linear regression is to assume that the errors follow an exponential power (EP) distribution. Specifically, assuming an EP distribution decreases the influence of outliers and, as a result, increases the robustness of the analysis (Box and Tiao 1962; Liang et al. 2007; Salazar et al. 2012; West 1984). In addition, the EP distribution includes the Gaussian distribution as a particular case. Further, the EP distribution may have tails either lighter (platykurtic) or heavier (leptokurtic) than Gaussian. Platykurtic distributions may be a result of truncation, whereas leptokurtic distributions provide protection against outliers. Salazar et al. (2012) have developed three types of Jeffreys priors for linear regression models with independent EP errors. Unfortunately, two of those priors lead to useless improper posterior distributions and only one leads to a proper posterior distribution. Here we develop explicit expressions for reference priors for all the six possible orderings of the model parameters.

We show that the six parameters orderings lead to two distinct reference priors. The parameter ordering corresponds to the order of importance of each parameter in the analysis, with the most important parameter appearing first and the least important appearing last (Berger and Bernardo 1992a,b). In addition to the two formally obtained reference priors, we propose an approximate reference prior that shares the same tail behavior but is much more straightforward to implement in practice. Finally, we show that the two reference priors lead to useful proper posterior distributions.

To make sure that Bayesian reference procedures do not bias the data analysis in an undesirable manner, it is important to study their frequentist properties. To study the frequentist properties of our proposed procedures, we have performed a Monte Carlo study that shows that our proposed Bayesian reference approaches compare favorably to a posterior analysis based on a competing prior in terms of coverage of credible intervals, relative mean squared error, and mean length of credible intervals. While the relative mean squared error and the mean length of credible intervals should be judged in comparison with those yielded by competing priors, the coverage of credible intervals should be as close as possible to the nominal level.

Coverage of credible intervals close to nominal provides a guarantee of level of performance of the procedure when used automatically and independently by many researchers in their problems. In our Monte Carlo study, we have found that the Bayesian reference credible intervals that we have obtained have frequentist coverage close to nominal. These good frequentist properties results agree with previous literature on Bayesian reference analyses for other models such as, for example, Gaussian random fields (Berger et al. 2001), Markov random fields (Ferreira and De Oliveira 2007), multivariate normal models (Sun and Berger 2007), and elapsed times in continuous-time Markov chains (Ferreira and Suchard 2008).

The EP density is given by

$$f(y|\mu,\sigma_p,p) = \left[2p^{1/p}\sigma_p\Gamma(1+1/p)\right]^{-1}\exp\left[-\left(p\sigma_p^p\right)^{-1}|y-\mu|^p\right],\ -\infty < y < \infty, \quad (1)$$

where $p > 1$, $-\infty < \mu < \infty$ and $\sigma_p > 0$. The EP distribution has three parameters: the location parameter $\mu = E(y)$, the scale parameter $\sigma_p = [E(|y - \mu|^p)]^{1/p}$, and the shape parameter p. The scale parameter σ_p can be seen as a variability index that generalizes the standard deviation. Moreover, σ_p is also known as power deviation of order p (Vianelli 1963). In addition, the kurtosis is $\kappa = \Gamma(1/p)\Gamma(5/p)/(\Gamma(3/p))^2$, implying that the shape parameter p determines the thickness of the tails of the EP density. Specifically, the EP distribution is leptokurtic if $p < 2$ ($\kappa > 3$) and platykurtic if $p > 2$ ($\kappa < 3$). Finally, the EP distribution has several important especial cases such as the Laplace distribution ($p = 1$), the normal distribution ($p = 2$) and, when $p \to \infty$, the uniform distribution on the interval $(\mu - \sigma_p, \mu + \sigma_p)$ (e.g., see Box and Tiao 1992).

There are just some few Bayesian procedures for the analysis of EP regression models published to date. Moreover, there are no published reference priors for EP regression models. Existing literature has considered the use of EP errors in a number of contexts such as, for example, EP errors to robustify linear models (Box and Tiao 1992; Salazar et al. 2012), and mixtures of regression models with EP errors (Achcar and Pereira 1999). In addition, the EP distribution has been used as a prior for a Gaussian model location parameter (Choy and Smith 1997). To implement simulation-based computation for models with EP errors, one may use representations of the EP distribution as a scale mixture of normals (West 1987) or as a scale mixture of uniforms (Walker and Gutiérrez-Peña 1999). As an alternative, Salazar et al. (2012) have developed fast analysis for EP regression models using Laplace approximations and Newton-Cotes integration. Here we use these latter fast computational methods.

The remainder of the paper is organized as follows. Section 2 presents the linear model with exponential power errors and the associated likelihood function. Section 3 derives the two reference priors and shows that both of these priors lead to proper posterior distributions. Section 4.1 presents a simulation study of the frequentist properties of the reference-priors-based Bayesian procedures and those of a competing noninformative prior. Section 4.2 presents applications of Bayesian reference analysis to two datasets. Section 5 concludes with a discussion of major findings and possible future research directions.

2 EP linear model

Let $y = (y_1, \ldots, y_n)'$ be the vector of observations and $x = (x_1, \ldots, x_n)'$ be the $n \times k$ design matrix of explanatory variables. We consider the linear model

$$y = x\beta + \epsilon, \tag{2}$$

where $\beta = (\beta_1, \ldots, \beta_k)' \in \mathbb{R}^k$ is a vector of regression coefficients, and $\epsilon = (\epsilon_1, \ldots, \epsilon_n)'$ is a vector of errors such that $\epsilon_1, \ldots, \epsilon_n$ are independent and identically distributed and follow the exponential power distribution with location parameter equal to zero, scale parameter σ_p, and shape parameter p. We reparameterize the model by defining $\sigma = p^{1/p}\sigma_p\Gamma(1 + 1/p)$. This reparametrization has also been considered by Zhu and Zinde-Walsh (2009) and Salazar et al. (2012). Let us denote the parameter vector by $\theta = (\beta, \sigma, p) \in \mathbb{R}^k \times (0, \infty) \times (1, \infty)$. Then, the log-likelihood function for the model given in Equation (2) is

$$l(\theta; y, x) = -n \log 2 - n \log \sigma - \sum_{i=1}^{n} \left[\frac{\Gamma(1 + 1/p)|y_i - x_i'\beta|}{\sigma} \right]^p. \tag{3}$$

We use the log-likelihood function to develop reference priors for the EP regression model.

3 Methods

In this section, we obtain explicit expressions for reference priors for all the six possible orderings of the parameters of the EP linear model, and show that associated with these six parameters orderings there are only two reference priors. Finally, we show that both of these reference priors lead to proper posterior distributions.

Specifically, we consider here the Bernardo reference priors (Bernardo 1979) that take into account the Kulback-Leibler divergency between the prior distribution and the posterior distribution. In a nutshell, the reference priors proposed by Bernardo maximize the expected value of perfect information about the model parameters (p. 300, Bernardo and Smith 1994). When the parameter space is one-dimensional and asymptotic normality of the posterior distribution holds, the reference prior coincides with Jeffreys prior (Jeffreys 1961). However, when the parameter space is multidimensional Jeffreys prior is known to lead to Bayesian procedures that may have undesirable frequentist properties, such as for example frequentist coverage of credible intervals far away from the desired nominal level.

For the multidimensional parameter case when the parameters may be partitioned in a block of parameters of interest and another block of nuisance parameters, Bernardo

(1979) suggested an approach in three stages. The first stage obtains the conditional distribution of the nuisance parameter conditional on the parameter of interest. The second stage integrates out the nuisance parameter with respect to that conditional distribution to obtain a marginal likelihood. Finally, the third stage applies the reference prior approach to the marginal likelihood to obtain the reference prior for the parameter of interest. This idea can be naturally extended to partitions of the parameter vector with more than two components. The resulting reference prior will then depend on the ordering of the parameter vector components. This multiparameter case has been developed in a series of papers by Berger and Bernardo (1992a,b,c). Here we use the Berger-Bernardo approach to develop reference priors for the parameters of the EP regression model.

As we show below, the reference priors obtained here are of the form

$$\pi(\theta) \propto \frac{\pi(p)}{\sigma^a}, \tag{4}$$

where $a \in \mathbb{R}$ is a hyperparameter and $\pi(p)$ is the 'marginal' prior of the shape parameter p. As shown by Salazar et al. (2012), the Jeffreys-rule prior and two independence Jeffreys priors also have the functional form (4). Specifically, using the same notation as in Salazar et al. (2012), the two independence Jeffreys priors have $a = 1$ and their marginal priors for p are respectively given by

$$\pi^{I_1}(p) \propto p^{-1}\left[\left(1+p^{-1}\right)\Psi'\left(1+p^{-1}\right) - 1\right]^{1/2}, \tag{5}$$

and

$$\pi^{I_2}(p) \propto p^{-3/2}\left[\left(1+p^{-1}\right)\Psi'\left(1+p^{-1}\right)\right]^{1/2}. \tag{6}$$

Meanwhile, the Jeffreys-rule prior is such that $a = k+1$ and its marginal prior for p is

$$\pi^{J}(p) \propto \left[\Gamma\left(p^{-1}\right)\Gamma\left(2-p^{-1}\right)\right]^{k/2}\pi^{I_1}(p). \tag{7}$$

In what follows we find that the reference priors for the EP regression model are related to the independence Jeffreys priors given in Equations (5) and (6). When developing noninformative priors, it is crucial to study whether the resulting posterior distribution is proper. Salazar et al. (2012) have shown that the Independence Jeffreys prior $\pi^{I_2}(p)$ yields a proper posterior distribution. Unfortunately, both the independence Jeffreys prior $\pi^{I_1}(p)$ and the Jeffreys-rule prior $\pi^{J}(p)$ yield improper posterior distributions.

The Berger-Bernardo approach to develop reference priors requires the Fisher information matrix. Specifically, for the EP regression model the Fisher information matrix $H(\theta)$, with elements ϕ_{ij} given by $\phi_{ij} = E_{y|\theta}\left[-\frac{\partial^2}{\partial\theta_i\partial\theta_j'}l(\theta; y, x)\right]$ with $\phi_{ij} = \phi_{ji}$ and θ_j the jth element of $\theta = (\beta, \sigma, p)$, is:

$$H(\theta) = \begin{bmatrix} \sigma^{-2}\Gamma(p^{-1})\Gamma(2-p^{-1})\sum_{i=1}^{n} x_i x_i' & 0 & 0 \\ 0 & np\sigma^{-2} & -n\sigma^{-1}p^{-1} \\ 0 & -n\sigma^{-1}p^{-1} & np^{-3}\left(1+p^{-1}\right)\Psi'\left(1+p^{-1}\right) \end{bmatrix}, \tag{8}$$

where $\Psi(\alpha) \equiv \Gamma'(\alpha)/\Gamma(\alpha)$ and $\Psi'(\alpha) \equiv \partial\Psi(\alpha)/\partial\alpha$ are the digamma and trigamma functions, respectively.

The Fisher information matrix is block diagonal, with one block corresponding to β and another block corresponding to (σ, p). One of the consequences of this structure is that reference priors that consider β, σ, and p as three separate groups will depend on the ordering of the groups only with respect to whether σ or p appears first in the ordering. The following theorem provides reference priors for the parameters of the EP regression model.

Theorem 1. *Consider the EP regression model with log-likelihood function given in Equation (3). Then, there are two reference priors for all six possible orderings of the model parameters. Moreover, these two reference priors are of the form (4) with* $a = 1$. *For the orderings* (β, σ, p), (σ, β, p), *and* (σ, p, β) *the 'marginal' reference prior for* p *is*

$$\pi^{r_1}(p) \propto p^{-3/2}\left[\left(1+p^{-1}\right)\Psi'\left(1+p^{-1}\right)\right]^{1/2}, \tag{9}$$

whereas for the orderings (β, p, σ), (p, β, σ), *and* (p, σ, β) *the 'marginal' reference prior for* p *is*

$$\pi^{r_2}(p) \propto p^{-3/2}\left[\left(1+p^{-1}\right)\Psi'\left(1+p^{-1}\right) - 1\right]^{1/2}. \tag{10}$$

Proof. See the Appendix. □

While reference prior π^{r_2} is a new prior that has not appeared before in the literature, there are similarities between the reference priors given in Theorem 1 and the independence Jeffreys priors given in Equations (5) and (6). Reference prior π^{r_1} coincides with the independence Jeffreys prior π^{I_2} given in Equation (6). Moreover, it is important to point out that reference prior π^{r_2} is somewhat similar to the independence Jeffreys prior π^{I_1} given in Equation (5), differing only by a factor of $p^{-1/2}$. However, as we show below this difference between π^{I_1} and π^{r_2} is enough to make π^{I_1} yield a useless improper posterior distribution while the reference prior π^{r_2} yields a useful proper posterior distribution.

Consider a prior of the form (4). Then the integrated likelihood for p is given by

$$L^I(p; y) \propto \int_{\mathbb{R}^k}\int_0^\infty L(\beta, \sigma, p; y)\sigma^{-a}d\sigma d\beta.$$

Then the prior leads to a proper posterior distribution if and only if

$$\int_1^\infty L^I(p; y)\pi(p)dp < \infty, \tag{11}$$

Thus, in order to determine whether a prior of the form (4) leads to a proper posterior distribution, one needs to investigate the tail behavior of both the marginal prior and the integrated likelihood for p. The tail behavior of the marginal reference priors for p given in Theorem 1 is given in the following lemma.

Lemma 1. *The marginal priors for* p *given in Theorem 1 are continuous functions in* $[1, \infty)$ *and are such that* $\pi^{r_1}(p) = O\left(p^{-3/2}\right)$ *and* $\pi^{r_2}(p) = O\left(p^{-3/2}\right)$ *as* $p \to \infty$.

Proof. Direct inspection shows that $\pi^{r_1}(p)$ and $\pi^{r_2}(p)$ are continuous functions in $[1, \infty)$. Their tail behavior when $p \to \infty$ follows from the fact that $\Psi'\left(1+p^{-1}\right) \to 1.6449$ and $\Gamma\left(p^{-1}\right) = O(p)$ as $p \to \infty$. □

Theorem 1 and Lemma 1 suggest the definition of an approximate reference prior inspired by priors π^{r_1} and π^{r_2} that has the same value for the hyperparameter $a = 1$ and share their tail behavior with respect to p. We define such an approximate reference prior in Definition 1.

Definition 1. *We define an approximate reference prior* π^{r_3} *to be of the form (4) with* $a = 1$ *and marginal prior for* p *equal to* $\pi^{r_3}(p) \propto p^{-3/2}$.

Computation of prior π^{r_3} is faster and more straightforward than that of priors π^{r_1} and π^{r_2}. In addition, Section 4.1 shows that the frequentist properties of procedures based on π^{r_3} are similar to those based on π^{r_1} and π^{r_2}. As a consequence, the approximate reference prior π^{r_3} may become more widely used than the reference priors π^{r_1} and π^{r_2}. Therefore, henceforth we drop the term "approximate" and simply refer to π^{r_3} as a reference prior.

The following lemma, that was proved by Salazar et al. (2012), provides the tail behavior for the integrated likelihood for p.

Lemma 2 (Salazar et al.2012). *Provided that* $n > k + 1 - a$, *the integrated likelihood for* p *under the class of priors (4) is a continuous function in* $[1, \infty)$ *and is such that* $L^I(p; y) = O(1)$ *as* $p \to \infty$.

The following proposition establishes that the two reference priors that we have obtained yield proper posterior distributions.

Proposition 1. *Provided that* $n > k + 1 - a$, *the two reference priors* π^{r_1} *and* π^{r_2} *given in Theorem 1 yield proper posterior distributions.*

Proof. This proposition follows directly from condition (11), and Lemmas 1 and 2. □

To implement posterior analysis for the parameters of the EP regression model based on the reference priors developed here, we use an approach proposed by Salazar et al. (2012) that combines Laplace approximations and Newton-Cotes integration.

4 Results and discussion

4.1 Frequentist properties

In this section we perform a simulation study to access the frequentist properties of Bayesian procedures based on the reference priors π^{r_1}, π^{r_2}, and π^{r_3}. In addition, we compare the performance of these reference priors to that of a competing noninformative prior π^U that takes the form (4) with $a = 1$ and $\pi^U(p) \propto 1$ for $1 < p < 10$ and $\pi^U(p) = 0$ otherwise. The joint prior $\pi^U(\theta)$ leads to a proper posterior distribution, however as we see below the uniform prior $\pi^U(p)$ is a naïve way to express lack of information about p. The Bayesian procedures we consider are the posterior modes and posterior medians for point estimation, and the 95% highest posterior density (HPD) credible intervals for interval estimation. Finally, we consider three frequentist measures of quality. For evaluating the quality of point estimation, we consider the square root of the frequentist relative mean squared error. For evaluating the performance of interval estimation, we consider two frequentist measures: the frequentist coverage and the mean length of the credible intervals.

We have considered several combinations of sample sizes and parameters. Specifically, we have considered three sample sizes: $n = 30$, $n = 50$ and $n = 100$. Moreover, we have considered a grid of values for p on the interval from 1 to 3. Further, for each simulated dataset we have used $k = 2$, $x_i = (1, x_{1i})$, $x_{1i} \sim N(2, 1)$, $\beta = (1.5, -3)$, and $\sigma = 1$.

Finally, for each combination of parameter values and sample sizes, we have simulated 1,500 datasets to estimate the frequentist properties of the several procedures.

The square root of the relative mean squared error (RMSE), $\sqrt{MSE(\widehat{\theta})}/\theta$, for estimators of p and σ is shown as a function of p in Figure 1. As intuitively expected, for all priors and for both posterior mode and median, as the sample size increases the RMSE decreases. The most substantial differences are between the performances of the posterior mode and posterior median, and between the performances of the reference priors when compared with the π^U prior. First, we compare the performance of the posterior median and the posterior mode. For each prior, for the estimation of p, the posterior median provides smaller RMSE than the posterior mode for most values of p considered except for p close to one. And this advantage of the posterior mode becomes less pronounced as the sample size increases. For each prior, for the estimation of σ, the posterior median provides smaller RMSE than the posterior mode. Therefore, for the reference analysis of the EP regression model we recommend the use of the posterior median.

Second, we compare the RMSE performance of the different priors. For each type of point estimator considered here, in terms of RMSE the reference priors π^{r_1}, π^{r_2}, and π^{r_3} provide qualitatively similar results, with π^{r_1} and π^{r_3} being slightly better for smaller values of p and π^{r_2} being slightly better for larger values of p. In addition, the difference in performance of the three reference priors becomes smaller as the sample size increases. In contrast, the performance of the reference priors differs dramatically from that of the π^U prior. For each class of estimators of p and for all values of p considered, when compared to the π^U prior the reference priors lead to smaller RMSE. For the estimation of σ, the results are mixed; for small sample sizes while the reference priors lead to smaller RMSE when p is small and π^U leads to better results when p is larger. But for larger sample

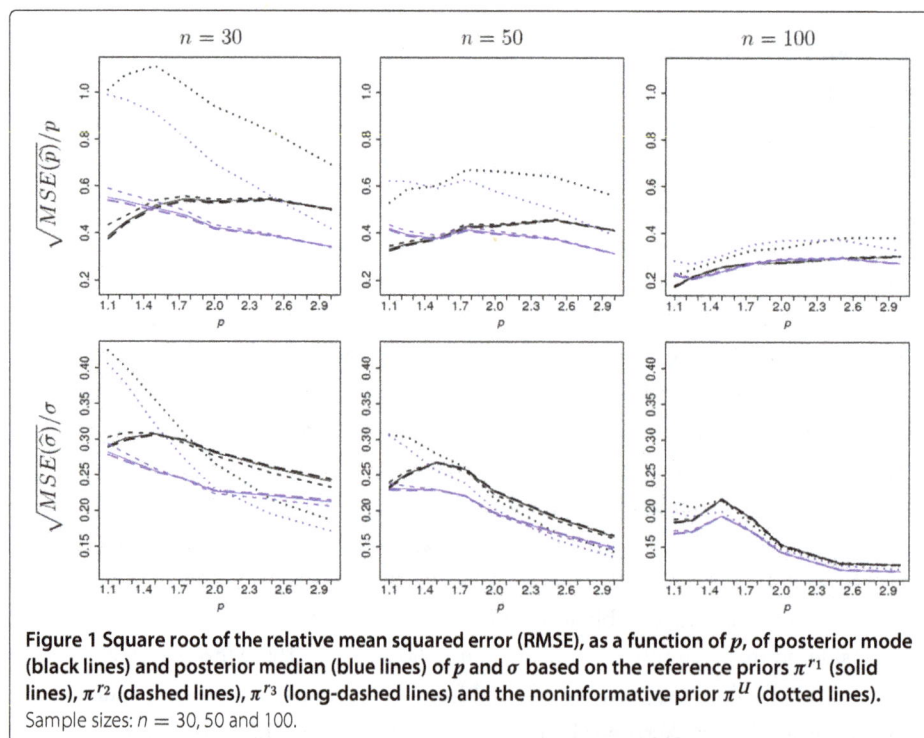

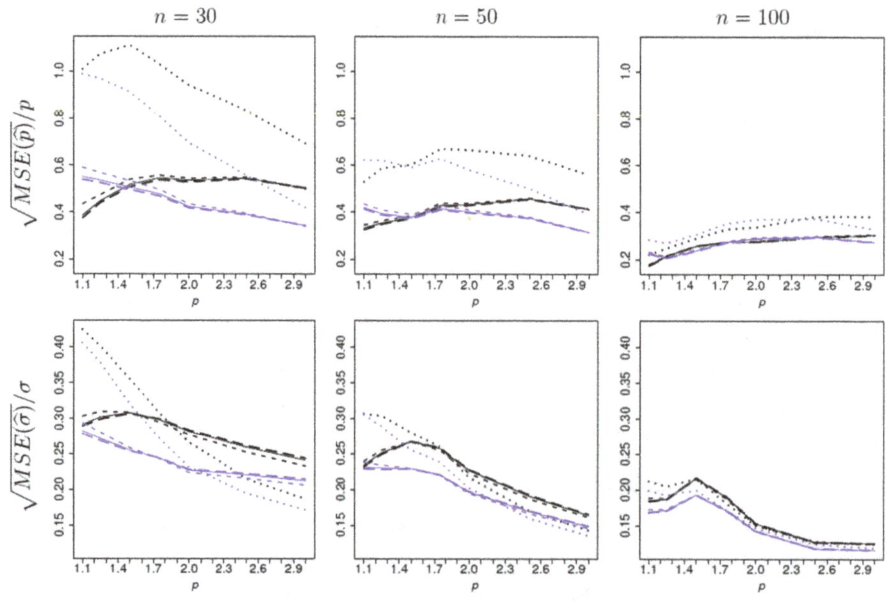

Figure 1 Square root of the relative mean squared error (RMSE), as a function of p, of posterior mode (black lines) and posterior median (blue lines) of p and σ based on the reference priors π^{r_1} (solid lines), π^{r_2} (dashed lines), π^{r_3} (long-dashed lines) and the noninformative prior π^U (dotted lines). Sample sizes: $n = 30, 50$ and 100.

sizes the reference priors-based posterior medians have smaller RMSE for all considered values of p.

The frequentist coverage (FC) of 95% HPD credible intervals for p and σ is shown, as a function of p, in Figure 2. As the sample size increases, the FC of the credible intervals based on the four priors becomes more similar. For both parameters, the π^{r_1}-, π^{r_2}-, and π^{r_3}-based credible intervals have frequentist coverage closer to the nominal level. This superiority of the Bayesian reference analysis is particularly pronounced for sample sizes equal to 30 or 50 and when $p < 2$.

The mean length of the 95% HDP credible intervals for p and σ is shown, as a function of p, in Figure 3. For the credible intervals based on the three reference priors, the mean lengths of the credible intervals are similar with slightly better results for π^{r_1}. For interval estimation for σ, the mean lengths of the credible intervals based on the three reference priors are smaller than the mean lengths of the credible intervals based on the π^{U} when $p < 2$ and are larger when $p > 2$. For interval estimation of p, in the range of values that we consider the π^{r_1}-, π^{r_2}-, and π^{r_3}-based credible intervals are on average shorter that those based on π^{U}. Therefore, for the interval estimation of p, in the range of values we consider, the credible intervals based on π^{r_1}, π^{r_2} and π^{r_3} provide uniformly superior results.

In summary, the reference priors π^{r_1}, π^{r_2}, and π^{r_3} lead to procedures that have similar frequentist properties. In addition, when compared to the competing noninformative prior π^{U}, the reference priors π^{r_1}, π^{r_2}, and π^{r_3} lead to overall superior results. Finally, the reference prior π^{r_3} has a simpler functional form and is more straightforward to be implemented. Therefore, in cases when there is no prior information for the analysis of EP linear regression models, we recommend the use of the reference prior π^{r_3}.

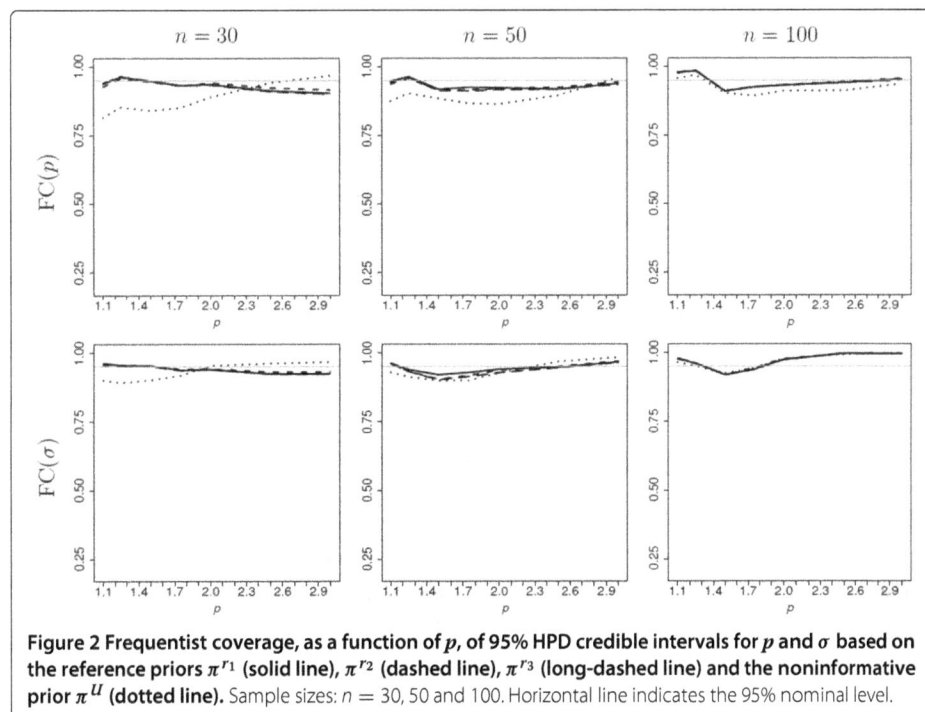

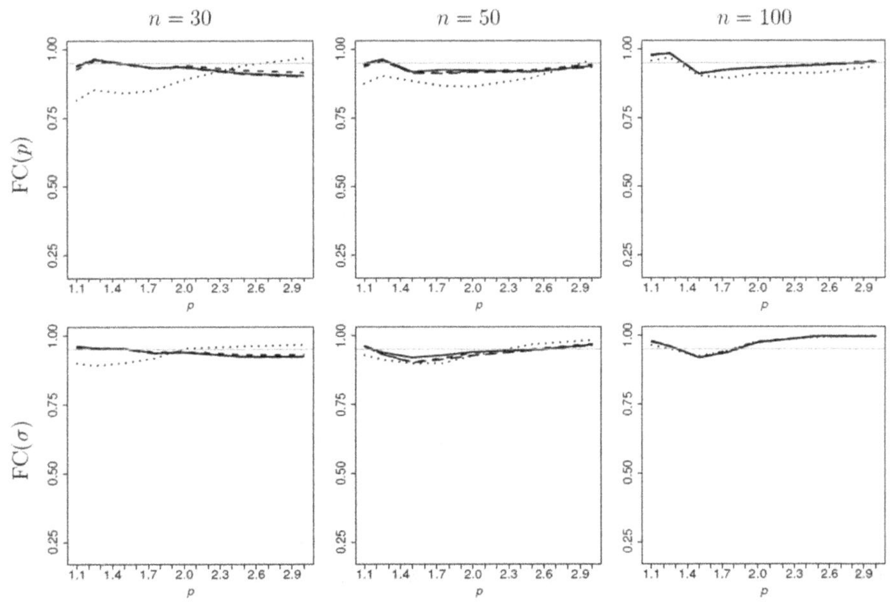

Figure 2 Frequentist coverage, as a function of p, of 95% HPD credible intervals for p and σ based on the reference priors π^{r_1} (solid line), π^{r_2} (dashed line), π^{r_3} (long-dashed line) and the noninformative prior π^{U} (dotted line). Sample sizes: $n = 30, 50$ and 100. Horizontal line indicates the 95% nominal level.

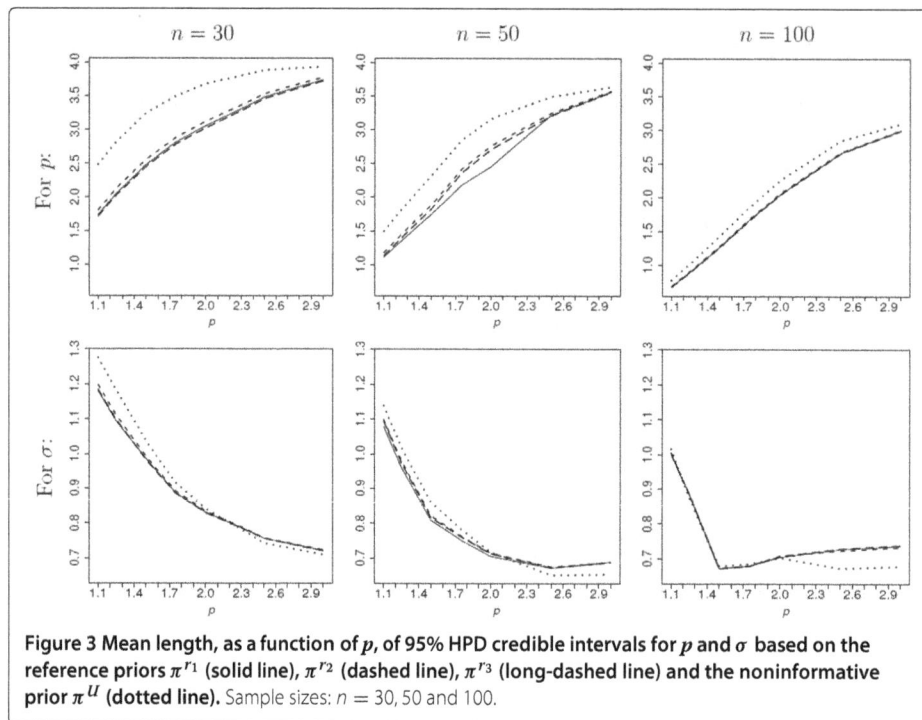

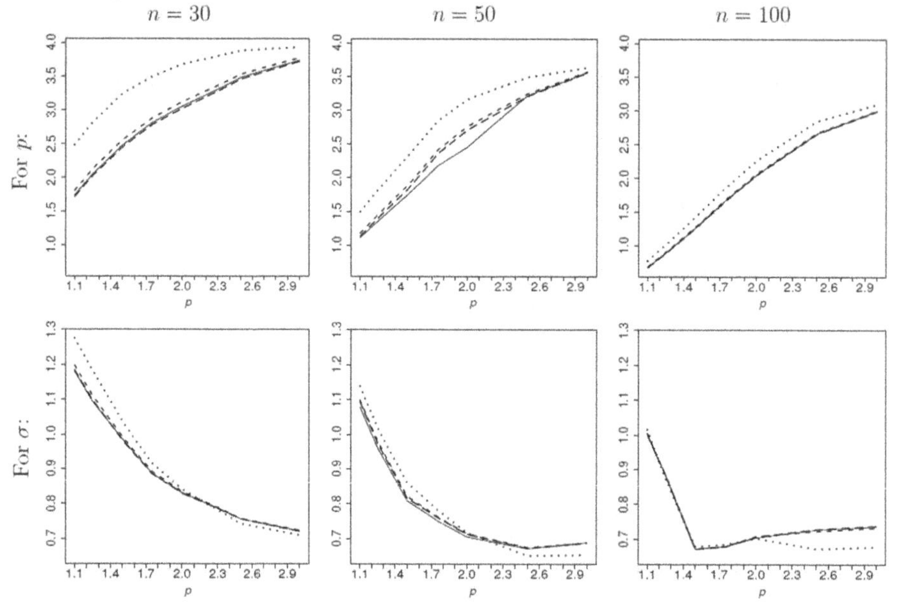

Figure 3 Mean length, as a function of p, of 95% HPD credible intervals for p and σ based on the reference priors π^{r_1} (solid line), π^{r_2} (dashed line), π^{r_3} (long-dashed line) and the noninformative prior π^{U} (dotted line). Sample sizes: $n = 30, 50$ and 100.

4.2 Applications

This section illustrates the use of the Bayesian reference analysis we propose for exponential power regression models with applications to two real world datasets. The first dataset illustrates leptokurtic errors and the second dataset illustrates platykurtic errors. Because the results based on the reference priors π^{r_1} and π^{r_3} are extremely similar, we show only the results for priors π^{r_1}, π^{r_2}, and π^{U}.

In both applications, we use the same truncation point at $p = 10$ used for $\pi^{U}(p)$ in Section 4.1 and assume $\pi^{U}(p) \propto 1$ for $1 < p < 10$ and $\pi^{U}(p) = 0$ otherwise. We have chosen the truncation point at $p = 10$ because datasets generated with $p = 10$ or with p close to 10 have similar statistical behavior. Hence, to distinguish whether a process follows an EP distribution with $p = 10$ or, say, $p = 10.1$ we would need an extremely large data set. Moreover, the choice of truncation should be made before the analyst looks at the data. For example, for the first application below, after looking at the scatterplot one may think about truncating the prior for values of p that correspond to leptokurtic distributions, that is, $1 < p < 2$. However, doing that would mean to use the data twice in the Bayes Theorem formula: once through the prior, and another time through the likelihood. Usually, such double use of the data leads to underestimation of the uncertainty. Therefore, we prefer to decide the truncation of the prior before looking at the data.

4.2.1 School spending

We analyze the relationship between per capita spending in public schools and per capita income by state in the United States. This dataset has been previously analyzed by Greene (1997), Cribari-Neto et al. (2000), and Fonseca et al. (2008). Specifically, Greene (1997) and Cribari-Neto et al. (2000) proposed analyses based on heuristic approaches to the so-called problem of heterocedasticity-of-unknown-form. In contrast, Fonseca et al. (2008)

have analyzed this dataset in the context of linear regression models with Student-t errors. Fonseca et al. (2008) found that when errors with distributions with heavy tails are assumed, a linear model is superior to a quadratic model. Here, we take a similar approach as that of Fonseca et al. (2008) in that we assume a linear model with errors that may have a heavy tail distribution. However, we assume that the errors follow an exponential power distribution.

Table 1 presents the posterior summaries based on the priors $\pi^U, \pi^{r_1}, \pi^{r_2}$. For all three priors both the posterior mode and the posterior median for p are smaller than one. In addition, both π^{r_1}- and π^{r_2}-based 95% credible intervals for p are contained in the interval $(1, 2)$ indicating evidence that the errors are leptokurtic. In contrast, the π^U-based 95% credible interval for p is not fully contained in the interval $(1, 2)$. However, from the results in Section 4.1 we know that for small true values of p, the use of the π^U prior leads to on average wider credible intervals for p that have lower coverage than nominal. Thus, this application provides an example when the superiority of the π^{r_1} and π^{r_2} priors matters to the conclusion that in this data set the errors distribution is leptokurtic.

Figure 4(a) shows the scatterplot for the school spending data set along with the fitted EP regression model based on π^{r_1} (solid line), π^{r_2} (dashed line), and π^U (dotted line). Figure 4(a) also shows the fitted Gaussian linear model (dot-dashed line). While the Gaussian model fit is clearly and strongly influenced by the outlier, the use of exponential power errors (with the four priors considered here) automatically makes the analysis robust against outliers. In particular, the model fits using the π^{r_1} and π^{r_2} priors (considering the posterior median) coincide and are equal to $\widehat{y} = -88.38 + 600.9x$. Another way to make the analysis robust against outliers is to use Student-t errors. Assuming Student-t errors, a model fitted by Fonseca et al. (2008) was $y = -75.3 + 583.2x$. We can see that both Student-t and exponential power errors fits are robust against outliers. However, the Student-t distribution cannot accommodate platykurtic errors and, therefore, the exponential power distribution provides more flexibility.

Figure 4(b) presents the marginal posterior densities for p based on π^{r_1} (solid line), π^{r_2} (dashed line), π^{r_3} (long-dashed line) and π^U (dotted line). In addition, the vertical lines indicate the limits of the 95% HPD credible intervals. The three reference priors lead to similar posterior densities for p, while the π^U prior leads to a substantially different posterior density for p. Figure 4(b) illustrates why the π^U leads to unnecessarily wider credible intervals. That combined with π^U-based credible intervals having coverage lower than nominal leads us to prefer the data analysis based on the reference priors.

4.2.2 Sold home videos vs. profits at the box office

We analyze a dataset about the relationship between the number of sold home videos in thousands (videos: y) and the profits at the box office in million of dollars (gross: x). This dataset has been previously analyzed by Levine et al. (2006) and Salazar et al. (2012) and comprises observations on 30 movies. A scatterplot of the variables of interest is shown in Figure 5(a). Using a linear model with EP errors and the independence Jeffreys prior π^{I_2} given in Equation (6), Salazar et al. (2012) found evidence of a platykurtic distribution for the errors. Here we compare three analyses of this home videos dataset with an EP linear regression model obtained by applying the reference priors π^{r_1} and π^{r_2}, and the noninformative prior π^U.

Table 1 School spending data set: Posterior summaries based on the noninformative prior π^{U} and the reference priors π^{r_1} and π^{r_2}

	π^{U}			π^{r_1}			π^{r_2}		
	Mode	**Median**	**95% C.I.**	**Mode**	**Median**	**95% C.I.**	**Mode**	**Median**	**95% C.I.**
p	1.18	1.33	(1.02, 2.03)	1.06	1.26	(1.00, 1.91)	1.08	1.27	(1.00, 1.92)
σ	52.73	54.75	(38.59, 74.44)	51.21	53.23	(38.08, 73.43)	51.72	53.23	(38.08, 73.43)
β_1	-89.37	-92.51	(-131.79, -37.85)	-91.51	-88.38	(-131.81, -36.86)	-91.51	-88.38	(-131.81, -36.86)
β_2	616.07	603.95	(525.16, 667.59)	609.99	600.90	(525.15, 667.57)	609.99	600.90	(525.15, 667.57)

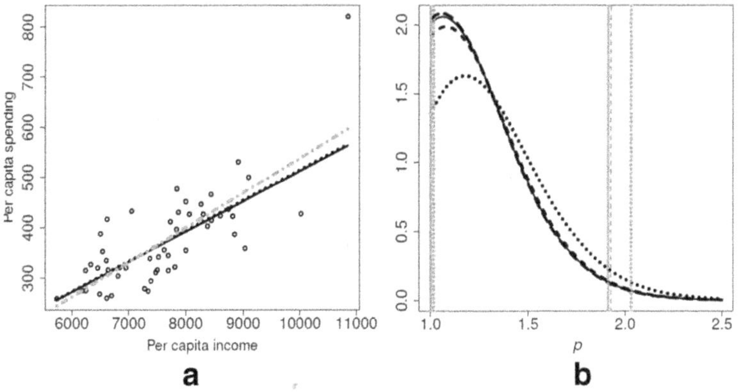

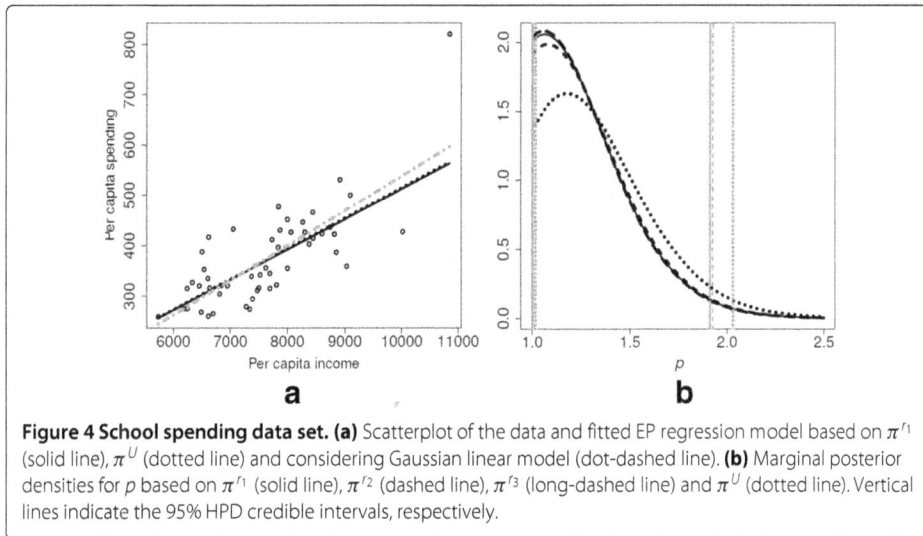

Figure 4 School spending data set. (a) Scatterplot of the data and fitted EP regression model based on π^{r_1} (solid line), π^U (dotted line) and considering Gaussian linear model (dot-dashed line). **(b)** Marginal posterior densities for p based on π^{r_1} (solid line), π^{r_2} (dashed line), π^{r_3} (long-dashed line) and π^U (dotted line). Vertical lines indicate the 95% HPD credible intervals, respectively.

Figure 5(a) shows the model fit for each of the priors we consider. The fits based on the reference priors visually coincide, whereas the fit based on π^U is slightly different. This is confirmed by Table 2, that shows that the slopes for the three fits are similar and around 4.33, whereas the intercept for the π^U-based fit is about 4.5% larger than the intercept for the π^{r_1}- and π^{r_2}-based fits. Even more striking are the differences between the reference analyses and the π^U-based analysis for σ and p. For σ, both posterior medians based on π^{r_1} and π^{r_2} are very similar and equal to 67.37 and 68.38 respectively, while the posterior median based on π^U is 77.50. Moreover, the 95% credible intervals for σ based on π^{r_1} and π^{r_2} are very similar and equal to (38.08, 93.64) and (38.10, 93.64) respectively, while the interval based on π^U is substantially different and equal to (47.17, 98.69).

The reference analyses for p are also strikingly distinct from the π^U-based analysis for p. First, the posterior medians for p based on π^{r_1} and π^{r_2} coincide and are equal to 2.64 while the π^U-based posterior median differs tremendously and is equal to 4.36. Second, the 95% credible intervals for p based on π^{r_1} and π^{r_2} are similar and equal to (1.00, 7.01) and (1.00, 7.18) respectively, while the π^U-based interval for p differs tremendously from

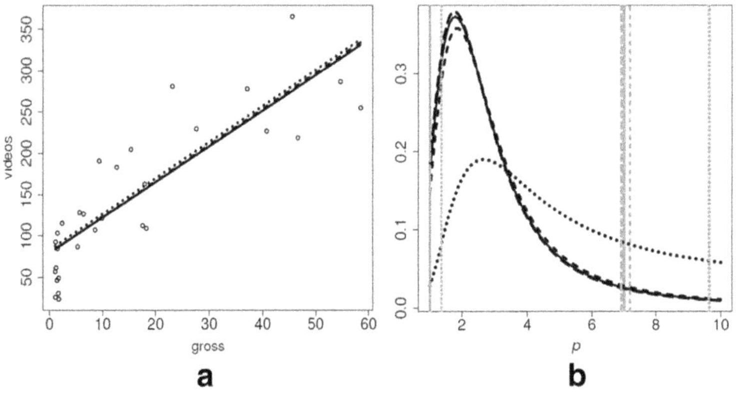

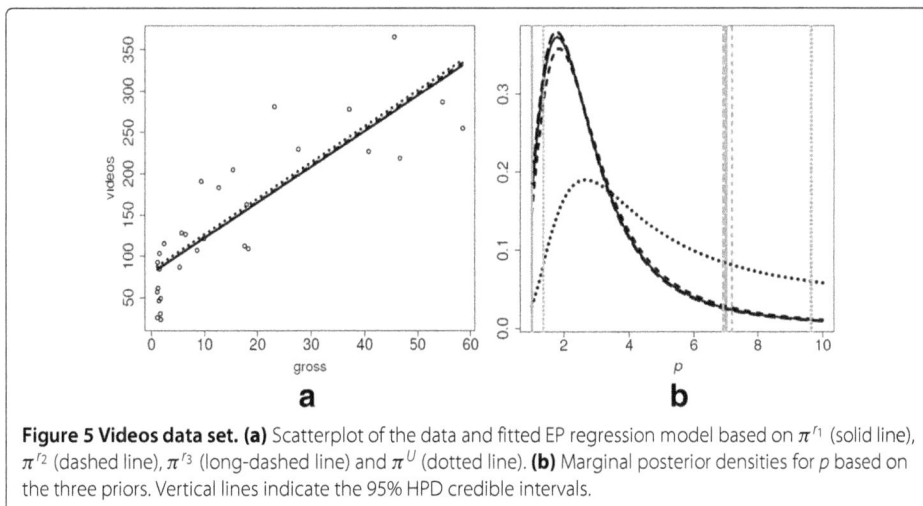

Figure 5 Videos data set. (a) Scatterplot of the data and fitted EP regression model based on π^{r_1} (solid line), π^{r_2} (dashed line), π^{r_3} (long-dashed line) and π^U (dotted line). **(b)** Marginal posterior densities for p based on the three priors. Vertical lines indicate the 95% HPD credible intervals.

Table 2 Videos data set: Posterior summaries based on the noninformative prior π^{U} and the reference priors π^{r_1} and π^{r_2}

	π^{U}			π^{r_1}			π^{r_2}		
	Mode	Median	95% C.I.	Mode	Median	95% C.I.	Mode	Median	95% C.I.
p	2.64	4.36	(1.36, 9.64)	1.82	2.64	(1.00, 7.01)	1.83	2.64	(1.00, 7.18)
σ	80.51	77.50	(47.17, 98.69)	67.37	67.37	(38.08, 93.64)	68.38	68.38	(38.10, 93.64)
β_1	83.11	83.11	(54.92, 107.65)	79.39	79.40	(53.03, 105.76)	79.53	79.53	(53.16, 104.98)
β_2	4.31	4.35	(3.42, 5.24)	4.32	4.32	(3.31, 5.33)	4.33	4.33	(3.32, 5.34)

the reference CIs and is equal to (1.36, 9.64). Hence, the π^{U}-based CI is more than 30% wider than the reference CIs. This undesirable feature of π^{U}-based CIs coincides with the results from the simulation study presented in Section 4.1.

Finally, Figure 5(b) presents the marginal posterior densities for p based on π^{r_1}, π^{r_2}, and π^{U}. This figure sheds light on the reason for the striking difference between the π^{r_1}- and π^{r_2}-based CIs and the π^{U}-based CI. The problem with the π^{U}-based analysis is that the right tail of the marginal posterior density for p decays too slowly. As a result, for the home video dataset the π^{U}-based CI depends dramatically on the right side truncation of the prior, which in this manuscript has been fixed at 10. Figure 5(b) makes it really clear that a larger truncation point would have a huge impact in the resulting π^{U}-based CI for p. This dataset clearly illustrates the superiority of the Bayesian reference analyses.

5 Conclusions

We have developed Bayesian reference analysis for linear models with exponential power errors. Specifically, we have developed three reference priors that lead to useful proper posterior distributions. In addition, we have shown through a simulation study that both priors yield procedures that have better frequentist properties than procedures resulting from a competing noninformative prior. Finally, we have illustrated our Bayesian reference analysis methodology with two real world applications that highlight the flexibility of the exponential power distribution to accommodate both cases when there are outliers in the dataset and also cases when the errors follow a platykurtic distribution.

The fact that the reference priors we have obtained for the EP regression model lead to proper posterior distributions is of substantial theoretical interest. The propriety of these reference posterior distributions contrasts with the impropriety of the posterior distribution associated with the Jeffreys-rule prior found by Salazar et al. (2012). Moreover, Salazar et al. (2012) found two independence Jeffreys priors, one of which leads to an improper posterior distribution whereas the other leads to a proper posterior distribution. We have found that the independence Jeffreys prior that yields a proper posterior distribution coincides with our reference prior π^{r_1}. Further, the independence Jeffreys prior that yields a useless improper posterior distribution differs only by a factor of $p^{-1/2}$ from the reference prior π^{r_2}. However, this difference is enough to make our reference prior π^{r_2} yield a useful proper posterior distribution.

Our results motivate many possible directions for future research. First, an open question is whether there exist general conditions under which reference priors yield proper posterior distributions. In addition, the existence of general conditions for posterior propriety may be investigated for Jeffreys-rule and independence Jeffreys priors. The search of general conditions for posterior propriety may benefit from our present work on EP regression and previous literature on examples of impropriety of posterior distributions for distinct objective Bayes priors (Berger et al. 2001; Ferreira and De Oliveira 2007; Salazar et al. 2012; Wasserman 2000).

We have considered the frequentist properties of the proposed Bayesian approaches via a simulation study. In particular, we have shown that credible intervals based on π^{r_1}, π^{r_2}, and π^{r_3} have similar frequentist properties with coverage close to nominal for p and σ. This is a reflection of the fact that for any prior satisfying some regularity conditions the frequentist coverage of credible intervals and the nominal level agree up to $O(n^{-1/2})$ (for a discussion and conditions, see Ghosh et al. 2006). A prior that leads to a more stringent agreement of order $O(n^{-1})$ is called a first-order probability matching prior. Such priors have to be derived with a specific parameter of interest in mind, and their derivation is far from trivial. Therefore, promising directions for future research for the EP regression model would be the derivation of priors that lead to Bayesian predictions that have approximate frequentist validity (Datta et al. 2000b) and the derivation of first-order probability matching priors (Datta and Ghosh 1995; Datta et al. 2000a).

Appendix

Proof of Theorem 1. To prove Theorem 1, we follow the methodology to obtain reference priors proposed by Berger and Bernardo (1992a). In particular, we assume that the reader is familiar with both the notation and the methodology of Berger and Bernardo (1992a). This proof is divided in two parts. In the first part, we obtain the reference prior for the orderings (β,σ,p), (σ,β,p), and (σ,p,β). Because the proofs are analogous for each of these three orderings, in the first part we obtain the reference prior for the ordering (σ,β,p). In the second part, we obtain the reference prior for the orderings (β,p,σ), (p,β,σ), and (p,σ,β). Because the proofs are analogous for each of these three orderings, in the second part we obtain the reference prior for the ordering (p,β,σ).

Part 1. Consider the ordering $\theta = (\sigma,\beta,p)$.

After rearranging the Fisher information matrix $H(\theta)$ given in Equation (8) to conform to this ordering, the inverse of the Fisher information matrix becomes

$$S(\theta)=H^{-1}(\theta)=\begin{bmatrix} \frac{\sigma^2}{np}\frac{(1+p^{-1})\Psi'(1+p^{-1})}{(1+p^{-1})\Psi'(1+p^{-1})-1} & 0 & \frac{\sigma p}{n}\frac{1}{(1+p^{-1})\Psi'(1+p^{-1})-1} \\ 0 & \sigma^2\left\{\Gamma\left(p^{-1}\right)\Gamma\left(2-p^{-1}\right)\sum_{i=1}^{n} x_i x_i'\right\}^{-1} & 0 \\ \frac{\sigma p}{n}\frac{1}{(1+p^{-1})\Psi'(1+p^{-1})-1} & 0 & \frac{p^3}{n}\frac{1}{(1+p^{-1})\Psi'(1+p^{-1})-1} \end{bmatrix}.$$

Thus,

$$S_1 = \frac{\sigma^2}{np}\frac{\left(1+p^{-1}\right)\Psi'\left(1+p^{-1}\right)}{\left(1+p^{-1}\right)\Psi'\left(1+p^{-1}\right)-1},$$

$$S_2 = \begin{bmatrix} \frac{\sigma^2}{np}\frac{(1+p^{-1})\Psi'(1+p^{-1})}{(1+p^{-1})\Psi'(1+p^{-1})-1} & 0 \\ 0 & \sigma^2\left\{\Gamma\left(p^{-1}\right)\Gamma\left(2-p^{-1}\right)\sum_{i=1}^{n} x_i x_i'\right\}^{-1} \end{bmatrix},$$

and $S_3 = S(\theta)$. Moreover, let $H_j = S_j^{-1}$. Thus,

$$H_1 = \frac{np}{\sigma^2}\frac{\left(1+p^{-1}\right)\Psi'\left(1+p^{-1}\right)-1}{\left(1+p^{-1}\right)\Psi'\left(1+p^{-1}\right)},$$

$$H_2 = \begin{bmatrix} \frac{np}{\sigma^2}\frac{(1+p^{-1})\Psi'(1+p^{-1})-1}{(1+p^{-1})\Psi'(1+p^{-1})} & 0 \\ 0 & \sigma^{-2}\Gamma\left(p^{-1}\right)\Gamma\left(2-p^{-1}\right)\sum_{i=1}^{n} x_i x_i' \end{bmatrix},$$

and $H_3 = H(\theta)$.

Let h_j be the $n_j \times n_j$ lower right corner of H_j. Thus,

$$h_1 = \frac{np}{\sigma^2}\frac{\left(1+p^{-1}\right)\Psi'\left(1+p^{-1}\right)-1}{\left(1+p^{-1}\right)\Psi'\left(1+p^{-1}\right)},$$

$$h_2 = \sigma^{-2}\Gamma\left(p^{-1}\right)\Gamma\left(2-p^{-1}\right)\sum_{i=1}^{n} x_i x_i', \quad \text{and}$$

$$h_3 = np^{-3}\left(1+p^{-1}\right)\Psi'\left(1+p^{-1}\right).$$

Let $\theta_{(1)} = \sigma$, $\theta_{(2)} = \beta$, and $\theta_{(3)} = p$. In addition, let $\theta_{[1]} = \theta_{(1)} = \sigma$, $\theta_{[2]} = (\theta_{(1)}, \theta_{(2)}) = (\sigma, \beta)$, and $\theta_{[3]} = (\theta_{(1)}, \theta_{(2)}, \theta_{(3)}) = (\sigma, \beta, p)$. Moreover, let $\theta_{[\sim 1]} = (\theta_{(2)}, \theta_{(3)}) = (\beta, p)$ and $\theta_{[\sim 2]} = (\theta_{(3)}) = p$. Further, consider the following compact sets: for σ, $\Theta^l_{(1)} = [l^{-1}, l]$; for β, $\Theta^l_{(2)} = [-l, l]^k$; for p, $\Theta^l_{(3)} = [1, l]$.

Then,

$$\begin{aligned}
\pi^l_3(p \mid \sigma, \beta) &= \pi^l_3\left(\theta_{[\sim 2]} \mid \theta_{[2]}\right) \\
&= \frac{|h_3(\theta)|^{1/2} \mathbf{1}_{\Theta^l_{(3)}}\left(\theta_{(3)}\right)}{\int_{\Theta^l_{(3)}} |h_3(\theta)|^{1/2} d\theta_{(3)}} \\
&= \frac{\left\{np^{-3}\left(1+p^{-1}\right)\Psi'\left(1+p^{-1}\right)\right\}^{1/2} \mathbf{1}_{[1,l]}(p)}{\int_1^l \left\{np^{-3}\left(1+p^{-1}\right)\Psi'\left(1+p^{-1}\right)\right\}^{1/2} dp} \\
&= \{c_1(l)\}^{-1} p^{-3/2}\left(1+p^{-1}\right)^{1/2}\left\{\Psi'\left(1+p^{-1}\right)\right\}^{1/2} \mathbf{1}_{[1,l]}(p),
\end{aligned}$$

where $c_1(l) = \int_1^l p^{-3/2}(1+p^{-1})^{1/2}\left\{\Psi'(1+p^{-1})\right\}^{1/2} dp$.

Now,

$$\begin{aligned}
\pi^l_2(\beta, p \mid \sigma) &= \pi^l_2\left(\theta_{[\sim 1]} \mid \theta_{[1]}\right) \\
&= \frac{\pi^l_3\left(\theta_{[\sim 2]} \mid \theta_{[2]}\right) \exp\left\{0.5E^l_2\left[\log|h_2(\theta)| \middle| \theta_{[2]}\right]\right\} \mathbf{1}_{\Theta^l_{(2)}}\left(\theta_{(2)}\right)}{\int_{\Theta^l_{(2)}} \exp\left\{0.5E^l_2\left[\log|h_2(\theta)| \middle| \theta_{[2]}\right]\right\} d\theta_{(2)}},
\end{aligned}$$

where

$$\begin{aligned}
E^l_2\left[\log|h_2(\theta)| \middle| \theta_{[2]}\right] &= \int_{\Theta^l_{(3)}} \log|h_2(\theta)| \pi^l_3(\theta_{[\sim 2]} \mid \theta_{[2]}) d\theta_{[\sim 2]} \\
&= \int_1^l \left\{-2k\log\sigma + k\log\Gamma\left(p^{-1}\right) + k\log\Gamma\left(2-p^{-1}\right) + \log\left|\sum_{i=1}^n x_i x_i'\right|\right\} \\
&\qquad \{c_1(l)\}^{-1} p^{-3/2}\left(1+p^{-1}\right)^{1/2}\left\{\Psi'\left(1+p^{-1}\right)\right\}^{1/2} dp \\
&= -2k\log\sigma + c_2(l),
\end{aligned}$$

with

$$\begin{aligned}
c_2(l) = \{c_1(l)\}^{-1} \int_1^l &\left\{k\log\Gamma\left(p^{-1}\right) + k\log\Gamma\left(2-p^{-1}\right) + \log\left|\sum_{i=1}^n x_i x_i'\right|\right\} \\
&p^{-3/2}\left(1+p^{-1}\right)^{1/2}\left\{\Psi'\left(1+p^{-1}\right)\right\}^{1/2} dp.
\end{aligned}$$

Hence,

$$\begin{aligned}
\pi^l_2(\beta, p \mid \sigma) &= \frac{\pi^l_3(p|\sigma,\beta) \exp\left\{0.5[-2k\log\sigma + c_2(l)]\right\} \mathbf{1}_{[-l,l]^k}(\beta)}{\int_{[-l,l]^k} \exp\left\{0.5[-2k\log\sigma + c_2(l)]\right\} d\beta} \\
&= \pi^l_3(p|\sigma,\beta)(2l)^{-k} \mathbf{1}_{[-l,l]^k}(\beta).
\end{aligned}$$

Finally,

$$\begin{aligned}
\pi^l_1(\sigma, \beta, p) &= \pi^l_1\left(\theta_{[\sim 0]} \mid \theta_{[0]}\right) \\
&= \frac{\pi^l_2\left(\theta_{[\sim 1]} \mid \theta_{[1]}\right) \exp\left\{0.5E^l_1\left[\log|h_1(\theta)| \middle| \theta_{[1]}\right]\right\} \mathbf{1}_{\Theta^l_{(1)}}\left(\theta_{(1)}\right)}{\int_{\Theta^l_{(1)}} \exp\left\{0.5E^l_1\left[\log|h_1(\theta)| \middle| \theta_{[1]}\right]\right\} d\theta_{(1)}},
\end{aligned}$$

with

$$E_1^l\left[\log|h_1(\theta)||\theta_{[1]}\right] = \int_{[-l,l]^k}\int_1^l \log\left\{\frac{np}{\sigma^2}\frac{\left(1+p^{-1}\right)\Psi'\left(1+p^{-1}\right)-1}{\left(1+p^{-1}\right)\Psi'\left(1+p^{-1}\right)}\right\}\pi_2^l(\beta,p\mid\sigma)dpd\beta$$
$$= -2\log\sigma + c_3(l),$$

where

$$c_3(l) = \int_{[-l,l]^k}\int_1^l \log\left\{np\frac{\left(1+p^{-1}\right)\Psi'\left(1+p^{-1}\right)-1}{\left(1+p^{-1}\right)\Psi'\left(1+p^{-1}\right)}\right\}\pi_2^l(\beta,p\mid\sigma)dpd\beta$$

does not depend on $\theta = (\sigma, \beta, p)$.

Hence,

$$\pi_1^l(\sigma,\beta,p) = \frac{\pi_2^l(\beta,p|\sigma)\exp\{0.5[2\log\sigma + c_3(l)]\}\mathbf{1}_{(l^{-1},l)}(\sigma)}{\int_{l^{-1}}^l \exp\{0.5[2\log\sigma + c_3(l)]\}d\sigma}$$
$$= \frac{\pi_2^l(\beta,p|\sigma)\sigma^{-1}\mathbf{1}_{(l^{-1},l)}(\sigma)}{2\log l}$$

Thus,

$$\pi_1^l(\sigma,\beta,p) = \sigma^{-1}p^{-3/2}\left(1+p^{-1}\right)^{1/2}\left\{\Psi'\left(1+p^{-1}\right)\right\}^{1/2}$$
$$\times\ \{c_1(l)\}^{-1}(2l)^{-k}(2\log l)^{-1}\mathbf{1}_{[l^{-1},l]}(\sigma)\mathbf{1}_{[-l,l]^k}(\beta)\mathbf{1}_{[1,l]}(p).$$

Now take any point $\theta^* = (\sigma^*, \beta^*, p^*) \in [l^{-1}, l] \times [-l, l]^k \times [1, l]$. Then, the reference prior for the ordering (σ, β, p) is

$$\pi(\sigma,\beta,p) \propto \lim_{l\to\infty}\frac{\pi_1^l(\sigma,\beta,p)}{\pi_1^l(\sigma^*,\beta^*,p^*)}$$
$$= \sigma^{-1}p^{-3/2}\left(1+p^{-1}\right)^{1/2}\left\{\Psi'\left(1+p^{-1}\right)\right\}^{1/2},$$

which is of the form (4).

Part 2. Consider the ordering $\boldsymbol{\theta} = (p, \beta, \sigma)$.

After rearranging the Fisher information matrix $H(\theta)$ given in Equation (8) to conform to this ordering, the inverse of the Fisher information matrix becomes

$$S(\theta)=H^{-1}(\theta)=\begin{bmatrix}\frac{p^3}{n}\frac{1}{(1+p^{-1})\Psi'(1+p^{-1})-1} & 0 & \frac{\sigma p}{n}\frac{1}{(1+p^{-1})\Psi'(1+p^{-1})-1}\\ 0 & \sigma^2\left\{\Gamma\left(p^{-1}\right)\Gamma\left(2-p^{-1}\right)\sum_{i=1}^n x_ix_i'\right\}^{-1} & 0\\ \frac{\sigma p}{n}\frac{1}{(1+p^{-1})\Psi'(1+p^{-1})-1} & 0 & \frac{\sigma^2}{np}\frac{(1+p^{-1})\Psi'(1+p^{-1})}{(1+p^{-1})\Psi'(1+p^{-1})-1}\end{bmatrix},$$

Thus,

$$S_1 = \frac{p^3}{n}\frac{1}{\left(1+p^{-1}\right)\Psi'\left(1+p^{-1}\right)-1},$$

$$S_2 = \begin{bmatrix}\frac{p^3}{n}\frac{1}{(1+p^{-1})\Psi'(1+p^{-1})-1} & 0\\ 0 & \sigma^2\left\{\Gamma(p^{-1})\Gamma\left(2-p^{-1}\right)\sum_{i=1}^n x_ix_i'\right\}^{-1}\end{bmatrix},$$

and $S_3 = S(\theta)$. Moreover, let $H_j = S_j^{-1}$. Thus,

$$H_1 = np^{-3}\{(1+p^{-1})\Psi'(1+p^{-1})-1\},$$

$$H_2 = \begin{bmatrix}np^{-3}\{\left(1+p^{-1}\right)\Psi'\left(1+p^{-1}\right)-1\} & 0\\ 0 & \sigma^{-2}\Gamma\left(p^{-1}\right)\Gamma\left(2-p^{-1}\right)\sum_{i=1}^n x_ix_i'\end{bmatrix},$$

and $H_3 = H(\theta)$.

Let h_j be the $n_j \times n_j$ lower right corner of H_j. Thus,

$$h_1 = np^{-3}\{(1+p^{-1})\Psi'(1+p^{-1}) - 1\},\ h_2 = \sigma^{-2}\Gamma\left(p^{-1}\right)\Gamma\left(2-p^{-1}\right)\sum_{i=1}^{n} x_i x_i', \text{ and } h_3 = np\sigma^{-2}.$$

Let $\theta_{(1)} = p$, $\theta_{(2)} = \beta$, and $\theta_{(3)} = \sigma$. In addition, let $\theta_{[1]} = \theta_{(1)} = p$, $\theta_{[2]} = (\theta_{(1)}, \theta_{(2)}) = (p, \beta)$, and $\theta_{[3]} = (\theta_{(1)}, \theta_{(2)}, \theta_{(3)}) = (p, \beta, \sigma)$. Moreover, let $\theta_{[\sim 1]} = (\theta_{(2)}, \theta_{(3)}) = (\beta, \sigma)$ and $\theta_{[\sim 2]} = (\theta_{(3)}) = \sigma$. Further, consider the following compact sets: for p, $\Theta^l_{(1)} = [1, l]$; for β, $\Theta^l_{(2)} = [-l, l]^k$; for σ, $\Theta^l_{(3)} = [l^{-1}, l]$.

Then,

$$\begin{aligned}
\pi_3^l(\sigma \mid p, \beta) &= \pi_3^l\left(\theta_{[\sim 2]} \mid \theta_{[2]}\right) \\
&= \frac{|h_3(\theta)|^{1/2}\mathbf{1}_{\Theta^l_{(3)}}\left(\theta_{(3)}\right)}{\int_{\Theta^l_{(3)}} |h_3(\theta)|^{1/2} d\theta_{(3)}} \\
&= \frac{\left\{np\sigma^{-2}\right\}^{1/2}\mathbf{1}_{[l^{-1},l]}(\sigma)}{\int_{l^{-1}}^{l}\left\{np\sigma^{-2}\right\}^{1/2} d\sigma} \\
&= \sigma^{-1}(2\log l)^{-1}\mathbf{1}_{[l^{-1},l]}(\sigma).
\end{aligned}$$

Moreover,

$$\begin{aligned}
\pi_2^l(\beta, p \mid \sigma) &= \pi_2^l(\theta_{[\sim 1]} \mid \theta_{[1]}) \\
&= \frac{\pi_3^l(\theta_{[\sim 2]} \mid \theta_{[2]}) \exp\left\{0.5E_2^l\left[\log|h_2(\theta)|\middle|\theta_{[2]}\right]\right\}\mathbf{1}_{\Theta^l_{(2)}}(\theta_{(2)})}{\int_{\Theta^l_{(2)}} \exp\left\{0.5E_2^l\left[\log|h_2(\theta)|\middle|\theta_{[2]}\right]\right\} d\theta_{(2)}},
\end{aligned}$$

where

$$\begin{aligned}
E_2^l\left[\log|h_2(\theta)|\middle|\theta_{[2]}\right] &= \int_{\Theta^l_{(3)}} \log|h_2(\theta)|\pi_3^l(\theta_{[\sim 2]} \mid \theta_{[2]}) d\theta_{[\sim 2]} \\
&= \int_{l^{-1}}^{l}\left\{-2k\log\sigma + k\log\Gamma(p^{-1}) + k\log\Gamma(2-p^{-1}) + \log\left|\sum_{i=1}^{n} x_i x_i'\right|\right\} \\
&\qquad \sigma^{-1}(2\log l)^{-1} d\sigma \\
&= c_1(l, p), \text{ which does not depend on } \beta.
\end{aligned}$$

Hence,

$$\begin{aligned}
\pi_2^l(\beta, \sigma \mid p) &= \frac{\pi_3^l(\sigma|p, \beta)\exp\left\{0.5c_1(l,p)\right\}\mathbf{1}_{[-l,l]^k}(\beta)}{\int_{[-l,l]^k}\exp\left\{0.5c_1(l,p)\right\} d\beta} \\
&= \pi_3^l(\sigma|p,\beta)(2l)^{-k}\mathbf{1}_{[-l,l]^k}(\beta).
\end{aligned}$$

Further,

$$\begin{aligned}
\pi_1^l(p, \beta, \sigma) &= \pi_1^l(\theta_{[\sim 0]} \mid \theta_{[0]}) \\
&= \frac{\pi_2^l(\theta_{[\sim 1]} \mid \theta_{[1]}) \exp\left\{0.5E_1^l\left[\log|h_1(\theta)|\middle|\theta_{[1]}\right]\right\}\mathbf{1}_{\Theta^l_{(1)}}(\theta_{(1)})}{\int_{\Theta^l_{(1)}} \exp\left\{0.5E_1^l\left[\log|h_1(\theta)|\middle|\theta_{[1]}\right]\right\} d\theta_{(1)}},
\end{aligned}$$

with

$$\begin{aligned}
E_1^l\left[\log|h_1(\theta)|\middle|\theta_{[1]}\right] &= \int_{[-l,l]^k}\int_{l^{-1}}^{l} \log\left[np^{-3}\{(1+p^{-1})\Psi'(1+p^{-1}) - 1\}\right]\pi_2^l(\beta, \sigma \mid p) d\sigma d\beta \\
&= \log\left[np^{-3}\{(1+p^{-1})\Psi'(1+p^{-1}) - 1\}\right].
\end{aligned}$$

Hence,

$$\pi_1^l(\sigma,\beta,p) = \frac{\pi_2^l(\beta,\sigma|p)\exp\{0.5\log\left[np^{-3}\{(1+p^{-1})\Psi'(1+p^{-1})-1\}\right]\}\mathbf{1}_{(1,l)}(p)}{\int_1^l \exp\{0.5\log\left[np^{-3}\{(1+p^{-1})\Psi'(1+p^{-1})-1\}\right]\}dp}$$
$$= \pi_2^l(\beta,p|\sigma)p^{-3/2}\{(1+p^{-1})\Psi'(1+p^{-1})-1\}^{1/2}c_2(l)\mathbf{1}_{[1,l]}(p),$$

where

$$\{c_2(l)\}^{-1} = \int_1^l \exp\{0.5\log\left[np^{-3}\{(1+p^{-1})\Psi'(1+p^{-1})-1\}\right]\}dp.$$

Thus,

$$\pi_1^l(p,\beta,\sigma) = \sigma^{-1}p^{-3/2}\{(1+p^{-1})\Psi'(1+p^{-1})-1\}^{1/2}c_2(l)(2l)^{-k}(2\log l)^{-1}\mathbf{1}_{[1,l]}(p)\mathbf{1}_{[-l,l]^k}(\beta)\mathbf{1}_{[l^{-1},l]}(\sigma).$$

Now take any point $\theta^* = (p^*,\beta^*,\sigma^*) \in [1,l] \times [-l,l]^k \times [l^{-1},l]$. Then, the reference prior for the ordering (p,β,σ) is

$$\pi(p,\beta,\sigma) \propto \lim_{l\to\infty} \frac{\pi_1^l(p,\beta,\sigma)}{\pi_1^l(p^*,\beta^*,\sigma^*)}$$
$$= \sigma^{-1}p^{-3/2}\{(1+p^{-1})\Psi'(1+p^{-1})-1\}^{1/2},$$

which is of the form (4).

Competing interests

The authors declare that they have no competing interests.

Authors' contributions

MARF proved Theorem 1, Lemma1, and Proposition 1, and wrote the manuscript. ES performed the computations for the simulation study and for the application, and wrote the manuscript. Both authors read and approved the final manuscript.

Acknowledgement

The work of Ferreira was supported in part by National Science Foundation Grant DMS-0907064. The authors gratefully acknowledge the constructive comments and suggestions made by three anonymous referees that led to a substantially improved article.

Author details

[1]Department of Statistics, University of Missouri, Columbia, USA. [2]Department of Electrical and Computer Engineering, Duke University, Durham, USA.

References

Achcar, JA, Pereira, GA: Use of exponential power distributions for mixture models in the presence of covariates. J. Appl. Stat. **26**(6), 669–679 (1999)

Berger, JO, Bernardo, JM: On the development of the reference prior method. In: Bernardo, JM, Berger, JO, Dawid, AP, Smith, AFM (eds.) Bayesian Statistics 4, pp. 35–60. Oxford University Press, London, (1992a)

Berger, JO, Bernardo, JM: Ordered group reference priors with applications to a multinomial problem. Biometrika **79**, 25–37 (1992b)

Berger, JO, Bernardo, JM: Reference priors in a variance components problem. In: Goel, PK, Iyengar, NS (eds.) *Bayesian Analysis in Statistics and Econometrics*, pp. 323–340. Springer, Berlin, (1992c)

Berger, JO, de Oliveira, V, Sansó, B: Objective Bayesian analysis of spatially correlated data. J. Am. Stat. Assoc. **96**(456), 1361–1374 (2001)

Bernardo, JM: Reference posterior distribution for Bayes inference. J. Roy. Stat. Soc. B. **41**, 113–147 (1979)

Bernardo, JM, Smith, AFM: Bayesian Theory. Wiley, New York (1994)

Box, GEP, Tiao, GC: A further look at robustness via Bayes's theorem. Biometrika. **49**, 419–432 (1962)

Box, GEP, Tiao, GC: Bayesian Inference in Statistical Analysis. Wiley-Interscience, Hoboken (1992)

Choy, STB, Smith, AFM: On robust analysis of a normal location parameter. J. Roy. Stat. Soc. B. **59**(2), 463–474 (1997)

Cribari-Neto, F, Ferrari, SLP, Cordeiro, GM: Improved heteroscedasticity-consistent covariance matrix estimators. Biometrika. **87**, 907–918 (2000)

Datta, GS, Ghosh, JK: Noninformative priors for maximal invariant parameter in group models. Test. **4**, 95–114 (1995)

Datta, GS, Ghosh, M, Mukerjee, R: Some new results on probability matching priors. Bull. Calcutta Stat. Assoc. **50**(199–200), 179–192 (2000a)

Datta, GS, Mukerjee, R, Ghosh, M, Sweeting, TJ: Bayesian prediction with approximate frequentist validity. Ann. Stat. **28**, 1414–1426 (2000b)

Ferreira, MAR, De Oliveira, V: Bayesian reference analysis for Gaussian Markov Random Fields. J. Multivariate Anal. **98**, 789–812 (2007)

Ferreira, MAR, Suchard, MA: Bayesian analysis of elapsed times in continuous-time Markov chains. Can. J. Stat. **36**, 355–368 (2008)

Fonseca, TCO, Ferreira, MAR, Migon, HS: Objective Bayesian analysis for the Student-t regression model. Biometrika. **95**(2), 325–333 (2008)

Greene, WH: Econometric Analysis. Prentice-Hall, Upper Saddle River (1997)

Ghosh, JK, Delampady, M, Samanta, T: An Introduction to Bayesian Statistics – Theory and Methods. Springer, New York (2006)

Jeffreys, H: Theory of Probability. Oxford University Press, Oxford (1961)

Levine, DM, Krehbiel, TC, Berenson, ML: Business Statistics: A First Course. Pearson Prentice Hall, Upper Saddle River (2006)

Liang, F, Liu, C, Wang, N: A robust sequential Bayesian method for identification of differentially expressed genes. Statistica Sinica. **17**, 571–597 (2007)

Salazar, E, Ferreira, MAR, Migon, HS: Objective Bayesian analysis for exponential power regression models. Sankhya - Series B. **74**, 107–125 (2012)

Sun, D, Berger, JO: Objective Bayesian analysis for the multivariate normal model. In: Bernardo, JM, Bayarri, MJ, Berger, JO, Dawid, AP, Heckerman, D, Smith, AFM, West, M (eds.) Bayesian Statistics 8, pp. 525–547. Oxford University Press, Oxford, (2007)

Vianelli, S: La misura della variabilità condizionata in uno schema generale delle curve normali di frequenza. Statistica. **23**, 447–474 (1963)

Walker, SG, Gutiérrez-Peña, E: Robustifying Bayesian procedures. In: Bayesian Statistics 6. Oxford University Press, New York, (1999)

Wasserman, L: Asymptotic inference for mixture models using data-dependent priors. J. Roy. Stat. Soc. B. **62**, 159–180 (2000)

West, M: Outlier models and prior distributions in Bayesian linear regression. J. Roy. Stat. Soc. B. **46**, 431–439 (1984)

West, M: On scale mixtures of normal distributions. Biometrika. **79**, 646–648 (1987)

Zhu, D, Zinde-Walsh, V: Properties and estimation of asymmetric exponential power distribution. J. Econometrics. **148**, 86–99 (2009)

3

A closed-form universal trivariate pair-copula

Werner Hürlimann

Correspondence:
whurlimann@bluewin.ch
Feldstrasse 145, CH-8004 Zürich,
Switzerland

Abstract

Based on the trivariate pair-copula construction for the bivariate linear circular copula by Perlman and Wellner (Symmetry 3:574-99, 2011) and the Theorem of Carathéodory, which states that any valid correlation matrix is a finite convex combination of extreme correlation matrices, we generate a class of closed-form analytical 3-universal copulas. We derive explicit product and lifting copula formulas for the set of all extremal correlation matrices. Our analytical proof makes use of a novel set of conditional copula inequalities, which are of independent interest.

Keywords: Pair-copula; Trivariate universality; Linear circular copula; Elliptical copula; Extremal correlation; Product copula; Lifting copula; Conditional product copula inequalities
MSC 2010: 60E15; 62E15; 62H05; 62H20

1 Introduction

In practice, the joint normal transform method for dependence modelling, which consists of transforming the margins to normal distributions, induce a dependence structure and transform back, is rather popular (e.g. Kurowicka and Cooke (2006), Section 4.2). Applied to a random vector $(X_1, X_2, ..., X_n)$ with invertible continuous marginal distributions $(F_1, F_2, ..., F_n)$ this method follows three steps:

1) Specify the rank correlation matrix $r = (r_{ij}),\ 1 \le i, j \le n$, of this random vector, i.e. $r_{ij} = Cov[F_i(X_i), F_j(X_j)],\ 1 \le i, j \le n$.
2) A random sample $(Y_1, Y_2, ..., Y_n)$ is drawn from a joint normal distribution with standard normal margins and the specified rank correlation matrix $r = (r_{ij})$.
3) Let $\Phi(x)$ denote the standard normal distribution. The random sample $(X_1, X_2, ..., X_n)$ with margins $(F_1, F_2, ..., F_n)$ and the rank correlation matrix $r = (r_{ij})$ is obtained from the inverse probability transform through $(X_1, X_2, ..., X_n) = (F_1^{-1}(\Phi(Y_1)), F_2^{-1}(\Phi(Y_2)), ..., F_n^{-1}(\Phi(Y_n)))$.

From a theoretical point of view this procedure suffers from several drawbacks. Concerning (1), it is known that every 3-dimensional valid correlation matrix can be realized as a rank correlation matrix, i.e. there exists a trivariate uniform distribution with this rank correlation structure, the so-called *3-universality* property of copulas. This result is first stated in Joe (1997), Exercise 4.17, pp. 137–138. Kurowicka and Cooke (2006), Section 4.4.6, p.102, produce such trivariate copulas using the pair-copula construction for the so-called "elliptical copula" (see also Kurowicka and Cooke (2001)).

Another recent proof is Devroye and Letac (2010). However, it is not yet known whether copulas are n -universal for $n \geq 6$ (see Letac (2010) for $n < 6$). Therefore (1) is in general not necessarily consistent. The statement (2) causes similar difficulties. Indeed, the *compatibility* conditions under which a joint normal distribution with a specified rank correlation matrix exist, imply that for any $n \geq 3$ there are compatibility counterexamples (see Hürlimann (2012a), Theorem 3.1 and Corollary 4.2). Moreover, the probability of incompatibility increases with the dimension (e.g. Kurowicka and Cooke (2001)), Section 4, Table 1, Ghosh and Henderson (2002), and Ghosh (2004), Figure 3.1, p. 67). On the other hand, the 3-universality of the elliptical pair-copula construction does not extend to the fourvariate case (e.g. Kurowicka and Cooke (2006), Example 4.9). Therefore, to enable a rigorous use of any valid rank correlation matrix, there is a need for a more comprehensive understanding of the class of universal copulas, at least in small dimensions.

In the present paper we offer an analytical approach to the pair-copula construction for the bivariate *linear circular copula* derived by Perlman and Wellner (2011), whose density obviously coincides with the "elliptical copula" introduced by Kurowicka et al. (2000). To the best of our knowledge Section 5 presents for the first time simple closed-form analytical 3-universal copula formulas. The possible extension to the multivariate case is a challenging topic for future research (see Section 6). A more detailed account of the content follows.

Section 2 recalls the trivariate pair-copula construction. We explain how the combined use of extremal correlation matrices and bivariate elliptical copulas lead to extremal lifting copulas whose bivariate product copula margins induce sharp bounds in the concordance order. The latter include the set of all feasible rank correlation matrices. Advocating the Theorem of Carathéodory, which states that any valid correlation matrix can be written as a finite convex combination of extreme correlation matrices, the generation of 3-universal copulas follows. Section 3 summarizes the needed bivariate linear circular copula formulas. Section 4 presents the analytical closed-form product copula formulas, and Section 5 states our main result about analytical 3-universal copulas. Section 6 contains a simple two-dimensional algorithm for generating random vectors with 3-universal linear circular copula, and many references to potential applications are provided. Detailed proofs follow in the Appendices 1 and 2.

2 Trivariate linear circular pair-copula construction

In response to the difficulties encountered with the multivariate normal copula, the pair-copula construction of multivariate copulas has become more and more popular as can be seen from the recent review by Czado (2010). The first pair-copula construction is due to Joe (1996), see also Joe (1997), Section 4.5. His construction is given in terms of distribution functions, while Bedford and Cooke (2001) Bedford and Cooke (2002) expressed these constructions in terms of densities, and Kurowicka and Cooke (2006), Section 6.4, designed various sampling algorithms for them. A trivariate restatement of these equivalent representations will play a major role.

As observed by Devroye and Letac (2010), any valid rank correlation matrix can be written as a finite convex combination of extreme correlation matrices, also called extreme points (Theorem of Carathéodory (1911) and Steinitz (1914)). By linearity it suffices to restrict the attention to the pair-copula construction for extreme points. This way a finite algorithm that permits the construction of 3-universal copulas in terms of the linear circular copula can be designed. The proposed approach is then made fully analytical in Section 3 through the derivation of simple closed-form analytical 3-universal copula formulas for the extreme points.

Recall the structure of the extreme points for the set of all positive semi-definite 3×3 correlation matrices. They have necessarily rank 1 or 2 and take the form (Ycart(1985), Corollary, p. 611):

$$\rho = \rho(a,b,c) = \left(\rho_{ij}\right) = \begin{pmatrix} 1 & \cos c & \cos b \\ \cos c & 1 & \cos a \\ \cos b & \cos a & 1 \end{pmatrix}, \quad a+b+c \equiv 0 \mod 2\pi. \tag{2.1}$$

One notes that there are four extreme points of rank one, namely

$$(a,b,c) \in \{(0,0,0),(0,\pi,\pi),(\pi,0,\pi),(\pi,\pi,0)\} \tag{2.2}$$

and the rank two extreme points are characterized by the condition

$$(\sin a, \ \sin b, \ \sin c) \neq (0,0,0) \tag{2.3}$$

The following immediate consequence is crucial.

Lemma 2.1

The absolute value of the partial correlation of a 3×3 extreme correlation matrix of the form (2.1) is always one. More precisely, one has

$$\rho_{12;3} = \frac{\rho_{12}-\rho_{13}\rho_{23}}{\sqrt{(1-\rho_{13}^2)\cdot(1-\rho_{23}^2)}} = -\operatorname{sgn}(\sin a \cdot \sin b) = \pm 1 \tag{2.4}$$

Proof

Since $a+b+c \equiv 0 \mod 2\pi$ we have from the cosine addition law that $\cos c = \cos(-(a+b) = \cos(a+b) = \cos a \cdot \cos b - \sin a \cdot \sin b$. Inserted in the defining relation for the partial correlation coefficient we get $\rho_{12;3} = -\frac{\sin a \cdot \sin b}{|\sin a \cdot \sin b|}$, which implies (2.4). ◊

Now, in a first step, let us assume that all joint, marginal and conditional distributions are absolutely continuous, and that the corresponding densities exist. Let (X_1, X_2, X_3) be a trivariate random vector from the Fréchet space $F(F_{13}, F_{23})$ of all trivariate distributions with given bivariate margins $F_{13}(x_1, x_3), F_{23}(x_2, x_3)$. Denote the marginal distributions by $F_1(x_1), F_2(x_2), F_3(x_3)$, the conditional distributions obtained from F_{13}, F_{23} by $F_{1|3}(x_1|x_3), F_{2|3}(x_2|x_3)$, and the marginal densities by $f_1(x_1), f_2(x_2), f_3(x_3)$. Let $C_{13}(u_1, u_3), C_{23}(u_2, u_3)$, respectively $C_{12|3}(u_1, u_2)$, be copulas associated to the random vectors $(X_1, X_3), (X_2, X_3)$, respectively to the conditional random vector $(X_1, X_2|X_3)$, which necessarily exist by the Theorem of Sklar (1959). The mixture of conditional distributions defined by (e.g. Joe (1997), equation (4.37))

$$\begin{aligned} F_{123}(x_1,x_2,x_3) &= \int_{-\infty}^{x_3} C_{12|3}(F_{1|3}(x_1|z_3), F_{2|3}(x_2|z_3))dF_3(z_3) \\ &= \int_{-\infty}^{x_3} C_{12|3}\left(\frac{\partial C_{13}(F_1(x_1),F_3(z_3))}{\partial u_3}, \frac{\partial C_{23}(F_2(x_2),F_3(z_3))}{\partial u_3}\right) f_3(z_3)dz_3, \end{aligned} \tag{2.5}$$

is a proper trivariate distribution in $F(F_{13}, F_{23})$. In terms of densities, the joint trivariate density corresponding to (2.5), if it exists, reads (e.g. Czado (2010) Section 2)

$$
\begin{aligned}
&f_{123}(x_1, x_2, x_3) \\
&= c_{13}(F_1(x_1), F_3(x_3))c_{23}(F_2(x_2), F_3(x_3))c_{12|3}(F_{1|3}(x_1|x_3), F_{2|3}(x_2|x_3))f_1(x_1)f_2(x_2)f_3(x_3),
\end{aligned} \tag{2.6}
$$

where $c_{13}(u_1, u_3), c_{23}(u_2, u_3), c_{12|3}(u_1, u_2)$ are the copula densities of $C_{13}(u_1, u_3), C_{23}(u_2, u_3)$, $C_{12|3}(u_1, u_2)$.

As we are looking for universal trivariate copulas we focus on uniform [0,1] marginal distributions and copulas for bivariate marginal and bivariate conditional distributions. Let (X_1, X_2, X_3) be a trivariate random vector with uniform [0,1] margins. A random sampling algorithm for the simulation of (X_1, X_2, X_3) corresponding to (2.5) reads (e.g. Kurowicka and Cooke (2006), equation (6.2)):

$$
X_3 = U_3, \quad X_2 = F_{2|3}^{-1}(U_2|U_3), \quad X_1 = F_{1|3}^{-1}(F_{12|3;U_2}^{-1}(U_1)|U_3) \tag{2.7}
$$

where U_1, U_2, U_3 are independent uniform [0,1] random variables, $F_{1|3}^{-1}(x_1|x_3), F_{2|3}^{-1}$ $(x_2|x_3)$ are the inverse cumulative distributions of $F_{1|3}(x_1|x_3), F_{2|3}(x_2|x_3)$, and $F_{12|3;U_2}^{-1}$ (x_1) denotes the inverse cumulative distribution for X_1 given U_2 under the conditional copula $C_{12|3}(u_1, u_2)$.

Kurowicka and Cooke (2006), Section 4.4.6, p.102, generate universal trivariate copulas with arbitrary valid rank correlation matrices using the sampling algorithm (2.7) based on the so-called "elliptical copula" for the involved copulas $C_{13}(u_1, u_3), C_{23}(u_2, u_3), C_{12|3}(u_1, u_2)$.

However, it seems that no attempt has been made so far to translate this Monte Carlo simulation procedure into closed-form analytical copula formulas. For this, it is appropriate to restate the representation (2.5) for the trivariate copula $C(u_1, u_2, u_3)$ associated to the uniform random vector (X_1, X_2, X_3) as follows:

$$
C(u_1, u_2, u_3) = \int_0^{u_3} C_{12|3}\Big(\frac{\partial}{\partial t}C_{13}(u_1, t), \frac{\partial}{\partial t}C_{23}(u_2, t)\Big)dt \tag{2.8}
$$

It is important to remark that (2.8) is a special case of the following more general trivariate copula construction. Let $A(u, v)$, $B(u, v)$ be two bivariate copulas, and let $C = \{C_t(u, v),\ t \in [0, 1]\}$ be a family of bivariate copulas. Then, the mapping $A *_C B : [0, 1]^3 \to [0, 1]$, defined by

$$
(A *_C B)(u_1, u_2, u_3) = \int_0^{u_3} C_t\Big(\frac{\partial}{\partial t}A(u_1, t), \frac{\partial}{\partial t}B(u_2, t)\Big)dt \tag{2.9}
$$

yields a trivariate copula, called the *C-lifting* of the copulas A and B (Durante et al. (2007a), Proposition 3.2). The bivariate copula margins of the trivariate copula (2.9) are $(A *_C B)(u_1, u_2)$, $A(u_1, u_3)$, $B(u_2, u_3)$, where the bivariate copula

$$
(A *_C B)(u_1, u_2) = \int_0^{1} C_t\Big(\frac{\partial}{\partial t}A(u_1, t), \frac{\partial}{\partial t}B(u_2, t)\Big)dt \tag{2.10}
$$

is called the *C-product* of the copulas A and B (see Durante et al. (2007b) for details). Clearly, the special instance (2.8) is obtained for the (constant) conditional copula family $C_t(u, v) \equiv C_{12|3}(u, v),\ t \in [0, 1]$.

Specializing further to bivariate elliptical copulas and extremal correlation matrices with the property $\rho_{12;3} = \pm 1$ (see Lemma 2.1), and finite convex combinations from these copulas, we claim that it suffices to consider the following two types of *C-lifting* copulas (2.8):

$$C^+(u_1,u_2,u_3) = \int_0^{u_3} \min\left\{\frac{\partial}{\partial t}C_{13}(u_1,t), \frac{\partial}{\partial t}C_{23}(u_2,t)\right\}dt, \quad \text{if } \rho_{12;3} = 1,$$
$$C^-(u_1,u_2,u_3) = \int_0^{u_3} \max\left\{\frac{\partial}{\partial t}C_{13}(u_1,t) + \frac{\partial}{\partial t}C_{23}(u_2,t) - 1, 0\right\}dt, \quad \text{if } \rho_{12;3} = -1, \tag{2.11}$$

where $C_{13}(u_1,u_3)$, $C_{23}(u_2,u_3)$ are bivariate elliptical copulas with correlation parameters $r_{13} = \rho_{13}$, $r_{23} = \rho_{23}$. Indeed, choosing the Hoeffding-Fréchet bounds as conditional copulas $C_{12|3}(u_1,u_2) = M(u_1,u_2) = \min\{u_1,u_2\}$, resp. $C_{12|3}(u_1,u_2) = W(u_1,u_2) = \max\{u_1 + u_2 - 1, 0\}$, corresponds to choosing the bivariate conditional elliptical copula for the extreme conditional correlations $r_{12|3} = \pm 1$ (note that the bivariate copula is a comprehensive family, which realizes any correlation value in the interval $(-1,1)$). But, for these elliptical copula choices, the conditional correlation is constant and coincides with the partial correlation (e.g. Kurowicka and Cooke (2006), Proposition 3.19, p.44). In particular, we have $r_{12|3} = \pm 1 \Leftrightarrow \rho_{12;3} = \pm 1$, which shows that the construction (2.11) realizes the extremal correlation matrices with the property

$$\rho_{12} = \rho_{13}\rho_{23} \pm \sqrt{(1-\rho_{13}^2)\cdot(1-\rho_{23}^2)} \tag{2.12}$$

In particular, this means that the correlations of the product copulas $C^+(u_1,u_2) = (C_{13} *_M C_{23})(u_1,u_2)$, $C^-(u_1,u_2) = (C_{13} *_W C_{23})(u_1,u_2)$ coincide with (2.12). Moreover, according to Durante et al. (2007a), Corollary 4.1, we have for all bivariate copulas C_{12} that are compatible with C_{13}, C_{23} the sharp bounds in the concordance order:

$$C^-(u_1,u_2) \le C_{12}(u_1,u_2) \le C^+(u_1,u_2) \tag{2.13}$$

But ρ_{13}, $\rho_{23} \in [-1,1]$ are arbitrary and the correlation ρ_{12} of C_{12} in (2.12) varies between the two extreme bounds (2.12), hence all rank correlation matrices are feasible (see also the special case $n = 3$ of Theorem 3.1 in Hürlimann (2012b)). The extreme lifting copulas (2.11) together with appropriate finite convex combinations generate a class of 3-universal copulas. In the next Sections, we show that the product and lifting copulas satisfy closed-form analytical expressions.

For reasons of symmetry it is more appropriate to work with uniform [–1,1] random margins. Then a copula is defined on the centred cube $C_n = [-1,1]^n$, $n \ge 2$, with uniform [–1,1] margins. In this setting and new notation, the lifting copulas (2.11) read:

$$C^+(x,y,z) = \int_{-1}^{z} \min\left\{\frac{\partial}{\partial t}C_{13}(x,t), \frac{\partial}{\partial t}C_{23}(y,t)\right\}dt,$$
$$C^-(x,y,z) = \int_{-1}^{z} \max\left\{\frac{\partial}{\partial t}C_{13}(x,t) + \frac{\partial}{\partial t}C_{23}(y,t) - \frac{1}{2}, 0\right\}dt, \quad (x,y,z) \in C_3. \tag{2.14}$$

In a first step, we compute the bivariate product copulas with extremal correlations (2.12), i.e.

$$C^{+}(x,y)=\int_{-1}^{1}\min\left\{\frac{\partial}{\partial t}C_{13}(x,t),\frac{\partial}{\partial t}C_{23}(y,t)\right\}dt,$$
$$C^{-}(x,y)=\int_{-1}^{1}\max\left\{\frac{\partial}{\partial t}C_{13}(x,t)+\frac{\partial}{\partial t}C_{23}(y,t)-\frac{1}{2},0\right\}dt,\quad (x,y)\in C_2. \tag{2.15}$$

For this we need the explicit formulas of the bivariate linear circular copula, which are summarized in the Section 3.

3 The bivariate linear circular copula

Starting point is the unique circular symmetric distribution on the unit disk B_2 in R^2 with uniform [−1,1] margins. The associated so-called *circular copula* on C_2 is given by (Perlman and Wellner (2011), Theorem 3.1):

$$C(u,v)=\tfrac{1}{4}(u+v+1)+\gamma_0(u,v),\quad \gamma_0(u,v)=\begin{cases}\alpha_0(u,v), & u^2+v^2<1,\\ \beta_0(u,v), & u^2+v^2\geq 1,\end{cases} \tag{3.1}$$

With

$$\alpha_0(u,v)=\frac{1}{2\pi}\cdot\left\{u\cdot\arcsin\left(\frac{v}{\sqrt{1-u^2}}\right)+v\cdot\arcsin\left(\frac{u}{\sqrt{1-v^2}}\right)-\arcsin\left(\frac{uv}{\sqrt{(1-u^2)(1-v^2)}}\right)\right\},$$
$$\beta_0(u,v)=\operatorname{sgn}(uv)\cdot\tfrac{1}{4}(|u|+|v|-1). \tag{3.2}$$

Taking partial derivatives one obtains the corresponding copula density

$$c(u,v)=\frac{1}{2\pi\sqrt{1-u^2-v^2}},\quad (u,v)\in B_2. \tag{3.3}$$

To obtain from this a one-parameter comprehensive family of copulas, let $(U,V)\sim c(u,v)$ be the unique circular symmetric random vector on B_2. For each angle $\varphi\in(-\frac{\pi}{2},\frac{\pi}{2})$, set $r=\sin(\varphi)\in(-1,1)$, and consider the linear transformed random vector (X,Y_r) defined by

$$X=U,\quad Y_r=r\cdot U+\sqrt{1-r^2}\cdot V \tag{3.4}$$

which by circular symmetry generates a copula on C_2. The support of (X,Y_r) is the *ellipse*

$$E_r=\{(x,y)\mid x^2+y^2-2rxy<1-r^2\} \tag{3.5}$$

Clearly, the correlation coefficient (that is equal to the rank correlation) of the random pair (X,Y_r) is $r\in(-1,1)$. The corresponding comprehensive one-parameter family of *linear circular copulas* has the explicit representation (Perlman and Wellner (2011), Theorem 5.1):

$$C_r(x,y)=\tfrac{1}{4}(x+y+1)+\gamma_r(x,y),\quad \gamma_r(x,y)=\begin{cases}\alpha_r(x,y), & (x,y)\in E_r,\\ \beta_r(x,y), & (x,y)\in C_2\setminus E_r\end{cases} \tag{3.6}$$

with

$$\alpha_r(x,y) = \frac{1}{2\pi}\cdot\left\{\begin{array}{l} x\cdot\arcsin\left(\frac{y-rx}{\sqrt{(1-r^2)(1-x^2)}}\right) + y\cdot\arcsin\left(\frac{x-ry}{\sqrt{(1-r^2)(1-y^2)}}\right) \\ -\arcsin\left(\frac{xy-r}{\sqrt{(1-x^2)(1-y^2)}}\right) \end{array}\right\}, \tag{3.7}$$

and

$$\beta_r(x,y) = \begin{cases} \frac{1}{4}(x+y-1), & (x,y)\in L_1(r) := (C_2\ E_r)\ \cap\ \{x+y\geq 1+r\} \\ \frac{1}{4}(x-y+1), & (x,y)\in L_2(r) := (C_2\ E_r)\ \cap\ \{y-x\geq 1-r\} \\ \frac{1}{4}(-x-y-1), & (x,y)\in L_3(r) := (C_2\ E_r)\ \cap\ \{x+y\leq -1-r\} \\ \frac{1}{4}(y-x+1), & (x,y)\in L_4(r) := (C_2\ E_r)\ \cap\ \{y-x\leq -1+r\} \end{cases} \tag{3.8}$$

It is important to remark that the copula density coincides with the "elliptical copula" density introduced in Kurowicka et al. (2000) and extensively used in the book by Kurowicka and Cooke (2006). It is given by (Perlman and Wellner (2011), Proposition 5.1):

$$c_r(x,y) = \frac{1}{2\pi\sqrt{(1-r^2)(1-x^2)-(y-rx)^2}}\cdot 1_{E_r}(x,y), \quad (x,y)\in C_2. \tag{3.9}$$

Since this family is obtained through linear transformation of the unique circular copula it is natural to use the name "linear circular copula". This new terminology clearly distinguishes it from the ubiquitous class of "elliptical copulas" that is generated by the multivariate elliptical distributions in any dimension. A further reason is that Perlman and Wellner (2011), Section 6, consider the possibility to generate *non-linear circular copulas* and provide an example.

For later use in Section 4, we also need expressions for the conditional linear circular copulas, i.e. the partial derivatives with respect to the arguments. For this, we first re-write (3.6) for the outer part $C_2\setminus E_r$ of the ellipse as follows:

$$C_r(x,y) = \begin{cases} \frac{1}{2}(x+y), & (x,y)\in L_1(r) \\ \frac{1}{2}(x+1), & (x,y)\in L_2(r) \\ 0, & (x,y)\in L_3(r) \\ \frac{1}{4}(y+1), & (x,y)\in L_4(r) \end{cases} \tag{3.10}$$

Through partial derivation we obtain the following formula

$$\frac{\partial C_r(x,y)}{\partial y} = \begin{cases} \frac{1}{4}+\frac{1}{2\pi}\arcsin\left(\frac{x-ry}{\sqrt{(1-r^2)(1-y^2)}}\right), & (x,y)\in E_r \\ \frac{1}{2}, & (x,y)\in L_1(r)\cup L_4(r) \\ 0, & (x,y)\in L_2(r)\cup L_3(r) \end{cases} \tag{3.11}$$

4 Analytical product copula formulas

Using the explicit formulas (3.6)-(3.8) for the linear circular copula, it is now possible to derive simple analytical expressions for the extreme linear circular product copulas (2.15). Given are extreme correlation coefficients $(\rho_{13},\rho_{23},\rho_{12}^{\pm})$ such that $\rho_{12}^{\pm} = \rho_{13}\rho_{23}\pm\sqrt{(1-\rho_{13}^2)\cdot(1-\rho_{23}^2)}$, $\rho_{13},\rho_{23}\in[-1,1]$. Consider the linear circular copulas $C_{13}(x,y) = C_{r_{12}}(x,y)$, $C_{23}(x,y) = C_{r_{23}}(x,y)$ obtained from (3.6) by setting $r_{13}=\rho_{13}$, $r_{23}=\rho_{23}$. The abbreviations $a = \sqrt{1-\rho_{13}^2}$, $b = \sqrt{1-\rho_{23}^2}$, $c^{\pm} = \rho_{23}a \mp b\rho_{13}$ are used throughout. Without loss of generality we assume that $a \geq b$.

Proposition 4.1

The upper extreme linear circular product copula $C^+(x,y) = \int_{-1}^{1} \min\left\{\frac{\partial}{\partial t} C_{13}(x,t), \frac{\partial}{\partial t} C_{23}(y,t)\right\} dt$, $(x,y) \in C_2$ satisfies the following analytical representation:

Case I: $c^+ = r_{23}a - br_{13} \geq 0$

$$\begin{aligned}
&(\text{I.a})\, C^+(x,y) = \tfrac{1}{2}(x+1), \quad ay - bx \geq c^+,\\
&(\text{I.b})\, C^+(x,y) = \tfrac{1}{2}(y+1) + C_{13}\left(x, \tfrac{ay-bx}{c^+}\right) - C_{23}\left(y, \tfrac{ay-bx}{c^+}\right), \quad -c^+ < ay-bx < c^+,\\
&(\text{I.c})\, C^+(x,y) = \tfrac{1}{2}(y+1), \quad ay-bx \leq -c^+.
\end{aligned} \tag{4.1}$$

Case II: $c^+ = r_{23}a - br_{13} < 0$

$$\begin{aligned}
&(\text{II.a})\, C^+(x,y) = \tfrac{1}{2}(x+1), \quad ay-bx \geq -c^+,\\
&(\text{II.b})\, C^+(x,y) = \tfrac{1}{2}(x+1) - C_{13}\left(x, \tfrac{ay-bx}{c^+}\right) + C_{23}\left(y, \tfrac{ay-bx}{c^+}\right), \quad c^+ < ay-bx < -c^+,\\
&(\text{II.c})\, C^+(x,y) = \tfrac{1}{2}(y+1), \quad ay-bx \leq c^+.
\end{aligned} \tag{4.2}$$

The special case $c^+ = 0$ occurs exactly when $\rho_{13} = \rho_{23} \in [-1,1]$, $\rho_{12}^+ = 1$, and (4.1) reduces to the formula $C^+(x,y) = \frac{1}{2}(x+1)$, $y \geq x$, $C^+(x,y) = \frac{1}{2}(y+1)$, $y \leq x$.

Proposition 4.2

The lower extreme linear circular product copula $C^-(x,y) = \int_{-1}^{1} \max\left\{\frac{\partial}{\partial t} C_{13}(x,t) + \frac{\partial}{\partial t} C_{23}\right.$ $\left.(y,t) - \frac{1}{2}, 0\right\} dt$, $(x,y) \in C_2$, satisfies the following analytical representation:

Case I: $c^- = r_{23}a + br_{13} \geq 0$

$$\begin{aligned}
&(\text{I.a})\, C^-(x,y) = \tfrac{1}{2}(x+y), \quad ay+bx \geq c^-,\\
&(\text{I.b})\, C^-(x,y) = C_{13}\left(x, \tfrac{ay+bx}{c^-}\right) + C_{23}\left(y, \tfrac{ay+bx}{c^-}\right) - \tfrac{1}{2}\left(1 + \tfrac{ay+bx}{c^-}\right), \quad -c^- < ay+bx < c^-,\\
&(\text{I.c})\, C^-(x,y) = 0, \quad ay+bx \leq -c^-.
\end{aligned} \tag{4.3}$$

Case II: $c^- = r_{23}a + br_{13} < 0$

$$\begin{aligned}
&(\text{II.a})\, C^-(x,y) = \tfrac{1}{2}(x+y), \quad ay+bx \geq -c^-,\\
&(\text{II.b})\, C^-(x,y) = \tfrac{1}{2}\left(x+y+1+\tfrac{ay+bx}{c^-}\right) - C_{13}\left(x, \tfrac{ay+bx}{c^-}\right) - C_{23}\left(y, \tfrac{ay+bx}{c^-}\right), \quad c^- < ay+bx < -c^-,\\
&(\text{II.c})\, C^-(x,y) = 0, \quad ay+bx \leq c^-.
\end{aligned} \tag{4.4}$$

The special case $c^- = 0$ occurs exactly when $\rho_{23} = -\rho_{13} \in [-1,1]$, $\rho_{12}^- = -1$, and (4.3) reduces to the formula $C^-(x,y) = \frac{1}{2}(x+y)$, $x+y \geq 0$, $C^-(x,y) = 0$, $x+y \leq 0$.

Remark 4.1

It has been stated after equation (2.12) that the (rank) correlations of the extreme linear circular product copulas coincide with $\rho_{12}^{\pm}$. The obtained formulas (4.1)-(4.4) can be verified indirectly by showing that the rank correlations of these copulas are the right ones, i.e.

$$\rho_{12}^{\pm} = 3 \cdot \int_{-1}^{1} \int_{-1}^{1} C^{\pm}(x,y)dxdy - 3 \tag{4.5}$$

This can be done through analytical calculation or numerical integral computation (the latter based on a computer algebra system, e.g. MATHCAD from MathSoft, Inc.). In fact, before conjectural statement and mathematical proof, we have first checked the simple case $c^{\pm} = 0$, for which $\rho_{12}^{\pm} = \pm 1$, and have verified numerically the formulas (4.5) in the general case.

A direct analytical proof of the obtained formulas is based on a set of (firstly conjectured) conditional copula inequalities, which are of independent interest, and are proved in detail in Appendix 1 and 2. These inequalities are also of primordial importance in Section 5.

Theorem 4.1

(*Upper conditional product copula or UPC inequalities*) In the above notations the following inequalities hold for all $x, y, t \in [-1, 1]$:

Case (I.a), $ay - bx \geq c^{+}$: $\frac{\partial}{\partial t} C_{13}(x,t) \leq \frac{\partial}{\partial t} C_{23}(y,t)$

Case (I.b), $-c^{+} < ay - bx < c^{+}$: $\frac{\partial}{\partial t} C_{13}(x,t) \leq \frac{\partial}{\partial t} C_{23}(y,t), \quad -1 \leq t \leq \frac{ay-bx}{c^{+}}$, $\frac{\partial}{\partial t} C_{13}(x,t) \geq \frac{\partial}{\partial t} C_{23}(y,t), \quad \frac{ay-bx}{c^{+}} \leq t \leq 1$

Case (I.c), $ay - bx \leq -c^{+}$: $\frac{\partial}{\partial t} C_{13}(x,t) \geq \frac{\partial}{\partial t} C_{23}(y,t)$

Case (II.a), $ay - bx \geq -c^{+}$: $\frac{\partial}{\partial t} C_{13}(x,t) \leq \frac{\partial}{\partial t} C_{23}(y,t)$

Case (II.b), $c^{+} < ay - bx < -c^{+}$: $\frac{\partial}{\partial t} C_{13}(x,t) \geq \frac{\partial}{\partial t} C_{23}(y,t), \quad -1 \leq t \leq \frac{ay-bx}{c^{+}}$, $\frac{\partial}{\partial t} C_{13}(x,t) \leq \frac{\partial}{\partial t} C_{23}(y,t), \quad \frac{ay-bx}{c^{+}} \leq t \leq 1$

Case (II.c), $ay - bx \leq c^{+}$: $\frac{\partial}{\partial t} C_{13}(x,t) \geq \frac{\partial}{\partial t} C_{23}(y,t)$

Theorem 4.2

(*Lower conditional product copula or LPC inequalities*) In the above notations the following inequalities hold for all $x, y, t \in [-1, 1]$:

Case (I.a), $ay + bx \geq c^{-}$: $\frac{\partial}{\partial t} C_{13}(x,t) + \frac{\partial}{\partial t} C_{23}(y,t) - \frac{1}{2} \geq 0$

Case (I.b), $-c^{-} < ay + bx < c^{-}$: $\frac{\partial}{\partial t} C_{13}(x,t) + \frac{\partial}{\partial t} C_{23}(y,t) - \frac{1}{2} \geq 0, \quad -1 \leq t \leq \frac{ay+bx}{c^{-}}$, $\frac{\partial}{\partial t} C_{13}(x,t) + \frac{\partial}{\partial t} C_{23}(y,t) - \frac{1}{2} \leq 0, \quad \frac{ay+bx}{c^{-}} \leq t \leq 1$

Case (I.c), $ay + bx \leq -c^{-}$: $\frac{\partial}{\partial t} C_{13}(x,t) + \frac{\partial}{\partial t} C_{23}(y,t) - \frac{1}{2} \leq 0$

Case (II.a), $ay + bx \geq -c^{-}$: $\frac{\partial}{\partial t} C_{13}(x,t) + \frac{\partial}{\partial t} C_{23}(y,t) - \frac{1}{2} \geq 0$

Case (II.b), $c^{-} < ay + bx < -c^{-}$: $\frac{\partial}{\partial t} C_{13}(x,t) + \frac{\partial}{\partial t} C_{23}(y,t) - \frac{1}{2} \geq 0, \quad \frac{ay+bx}{c^{-}} \leq t \leq 1$, $\frac{\partial}{\partial t} C_{13}(x,t) + \frac{\partial}{\partial t} C_{23}(y,t) - \frac{1}{2} \leq 0, \quad -1 \leq t \leq \frac{ay+bx}{c^{-}}$

Case (II.c), $ay + bx \leq c^{-}$: $\frac{\partial}{\partial t} C_{13}(x,t) + \frac{\partial}{\partial t} C_{23}(y,t) - \frac{1}{2} \leq 0$

Proof of the Propositions 4.1 and 4.2

Inserting the inequalities from Theorem 4.1 (4.2) case by case into the upper (lower) product copula (2.15) shows the desired formulas. ◊

5 Analytical closed-form 3-universal copulas

Based on the relevant conditional copula inequalities provided in the Theorems 4.1 and 4.2, we derive first analytical closed-form expressions for the extreme linear circular lifting copulas (2.14). We use the same notations, abbreviations and assumptions as in the preceding Section.

Theorem 5.1

The upper extreme linear circular lifting copula (2.14) satisfies the following analytical representation:

Case I: $c^+ = r_{23}a - br_{13} \geq 0$

$$\begin{aligned}
&(\text{I.a})\, C^+(x,y,z) = C_{13}(x,z), \quad ay{-}bx{\geq}c^+, \\
&(\text{I.b})\, C^+(x,y,z) = C_{13}\left(x, \tfrac{ay-bx}{c^+}\right) + C_{23}(y,z){-}C_{23}\left(y, \tfrac{ay-bx}{c^+}\right), \quad z > \tfrac{ay-bx}{c^+} > -1, \\
&\qquad C^+(x,y,z) = C_{13}(x,z), \quad z{\leq}\tfrac{ay-bx}{c^+} < 1, \\
&(\text{I.c})\, C^+(x,y) = C_{23}(y,z), \quad ay{-}bx{\leq}{-}c^+.
\end{aligned} \tag{5.1}$$

Case II: $c^+ = r_{23}a - br_{13} < 0$

$$\begin{aligned}
&(\text{II.a})\, C^+(x,y,z) = C_{13}(x,z), \quad ay{-}bx{\geq}{-}c^+, \\
&(\text{II.b})\, C^+(x,y,z) = C_{13}(x,z){-}C_{13}\left(x, \tfrac{ay-bx}{c^+}\right) + C_{23}\left(y, \tfrac{ay-bx}{c^+}\right), \quad z > \tfrac{ay-bx}{c^+} > -1, \\
&\qquad C^+(x,y,z) = C_{23}(y,z), \quad z{\leq}\tfrac{ay-bx}{c^+} < 1, \\
&(\text{II.c})\, C^+(x,y,z) = C_{23}(y,z), \quad ay{-}bx{\leq}c^+.
\end{aligned} \tag{5.2}$$

The special case $c^+ = 0$ occurs exactly when $r = \rho_{13} = \rho_{23} \in [-1,1],\ \rho_{12}^+ = 1$, and (5.1) reduces to the formula $C^+(x,y,z) = C_r(x,z),\ y \geq x,\ C^+(x,y,z) = C_r(y,z),\ y \leq x$.

Theorem 5.2

The lower extreme linear circular lifting copula (2.14) satisfies the following analytical representation:

Case I: $c^- = r_{23}a + br_{13} \geq 0$

$$\begin{aligned}
&(\text{I.a})\, C^-(x,y,z) = C_{13}(x,z) + C_{23}(y,z){-}\tfrac{1}{2}(1+z), \quad ay+bx{\geq}c^-, \\
&(\text{I.b})\, C^-(x,y,z) = C_{13}\left(x, \min\left\{\tfrac{ay+bx}{c^-}, z\right\}\right) + C_{23}\left(y, \min\left\{\tfrac{ay+bx}{c^-}, z\right\}\right) \\
&\qquad {-}\tfrac{1}{2}\left(1 + \min\left\{\tfrac{ay+bx}{c^-}, z\right\}\right), \quad -c^- < ay+bx < c^-, \\
&(\text{I.c})\, C^-(x,y,z) = 0, \quad ay+bx{\leq}{-}c^-.
\end{aligned} \tag{5.3}$$

Case II: $c^- = r_{23}a + br_{13} < 0$

$$\begin{aligned}
&(\text{II.a})\, C^-(x,y,z) = C_{13}(x,z) + C_{23}(y,z){-}\tfrac{1}{2}(1+z), \quad ay+bx{\geq}{-}c^-, \\
&(\text{II.b})\, C^-(x,y,z) = C_{13}(x,z){-}C_{13}\left(x, \tfrac{ay+bx}{c^-}\right) + C_{23}(y,z){-}C_{23}\left(y, \tfrac{ay+bx}{c^-}\right), \\
&\qquad z > \tfrac{ay+bx}{c^-} > -1, \quad C^-(x,y,z) = 0, \quad z{\leq}\tfrac{ay+bx}{c^-} < 1, \\
&(\text{II.c})\, C^-(x,y,z) = 0, \quad ay+bx{\leq}c^-.
\end{aligned} \tag{5.4}$$

The special case $c^- = 0$ occurs exactly when $\rho_{13} = \rho_{23} = 0,\ \rho_{12}^- = -1$, and (5.3) reduces to the formula $C^-(x,y,z) = C_0(x,z) + C_0(y,z){-}\tfrac{1}{2}(1+z),\ x+y{\geq}0, \quad C^-(x,y,z) = 0,\ x+y{\leq}0$.

Proof of the Theorems 5.1 and 5.2

Inserting the inequalities from Theorem 4.1 (4.2) case by case into the upper (lower) lifting copula (2.14) shows the desired formulas. ◊

By the Theorem of Carathéodory, the obtained results show the existence of a 3-universal linear circular based copula. However, instead of linearly combining four extreme points (Carathéodory's result), two such points suffice here, as shown in the next main result. Recall that the linear circular lifting copulas $C^{\pm}(x,y,z)$ belong to the extreme points $\rho^{\pm} = (\rho_{13}, \rho_{23}, \rho_{12}^{\pm})$.

Theorem 5.3

(*3-universal linear circular copula*) Let $\rho = (\rho_{13}, \rho_{23}, \rho_{12})$ belong to an arbitrary 3×3 positive semi-definite rank correlation matrix. Then, the following 3-copula is universal:

$$C_{\rho}(x,y,z) = \lambda^{-}C^{-}(x,y,z) + \lambda^{+}C^{+}(x,y,z), \quad (x,y,z) \in C_3,$$
$$\lambda^{-} = \frac{1}{2}\frac{\rho_{12}^{+}-\rho_{12}}{\sqrt{(1-\rho_{13}^2)\cdot(1-\rho_{23}^2)}} \geq 0, \quad \lambda^{+} = \frac{1}{2}\frac{\rho_{12}-\rho_{12}^{-}}{\sqrt{(1-\rho_{13}^2)\cdot(1-\rho_{23}^2)}} \geq 0, \quad \lambda^{-}+\lambda^{+} = 1.$$

Proof

This follows from the preceding results by noting that $\rho = \lambda^{-}\rho^{-} + \lambda^{+}\rho^{+}$. ◊

6 Applications and conclusions

As an immediate statistical application, we show that the sampling algorithm for trivariate canonical vine copulas (Kurowicka and Cooke (2006), Section 6.4.2) reduces to a *two-dimensional simulation algorithm* for the uniform [0,1] random vectors $X^{\pm} = (X_1^{\pm}, X_2^{\pm}, X_3^{\pm})$ with lower and upper extreme linear circular lifting copulas $C^{\pm}(u,v,w)$, $(u, v, w) \in [0,1]^3$. Invoking Theorem 5.3, this yields below a simple two-dimensional algorithm for generating random vectors $X = (X_1, X_2, X_3)$ with 3-universal linear circular copula. To show this, one needs some regression properties of the linear circular copula of Section 3.

Proposition 6.1

(Kurowicka and Cooke (2006), Proposition 3.18) Let U, V be uniform [0,1] random variables joined by the linear circular copula with correlation ρ.

Then one has

$$E\ [V|U] = \tfrac{1}{2}(1-\rho) + \rho U \tag{6.1}$$

$$Var\ [V|U] = \tfrac{1}{2}(1-\rho^2)U(1-U) \tag{6.2}$$

$$F_{V|U}(v|u) = \frac{1}{2} + \frac{1}{\pi}\arcsin\left(\frac{v-\rho u-\frac{1}{2}(1-\rho)}{\sqrt{(1-\rho^2)u(1-u)}}\right), \quad u,v \in [0,1] \tag{6.3}$$

$$F_{V|U}^{-1}(\xi|u) = \tfrac{1}{2}(1-\rho) + \rho u + \sqrt{(1-\rho^2)u(1-u)}\cdot\sin\left(\pi\xi-\tfrac{\pi}{2}\right), \quad u,\xi \in [0,1] \tag{6.4}$$

Of further relevance is the following result, which states that conditional correlations are constant and equal to partial correlations for the linear circular copula.

Proposition 6.2

Kurowicka and Cooke (2006), Proposition 3.19) Let U_1, U_2, U_3 be uniform [0,1] random variables, and let U_1, U_2 and U_2, U_3 be joined by the linear circular copula with correlations ρ_{12} and ρ_{13} respectively. Assume that the conditional copula for U_2, U_3 given U_1 does not depend on U_1. Then, the conditional correlation $\rho_{23|1} = \rho(U_2|U_1, U_3|U_1)$ is constant in U_1 and $\rho_{23|1} = \rho_{23;1} = \frac{\rho_{23}-\rho_{12}\rho_{13}}{\sqrt{(1-\rho_{12}^2)\cdot(1-\rho_{23}^2)}}$.

Now, let U_1, U_2, U_3 be uniform [0,1] random variables. To generate a random vector $X = (X_1, X_2, X_3)$ with valid rank correlation matrix $\rho = (\rho_{13}, \rho_{23}, \rho_{12})$, one uses the general sampling algorithm by Kurowicka and Cooke (2006), Section 6.4.2, which in the linear circular copula case reads as follows (use equation (6.4)):

$$X_1 = U_1, \quad X_2 = F^{-1}_{\rho_{12};U_1}(U_2) = \tfrac{1}{2}(1-\rho_{12}) + \rho_{12}U_1 + \sqrt{(1-\rho_{12}^2)U_1(1-U_1)}\cdot \sin(\pi U_2 - \tfrac{\pi}{2}),$$
$$X_3 = F^{-1}_{\rho_{13};U_1}(F^{-1}_{\rho_{23|1};U_2}(U_3)) = \tfrac{1}{2}(1-\rho_{13}) + \rho_{13}U_1 + \sqrt{(1-\rho_{13}^2)U_1(1-U_1)}\cdot \sin(\pi F^{-1}_{\rho_{23|1};U_2}(U_3) - \tfrac{\pi}{2}),$$
$$F^{-1}_{\rho_{23|1};U_2}(U_3) = \tfrac{1}{2}(1-\rho_{23|1}) + \rho_{23|1}U_2 + \sqrt{\left(1-\rho_{23|1}^2\right)U_2(1-U_2)}\cdot \sin(\pi U_3 - \tfrac{\pi}{2}).$$

For the extreme points $\rho^{\pm} = \left(\rho_{13}, \rho_{23}^{\pm}, \rho_{12}\right)$ with $\rho_{23}^{\pm} = \rho_{12}\rho_{13} \pm \sqrt{(1-\rho_{12}^2)\cdot(1-\rho_{13}^2)}$ one has $\rho_{23|1}^{\pm} = \rho_{23;1}^{\pm} = \pm 1$ by Proposition 6.2 (see also Kurowicka and Cooke (2006), p.102). To generate uniform [0,1] random vectors $X^{\pm} = \left(X_1^{\pm}, X_2^{\pm}, X_3^{\pm}\right)$ with correlation matrices $\rho^{\pm}$, one uses two times the preceding equations. In case $\rho_{23|1}^{+} = 1$ one has $F^{-1}_{\rho_{23|1}^{+};U_2}(U_3) = U_2$, and in case $\rho_{23|1}^{-} = -1$ one has $F^{-1}_{\rho_{23|1}^{-};U_2}(U_3) = 1-U_2$.

It follows that

$$X_1^{\pm} = U_1, \quad X_2^{\pm} = \tfrac{1}{2}(1-\rho_{12}) + \rho_{12}U_1 + \sqrt{(1-\rho_{12}^2)U_1(1-U_1)}\cdot \sin(\pi U_2 - \tfrac{\pi}{2}),$$
$$X_3^{\pm} = \tfrac{1}{2}\left(1-\rho_{13}\right) + \rho_{13}U_1 \pm \sqrt{(1-\rho_{13}^2)U_1(1-U_1)}\cdot \sin(\pi U_2 - \tfrac{\pi}{2}). \tag{6.5}$$

Theorem 6.1

Let U_1, U_2 be uniform [0,1] random variables, and let $\rho = (\rho_{13}, \rho_{23}, \rho_{12})$ be a valid rank correlation matrix. Then, the sampling procedure

$$X_1 = U_1, \quad X_2 = \tfrac{1}{2}(1-\rho_{12}) + \rho_{12}U_1 + \sqrt{(1-\rho_{12}^2)U_1(1-U_1)}\cdot \sin(\pi U_2 - \tfrac{\pi}{2}),$$
$$X_3 = \tfrac{1}{2}\left(1-\rho_{13}\right) + \rho_{13}U_1 + \rho_{23;1}\sqrt{(1-\rho_{13}^2)U_1(1-U_1)}\cdot \sin(\pi U_2 - \tfrac{\pi}{2}), \tag{6.6}$$

generates a random vector with 3-universal linear circular copula that has correlation matrix ρ.

Proof

With a permutation of indices, one has by Theorem 5.3 that $X = \lambda^- X^- + \lambda^+ X^+$, with $\lambda^- = \frac{1}{2}\frac{\rho_{23}^{+}-\rho_{23}}{\sqrt{(1-\rho_{12}^2)\cdot(1-\rho_{13}^2)}} \geq 0, \quad \lambda^+ = \frac{1}{2}\frac{\rho_{23}-\rho_{23}^{-}}{\sqrt{(1-\rho_{12}^2)\cdot(1-\rho_{13}^2)}} \geq 0, \quad \lambda^- + \lambda^+ = 1.$

The result follows by combining the equations (6.5) noting that $\lambda^+ - \lambda^- = \rho_{23;1}$. ◊

Let us conclude. Since simulation algorithms are almost always used in vine copula applications, any dimension reduction of such algorithms is a significant computational improvement. Vine copulas have received a lot of attention in recent years with applications in nearly all branches of mathematical sciences. Some further representative works on theory and applications of (vine) copulas include Cherubini et al. (2004),

Mauguis and Guégan (2010), Vaz de Melo Mendes et al. (2010), Hürlimann (2011), Joe and Kurowicka (2011), Bernard and Czado (2012), Czado et al. (2012), Allen et al. (2013), Brechmann and Czado (2013), etc. Vine copulas are especially useful in higher dimensions. Unfortunately, due to its complexity, the present work could not be so far extended beyond dimension three. This is an important challenge for future research.

For $n < 6$ Letac (2010) has announced some universal copulas, which use quite advanced mathematics, however. As stated by him, a difficulty lies in the lack of an easy characterization of the extreme points of the convex set of correlation matrices of order n. One can ask whether in higher dimensions the explicit recursive closed form correlation bounds for positive semi-definite correlation matrices derived in Hürlimann (2012b) are of help in this matter, as is the case in dimension three.

Explicit simpler constructions than ours are also open for investigation. Note that one of the most elementary approach, which is based on bivariate linear Spearman copulas, though completely classified, does not lead to 3-universal copulas, as shown in Hürlimann (2012c).

Appendix 1

Some preliminary identities and inequalities

We begin with some preliminary material that will be used throughout in Appendix 2. Due to the structure of the bivariate linear circular copula, the analysis of the conjectured inequalities requires a case by case analysis of whether or not the coordinates (x, t), respectively (y, t), belong to one of the following disjoint regions of the centred square C_2. For (x, t) these are the regions defined and denoted by

$E(r_{13}) = \{(x,t) \ : \ x^2 + t^2 - 2r_{13}xt < 1 - r_{13}^2\}$: inner of ellipse

$E^-(r_{13}) = \left\{(x,t) \ : \ t \le r_{13}x - \sqrt{(1-r_{13}^2)(1-x^2)}\right\}$: lower outer part of ellipse

$E^+(r_{13}) = \left\{(x,t) \ : \ t \ge r_{13}x + \sqrt{(1-r_{13}^2)(1-x^2)}\right\}$: upper outer part of ellipse

$L_1(r_{13}) = (C_2 \setminus E(r_{13})) \cap \{x + t \ge 1 + r_{13}\}$: outer part of ellipse above line in 1st quadrant

$L_2(r_{13}) = (C_2 \setminus E(r_{13})) \cap \{t - x \ge 1 - r_{13}\}$: outer part of ellipse above line in 2nd quadrant

$L_3(r_{13}) = (C_2 \setminus E(r_{13})) \cap \{x + t \le -1 - r_{13}\}$: outer part of ellipse below line in 3rd quadrant

$L_4(r_{13}) = (C_2 \setminus E(r_{13})) \cap \{t - x \le -1 + r_{13}\}$: outer part of ellipse below line in 4th quadrant

For (y, t) these are the disjoint regions defined and denoted by

$E(r_{23}) = \{(y,t) \ : \ y^2 + t^2 - 2r_{23}yt < 1 - r_{23}^2\}$: inner of ellipse

$E^-(r_{23}) = \left\{(y,t) \ : \ t \le r_{23}y - \sqrt{(1-r_{23}^2)(1-y^2)}\right\}$: lower outer part of ellipse

$E^+(r_{23}) = \left\{(y,t) \ : \ t \ge r_{13}y + \sqrt{(1-r_{23}^2)(1-y^2)}\right\}$: upper outer part of ellipse

$L_1(r_{23}) = (C_2 \setminus E(r_{23})) \cap \{y + t \ge 1 + r_{23}\}$: outer part of ellipse above line in 1st quadrant

$L_2(r_{23}) = (C_2 \setminus E(r_{23})) \cap \{t - y \ge 1 - r_{23}\}$: outer part of ellipse above line in 2nd quadrant

$L_3(r_{23}) = (C_2 \setminus E(r_{23})) \cap \{y + t \le -1 - r_{23}\}$: outer part of ellipse below line in 3rd quadrant

$L_4(r_{23}) = (C_2 \setminus E(r_{23})) \cap \{t - y \le -1 + r_{23}\}$: outer part of ellipse below line in 4th quadrant

We note that involved expressions of the type $uv \pm \sqrt{(1-u^2)(1-v^2)}$ necessarily belong to the interval $[-1, 1]$ in virtue of the following slightly more general property.

Lemma A1.1

For $x, y, z \in [-1, 1]$ the function defined by $w(x, y, z) = xy + z\sqrt{(1-x^2)(1-y^2)}$ takes values in [−1,1].

Proof

See the proof of Lemma 4.7 in Kurowicka and Cooke (2006), pp. 121–122. ◊

Some inequality properties of the function $w(x, y, \pm 1) = xy \pm \sqrt{(1-x^2)(1-y^2)}$ defined in Lemma A1.1 will be useful throughout the proofs.

Lemma A1.2

The following properties hold:

(P1) $w(x, y, -1) \le x + y - 1 \Leftrightarrow x + y \ge 0$

(P2) $w(x, y, 1) \ge x - y + 1 \Leftrightarrow x - y \le 0$

(P3) $\frac{\partial w(x,y,-1)}{\partial x} \ge 0 \Leftrightarrow x + y \ge 0, \quad \frac{\partial w(x,y,-1)}{\partial y} \ge 0 \Leftrightarrow x + y \ge 0$

(P4) $\frac{\partial w(x,y,1)}{\partial x} \ge 0 \Leftrightarrow x - y \le 0, \quad \frac{\partial w(x,y,1)}{\partial y} \ge 0 \Leftrightarrow x - y \le 0$

(P5) if $x, y < 1$ then $w(x, y, 1) > x + y - 1$

(P6) if $x, y < 1$ then $w(x, y, -1) < 1 + y - x, \; w(x, y, -1) < 1 + x - y$

Proof

This is obtained without difficulty through standard algebraic calculation. ◊

Lemma A1.3

The quantities $r_{13}, r_{23} \in [-1, 1], \; a = \sqrt{1-r_{13}^2} \ge b = \sqrt{1-r_{23}^2}$ and $c^{\pm} = ar_{23} \mp br_{13}$ satisfy the following identities:

$$\frac{b \pm c^{\pm} r_{13}}{a} = \frac{a - c^{\pm} r_{23}}{b}, \quad \left(\frac{a - c^{\pm} r_{23}}{b}\right)^2 = 1 - \left(c^{\pm}\right)^2 \tag{A1.1}$$

Proof

The first identity is shown by verifying the following equivalences:

$$b^2 \pm bc^{\pm} r_{13} = a^2 - ac^{\pm} r_{23} \quad \Leftrightarrow \quad c^{\pm}(ar_{23} \pm br_{13}) = a^2 - b^2$$
$$\Leftrightarrow \quad a^2 r_{23}^2 - b^2 r_{13}^2 = a^2 - b^2 \quad \Leftrightarrow \quad a^2\left(1-b^2\right) - b^2\left(1-a^2\right) = a^2 - b^2$$

The second one is shown similarly as follows:

$$a^2 - 2ac^{\pm} r_{23} + \left(c^{\pm}\right)^2 r_{23}^2 = b^2 - b^2 (c^{\pm})^2 \quad \Leftrightarrow \quad a^2 - 2ac^{\pm} r_{23} + (c^{\pm})^2 = b^2$$
$$\Leftrightarrow \quad c^{\pm}(2ar_{23} - c^{\pm}) = a^2 - b^2 \quad \Leftrightarrow \quad c^{\pm}\left(ar_{23} \pm b\rho_{13}\right) = a^2 - b^2$$

Since the last identity is true by the preceding equivalences, the result is shown. ◊

Lemma A1.4

The functions defined by

$$u_{\pm} = u_{\pm}(x) = \frac{c^{\pm} \cdot w(x, r_{13}, \pm 1) \pm bx}{a}, \quad v_{\pm} = v_{\pm}(y) = \pm\left(\frac{ay - c^{\pm} \cdot w(y, r_{23}, \pm 1)}{b}\right) \tag{A1.2}$$

have the following inverse functions

$$x = u_{\pm}^{*}(u_{\pm}) = \pm\left(\frac{a - c^{\pm} r_{23}}{b}\right) \cdot u_{\pm} \pm c^{\pm} \cdot \sqrt{1-u_{\pm}^{2}} = \left(\frac{c^{\pm} r_{13} \pm b}{a}\right) \cdot u_{\pm} \pm c^{\pm} \cdot \sqrt{1-u_{\pm}^{2}},$$
$$y = v_{\pm}^{*}(v_{-}) = \pm\left(\frac{a - c^{\pm} r_{23}}{b}\right) \cdot v_{\pm} \pm c^{\pm} \cdot \sqrt{1-v_{\pm}^{2}} = \left(\frac{c^{\pm} r_{13} \pm b}{a}\right) \cdot v_{\pm} \pm c^{\pm} \cdot \sqrt{1-v_{\pm}^{2}} \qquad \text{(A1.3)}$$

Proof

This is obtained through straightforward calculation using the identities (A1.1). ◊

Appendix 2

Proof of the conditional product copula inequalities

Starting point are the following preliminary remarks. According to (3.11), in case $(x, t) \in E(r_{13})$, $(y, t) \in E(r_{23})$, we have $\frac{\partial}{\partial t} C_{13}(x, t) \leq \frac{\partial}{\partial t} C_{23}(y, t)$ if, and only if

$$\arcsin\left(\frac{x - r_{13}t}{a\sqrt{1-t^2}}\right) \leq \arcsin\left(\frac{y - r_{23}t}{b\sqrt{1-t^2}}\right) \quad \Leftrightarrow \quad c^{+}t \leq ay - bx \qquad \text{(A2.1)}$$

and similarly, we have $\frac{\partial}{\partial t} C_{13}(x, t) + \frac{\partial}{\partial t} C_{23}(y, t) - \frac{1}{2} \geq 0$ if, and only if

$$\arcsin\left(\frac{x - r_{13}t}{a\sqrt{1-t^2}}\right) + \arcsin\left(\frac{y - r_{23}t}{b\sqrt{1-t^2}}\right) \geq 0 \quad \Leftrightarrow \quad c^{-}t \leq ay + bx \qquad \text{(A2.2)}$$

By symmetry it suffices to show the inequalities for Case I. The proof is done through exhaustive analysis of all possible disjoint regions that can contain the coordinates (x, t), respectively (y, t). We assume $c^{\pm} > 0$ and identify the cases $c^{\pm} = 0$ as limiting cases $c^{\pm} \to 0$. Items related to Theorem 4.1 (4.2) are distinguished using the abbreviation UPC (LPC) for upper (lower) conditional product copula inequalities unless quantities involving $c^{\pm}$ are used.

(1) $(x, t) \in E^{-}(r_{13})$ (y, t) arbitrary

(1.1) $x + r_{13} \geq 0$ (UPC), $x + r_{13} < 0$ (LPC)

First of all, we note that $\frac{\partial}{\partial t} C_{13}(x, t) = \frac{1}{2}$ (UPC), $\frac{\partial}{\partial t} C_{13}(x, t) = 0$ (LPC). To show this we use the properties of Lemma A1.2 and (3.11) as follows:

(UPC): (P1) $\Rightarrow$ $t \leq w(x, r_{13}, -1) \leq x + r_{13} - 1$ $\Rightarrow$ $(x, t) \in L_4(r_{13})$

(LPC): (i) using (P6) of Lemma 3.2: $t \leq w(x, r_{13}, -1) < 1 + r_{13} - x$ $\Rightarrow$ $(x, t) \notin L_1(r_{13})$

(ii) using $x + r_{13} < 0$ $t \geq -1 > x + r_{13} - 1$ $\Rightarrow$ $(x, t) \notin L_4(r_{13})$

Furthermore, we must have $y + r_{23} \geq 0$ in Cases (I.a) and (I.b), which is shown as follows:

(UPC) Case (I.a): if $y < -r_{23}$ then $0 \leq c^{+} \leq ay - bx < -ar_{23} - br_{13} = -c^{+} \leq 0$ a contradiction

(UPC) Case (I.b): $y > (-c^{+} + bx)/a \geq -(c^{+} + br_{13})/a = -r_{23}$

(LPC) Case (I.a): if $y < -r_{23}$, then $0 \leq c^{-} \leq ay + bx < -ar_{23} - br_{13} = -c^{-} \leq 0$ a contradiction

(LPC) Case (I.b): if $y < -r_{23}$ then $-c^{-} < ay + bx < -ar_{23} - br_{13} = -c^{-}$ a contradiction

If one assumes that $t \le w(y, r_{23}, -1)$, then $\frac{\partial}{\partial t} C_{23}(y,t) = \frac{1}{2}$ in virtue of (3.11) because (P1) $\Rightarrow$ $t \le w(y, r_{23}, -1) \le y + r_{23} - 1$ $\Rightarrow$ $(y, t) \in L_4(r_{23})$

We distinguish between three main cases:

(a) $c^{\pm}t \ge ay \mp bx$ (Case (I.c) and possibly Case (I.b))

Since $\frac{\partial}{\partial t} C_{23}(y,t) \le \frac{1}{2}$ we have necessarily $\frac{\partial}{\partial t} C_{13}(x,t) = \frac{1}{2} \ge \frac{\partial}{\partial t} C_{23}(y,t)$ (UPC), and $\frac{\partial}{\partial t} C_{13}(x,t) + \frac{\partial}{\partial t} C_{23}(y,t) - \frac{1}{2} \le 0$ (LPC), as in the desired inequalities.

(b) $c^{\pm}t \le ay \mp bx < c^{\pm} \cdot w(y, r_{23}, -1)$ (possible in Cases (I.a) and (I.b))

By the above we have $\frac{\partial}{\partial t} C_{13}(x,t) = \frac{\partial}{\partial t} C_{23}(y,t) = \frac{1}{2}$ (UPC) and $\frac{\partial}{\partial t} C_{13}(x,t) + \frac{\partial}{\partial t} C_{23}(y,t) - \frac{1}{2} = 0$ (LPC), which is consistent with the desired inequalities.

(c) $c^{\pm}t \le ay \mp bx$ and $c^{\pm} \cdot w(y, r_{23}, -1) \le ay \mp bx$ (possible in Cases (I.a) and (I.b))

We derive the inequality (C1.1) $t \le w(x, r_{13}, -1) \le w(y, r_{23}, -1)$ which again implies that $\frac{\partial}{\partial t} C_{13}(x,t) = \frac{\partial}{\partial t} C_{23}(y,t) = \frac{1}{2}$ (UPC) and $\frac{\partial}{\partial t} C_{13}(x,t) + \frac{\partial}{\partial t} C_{23}(y,t) - \frac{1}{2} = 0$ (LPC). We use the transformation of variable $z = z(y) = \pm(ay - c^{\pm} \cdot w(y, r_{23}, -1))/b$ and its inverse $y = y(z) = \pm(a - c^{\pm} r_{23})/b \cdot z \pm c^{\pm} \cdot \sqrt{1-z^2}$ in (A1.3), where the second "±" signs hold for both (UPC) and (LPC). We show (C1.1), that is $\Delta(x,y) := w(y, r_{23}, -1) - w(x, r_{13}, -1) \ge 0$ under the constraints $-r_{13} \le x \le z(y)$ (UPC), $z(y) \le x < -r_{13}$ (LPC), $-r_{23} \le y \le 1$. By (P3) of Lemma A1.2 the function $w(x, r_{13}, -1)$ is increasing in x (UPC), decreasing in x (LPC). It follows that $\Delta(x,y) \ge \Delta^*(y, z(y)) := w(y, r_{23}, -1) - w(z(y), r_{13}, -1)$. Using the transformation $z(y)$ we can write $w(y, r_{23}, -1) = (ay \mp bz(y))/c^{\pm}$. Making also use of the inverse $y = y(z)$ we see that (use the first identity in (A1.1)) $\Delta^*(z) := \Delta^*(y(z), z) = \dfrac{ay(z) \mp bz}{c^{\pm}} - w(z, r_{13}, -1) = \dfrac{a}{c^{\pm}}\left(\pm(a - c^{\pm} r_{23})/b \cdot z \pm c \cdot \sqrt{1-z^2}\right) \mp \dfrac{b}{c^{\pm}} z - r_{13} z + a\sqrt{1-z^2} = a\sqrt{1-z^2}(1 \pm 1)$, which implies that $\Delta(x,y) \ge \Delta^*(y, z(y)) = \Delta^*(z) \ge 0$ as desired.

(1.2) $x + r_{13} < 0$ (UPC), $x + r_{13} > 0$ (LPC)

We have $\frac{\partial}{\partial t} C_{13}(x,t) = 0$ $\quad \frac{\partial}{\partial t} C_{13}(x,t) = \frac{1}{2}$ (UPC), (LPC), which is shown as follows:

(UPC): (i) using (P6) of Lemma A1.2: $t \le w(x, r_{13}, -1) < 1 + r_{13} - x$ $\Rightarrow$ $(x, t) \notin L_1(r_{13})$

(ii) using $x + r_{13} < 0$: $t \ge -1 > x + r_{13} - 1$ $\Rightarrow$ $(x, t) \notin L_4(r_{13})$

(LPC): (P1) of Lemma A1.2 $\Rightarrow$ $t \le w(x, r_{13}, -1) \le x + r_{13} - 1$ $\Rightarrow$ $(x, t) \in L_4(r_{13})$

We distinguish between two main cases:

(a) $c^{\pm}t \le ay \mp bx$ (Case (I.a) and possibly Case (I.b))

We have $\frac{\partial}{\partial t} C_{13}(x,t) = 0 \le \frac{\partial}{\partial t} C_{23}(y,t)$ (UPC), $\frac{\partial}{\partial t} C_{13}(x,t) + \frac{\partial}{\partial t} C_{23}(y,t) - \frac{1}{2} \ge 0$ (LPC) as desired

(b) $ay \mp bx \le c^{\pm}t \le c^{\pm} \cdot w(x, r_{13}, -1)$ (Case (I.c) and possibly Case (I.b))

We show $\frac{\partial}{\partial t} C_{23}(y,t) = 0$, which is consistent with the desired results $\frac{\partial}{\partial t} C_{13}(x,t) \ge \frac{\partial}{\partial t} C_{23}(y,t)$ (UPC) and $\frac{\partial}{\partial t} C_{13}(x,t) + \frac{\partial}{\partial t} C_{23}(y,t) - \frac{1}{2} \le 0$ (LPC). It suffices to show the following inequalities:

Case 1: $x \ge -(b + c^{+} r_{13})/a$ (UPC), $x \le (b - c^{-} r_{13})/a$ (LPC)

(C1.2.1) $t \le w(x, r_{13}, -1) \le w(y, r_{23}, -1)$, and $y + r_{23} \le 0$

Case 2: $x \le -(b + c^{+} r_{13})/a$ (UPC), $x \ge (b - c^{-} r_{13})/a$ (LPC)

(C1.2.2) $c^{\pm} \cdot w(y, r_{23}, 1) \le ay \mp bx \le c^{\pm}t \le c^{\pm} \cdot w(x, r_{13}, -1)$, and $y - r_{23} \le 0$

Indeed, if these conditions hold, then $\frac{\partial}{\partial t}C_{23}(y,t) = 0$ in virtue of (3.11) as follows:

Case 1: (i) using (P6) of Lemma A1.2: $t \le w(y, r_{23}, -1) < 1 + r_{23} - y \Rightarrow (y, t) \notin L_1(r_{23})$
(ii) using $y + r_{23} \le 0$: $t \ge -1 \ge y + r_{23} - 1 \Rightarrow (y, t) \notin L_4(r_{23})$
Case 2: (i) using $y - r_{23} \le 0$: $t \le 1 \le 1 + r_{23} - y \Rightarrow (y, t) \notin L_1(r_{23})$
(ii) using (P5) of Lemma A1.2: $t \ge w(y, r_{23}, 1) > y + r_{23} - 1 \Rightarrow (y, t) \notin L_4(r_{23})$

To show the above inequalities we use the transformation $z = z(x) = (c^{\pm} \cdot w(x, r_{13}, -1) \pm bx)/a$.

Case 1: The function $z(x)$ is increasing in the interval $-(b + c^{+}r_{13})/a \le x < -r_{13}$ (UPC), respectively $z(x)$ is decreasing in the interval $-r_{13} \le x \le (b - c^{-}r_{13})/a$ (LPC):

$$z'(x) = (c^{\pm}r_{13} \pm b)/a + c^{\pm}\cdot x/\sqrt{1-x^2} \ge 0 \quad \Leftrightarrow \quad x/\sqrt{1-x^2} \ge \alpha := -(\pm b + c^{\pm}r_{13})/ac^{\pm}$$
$$\Leftrightarrow \quad x \ge \alpha/\sqrt{1+\alpha^2} = -(\pm b + c^{\pm}r_{13})/a$$

In particular, by the restriction defining case (b), we have necessarily $y \le z(x) \le z(-r_{13}) = (-c^{\pm} \mp br_{13})/a = -r_{23} \Rightarrow y + r_{23} \le 0$.

Now, we show that $\Delta(x,y) = w(y, r_{23}, -1) - w(x, r_{13}, -1) \ge 0$ under the constraints, $-1 \le y \le z(x)$, $-(b + c^{+}r_{13})/a \le x < -r_{13}$ (UPC), $-r_{13} \le x \le (b - c^{-}r_{13})/a$ (LPC). Since $w(y, r_{23}, -1)$ is decreasing in y by (P3) of Lemma A1.2, we have $\Delta(x,y) \ge \Delta^*(x, z(x)) := w(z(x), r_{23}, -1) - w(x, r_{13}, -1) = w(z(x), r_{23}, -1) - (az(x) \mp bx)/c^{\pm}$.

Since $z(x)$ is increasing and $z(-r_{13}) = -r_{23}$, $z(-(b + c^{+}r_{13})/a) = -1$ (UPC), respectively $z(x)$ is decreasing and $z(-r_{13}) = -r_{23}$, $z((b - c^{-}r_{13})/a) = -1$, its inverse in (A1.3) takes the "+" sign (UPC) respectively the "-" sign (LPC), that is $x = x(z) = \pm(b \pm c^{\pm}r_{13})/a\cdot z \pm c^{\pm}\cdot\sqrt{1-z^2}$. Insert it into the function $\Delta^*(z) = \Delta^*(x(z), z) = w(z, r_{23}, -1) - (az \mp bx(z))/c^{\pm}$ to see that $\Delta(x,y) \ge \Delta^*(x, z(x)) = \Delta^*(z) = 0$ as desired.

Case 2: Similarly to Case 1, one shows that $z(x)$ is decreasing in the interval $-1 \le x \le -(b + c^{+}r_{13})/a$ (UPC), respectively $z(x)$ is increasing in the interval $(b - c^{-}r_{13})/a \le x \le 1$ (LPC), and in particular $y \le z(x) \le z(-1) = -(\pm c^{\pm}r_{13} + b)/a \le r_{23}$, where the last inequality is true because $ar_{23} + b + c^{+}r_{13} = (b + c^{+}) \cdot (1 + r_{13}) \ge 0$ (UPC), $ar_{23} + b - c^{-}r_{13} \ge c^{-} \cdot (1 - r_{13}) \ge 0$ (LPC).

Now, we show that $\Delta(x,y) = (ay \mp bx)/c^{\pm} - w(y, r_{23}, 1) \ge 0$ under the constraints $-1 \le y \le z(x)$, $-1 \le x \le -(b + c^{+}r_{13})/a$ (UPC), $(b - c^{-}r_{13})/a \le x \le 1$ (LPC). The function $h(y) = \pm(ay/c^{\pm} - w(y, r_{23}, 1))$ is decreasing (UPC), respectively increasing (LPC), in the interval $-1 \le y \le -(b \pm c^{\pm}r_{13})/a$:

$$h'(y) = \pm(a - c^{\pm}r_{23})/c^{\pm} \pm b\cdot y/\sqrt{1-y^2} \le (\ge) 0 \quad \Leftrightarrow \quad y/\sqrt{1-y^2} \le \beta := (c^{\pm}r_{23} - a)/bc^{\pm}$$
$$\Leftrightarrow \quad y \le \beta/\sqrt{1+\beta^2} = (c^{\pm}r_{23} - a)/b = \mp(b \pm c^{\pm}r_{13})/a$$

It follows that $\Delta(x,y) \ge \Delta^*(x, z(x)) = (az(x) \mp bx)/c^{\pm} - w(z(x), r_{23}, 1)$. Since $z(x)$ is decreasing (UPC), respectively increasing (LPC), and $z(\mp(b \pm c^{\pm}r_{13})/a) = -1$, $z(\mp 1) = \mp(b \pm c^{\pm}r_{13})/a$, its inverse in (A1.3) takes the "-" sign (UPC) respectively the "+" sign (LPC), that is we have $x = x(z) = \pm(b \pm c^{\pm}r_{13})/a\cdot z \mp c^{\pm}\cdot\sqrt{1-z^2}$. Insert this expression into the function $\Delta^*(z) = \Delta^*(x(z), z) = (az \mp bx(z))/c^{\pm} - w(z, r_{23}, 1)$ to see that $\Delta(x,y) \ge \Delta^*(x, z(x)) = \Delta^*(z) = 0$.

(2) $(y, t) \in E^{-}(r_{23})$ (x, t) arbitrary

(2.1) $y + r_{23} < 0$

We have $\frac{\partial}{\partial t}C_{23}(y,t)=0$, which follows from Lemma A1.2 and (3.11) as follows:

(i) using (P6) of Lemma A1.2: $t \le w(y, r_{23}, -1) < 1 + r_{23} - y \Rightarrow (y,t) \notin L_1(r_{23})$
(ii) using $y + r_{23} < 0$: $t \ge -1 > y + r_{23} - 1 \Rightarrow (y,t) \notin L_4(r_{23})$

Furthermore, we must have $x + r_{13} < 0$ (UPC), respectively $x + r_{13} \ge 0$ (LPC), in Cases (I.a) and (I.b), which is shown as follows:

(UPC) Case (I.a): $ay - bx \ge c^+ \Rightarrow x \le (ay - c^+)/b < -(c^+ + ar_{13})/b \le -r_{13}$, where the last inequality follows from $ar_{23} - br_{13} + c^+ = 2c^+ \ge 0$
(UPC) Case (I.b): $ay - bx > -c^+ \Rightarrow x < (c^+ + ay)/b < (c^+ - ar_{23})/b = -r_{13}$
(LPC) Case (I.a): if $x < -r_{13}$, then $0 \le c^- \le ay + bx < -ar_{23} - br_{13} = -c^- \le 0$, a contradiction
(LPC) Case (I.b): if $x < -r_{13}$, then $-c^- < ay + bx < -ar_{23} - br_{13} = -c^-$, a contradiction

In these Cases, if $t \le w(x, r_{13}, -1)$, then $\frac{\partial}{\partial t}C_{13}(x,t)=0$ (UPC), $\frac{\partial}{\partial t}C_{13}(x,t)=\frac{1}{2}$ (LPC):

(UPC): (i) using (P6) of Lemma 3.2: $t \le w(x, r_{13}, -1) < 1 + r_{13} - x \Rightarrow (x,t) \notin L_1(r_{13})$
(ii) using $x + r_{13} < 0$: $t \ge -1 > x + r_{13} - 1 \Rightarrow (x,t) \notin L_4(r_{13})$

(LPC): (P1) of Lemma A1.2 $\Rightarrow x + r_{13} \ge 0 \Rightarrow w(x, r_{13}, -1) \le x + r_{13} - 1 \Rightarrow (x,t) \in L_4(r_{13})$
We have three main cases:

(a) $c^{\pm}t \ge ay \mp bx$ (Case (I.c) and possibly Case (I.b))
We have necessarily $\frac{\partial}{\partial t}C_{13}(x,t) \ge \frac{\partial}{\partial t}C_{23}(y,t)=0$ (UPC), $\frac{\partial}{\partial t}C_{13}(x,t)+\frac{\partial}{\partial t}C_{23}(y,t)-\frac{1}{2} \le 0$ (LPC)
(b) $c^{\pm}t \le ay \mp bx < c^{\pm} \cdot w(x, r_{13}, -1)$ (possible in Cases (I.a) and (I.b))
By the above we have $\frac{\partial}{\partial t}C_{13}(x,t)=\frac{\partial}{\partial t}C_{23}(y,t)=0$ (UPC), $\frac{\partial}{\partial t}C_{13}(x,t)+\frac{\partial}{\partial t}C_{23}(y,t)-\frac{1}{2}=0$ (LPC), which is consistent with the desired inequalities.
(c) $c^{\pm}t \le ay \mp bx$ and $c^{\pm} \cdot w(y, r_{23}, -1) \le ay \mp bx$ (possible in Cases (I.a) and (I.b))

We derive the inequality (C2.1) $t \le w(y, r_{23}, -1) \le w(x, r_{13}, -1)$, which again implies $\frac{\partial}{\partial t}C_{13}(x,t)=\frac{\partial}{\partial t}C_{23}(y,t)=0$ (UPC), $\frac{\partial}{\partial t}C_{13}(x,t)+\frac{\partial}{\partial t}C_{23}(y,t)-\frac{1}{2}=0$ (LPC). We use the transformation of variable $z = z(x) = (c^{\pm} \cdot w(x, r_{13}, -1) \pm bx)/a$ and its inverse $x = x(z) = \pm(a-c^{\pm}r_{23})/b \cdot z \pm c^{\pm} \cdot \sqrt{1-z^2}$ in (A1.3), where the second "±" signs hold for both (UPC) and (LPC). We show that $\Delta(x,y) := w(x, r_{13}, -1) - w(y, r_{23}, -1) \ge 0$ under the constraints $z(x) \le y < -r_{23}$, $-1 \le x < -r_{13}$ (UPC), $-r_{13} \le x \le 1$ (LPC). By (P3) of Lemma A1.2 the function $w(y, r_{23}, -1)$ is decreasing in y. It follows that $\Delta(x,y) \ge \Delta^*(x, z(x)) = w(x, r_{13}, -1) - w(z(x), r_{23}, -1) = (az(x) \mp bx)/c^{\pm} - w(z(x), r_{23}, -1)$.
Now, insert the inverse $x = x(z)$ into $\Delta^*(z) = \Delta^*(x(z), z) = (az \mp bx(z))/c^{\pm} - w(z, r_{23}, -1)$ to see that $\Delta(x,y) \ge \Delta^*(x, z(x)) = \Delta^*(z) = b\sqrt{1-z^2}(1 \pm 1) \ge 0$ as desired.
(2.2) $y + r_{23} \ge 0$
Property (P1) implies $t \le w(y, r_{23}, -1) \le y + r_{23} - 1 \Rightarrow (y,t) \in L_4(r_{23}) \Rightarrow \frac{\partial}{\partial t}C_{23}(y,t)=\frac{1}{2}$
We distinguish between two main cases:

(a) $c^{\pm}t \le ay \mp bx$ (Case (I.a) and possibly Case (I.b))

We have necessarily $\frac{\partial}{\partial t}C_{13}(x,t) \le \frac{\partial}{\partial t}C_{23}(y,t) = \frac{1}{2}$ (UPC), $\frac{\partial}{\partial t}C_{13}(x,t) + \frac{\partial}{\partial t}C_{23}(y,t) - \frac{1}{2} \ge 0$ (LPC).

(b)$ay \mp bx \le c^{\pm}t \le c^{\pm} \cdot w(x, r_{13}, -1)$ (Case (I.c) and possibly Case (I.b))

We show that $\frac{\partial}{\partial t}C_{13}(x,t) = \frac{1}{2}$ (UPC), $\frac{\partial}{\partial t}C_{13}(x,t) = 0$ (LPC), which is consistent with $\frac{\partial}{\partial t}C_{13}(x,t) \ge \frac{\partial}{\partial t}C_{23}(y,t)$ (UPC), $\frac{\partial}{\partial t}C_{13}(x,t) + \frac{\partial}{\partial t}C_{23}(y,t) - \frac{1}{2} \le 0$ (LPC). It suffices to show the following inequalities:

Case 1: $y \le (a - c^{\pm}r_{23})/b$
(C2.2.1) $t \le w(y, r_{23}, -1) \le w(x, r_{13}, -1$, and $x + r_{13} \ge 0$ (UPC), $x + r_{13} \le 0$ (LPC)
Case 2: $y \ge (a - c^{\pm}r_{23})/b$
(C2.2.2) $w(x, r_{13}, 1) \le (ay - bx)/c \le t \le w(y, r_{23}, -1)$, and $x - r_{13} \ge 0$ (UPC), $x - r_{13} \le 0$ (LPC)

Indeed, if these conditions hold, then $\frac{\partial}{\partial t}C_{13}(x,t) = \frac{1}{2}$ (UPC), $\frac{\partial}{\partial t}C_{13}(x,t) = 0$ (LPC) as follows:

(UPC) Case 1: Property (P1) implies $t \le w(x, r_{13}, -1) \le x + r_{13} - 1 \Rightarrow (x,t) \in L_4(r_{13})$
(LPC) Case 1:
(i) using (P6) of Lemma 3.2: $t \le w(x, r_{13}, -1) < 1 + r_{13} - x \Rightarrow (x,t) \notin L_1(r_{13})$
(ii)using $x + r_{13} \le 0$: $t \ge -1 \ge x + r_{13} - 1 \Rightarrow (x,t) \notin L_4(r_{13})$
(UPC) Case 2: Property (P2) implies $t \ge w(x, r_{13}, 1) \ge 1 + r_{13} - x \Rightarrow (x,t) \in L_1(r_{13})$
(LPC) Case 2:
(i) using $x - r_{13} \le 0$: $t \le 1 \le 1 + r_{13} - x \Rightarrow (x,t) \notin L_1(r_{13})$
(ii)using (P5) of Lemma 3.2: $t \ge w(x, r_{13}, 1) > x + r_{13} - 1 \Rightarrow (x,t) \notin L_4(r_{13})$

To show the above inequalities we use the transformation $z = z(y) = \pm(ay - c^{\pm} \cdot w(y, r_{23}, -1))/b$.

Case 1: The function $z(y)$ is increasing (UPC) (respectively decreasing (LPC)) in the interval $-r_{23} \le y \le (a - c^{\pm}r_{23})/b$: $z'(y) = \pm(a - c^{\pm}r_{23})/b \mp c^{\pm} \cdot y/\sqrt{1-y^2} \ge (\le) 0 \Leftrightarrow y \le (a - c^{\pm}r_{23})/b$, hence $x \ge z(y) \ge z(-r_{23}) = (c^{+} - ar_{23})/b = -r_{13}$ (UPC), $x \le z(y) \le z(-r_{23}) = (ar_{23} - c^{-})/b = -r_{13}$ (LPC). Now, we show that $\Delta(x,y) = w(x, r_{13}, -1) - w(y, r_{23}, -1) \ge 0$ under the constraints $z(y) \le x \le 1$ (UPC), $-1 \le x \le z(y)$ (LPC), $-r_{23} \le y \le (a - c^{\pm}r_{23})/b$.

Since $w(x, r_{13}, -1)$ is increasing in x (UPC) (decreasing in x (LPC)) by (P3) of Lemma A1.2, we have $\Delta(x,y) \ge \Delta^*(y, z(y)) = w(z(y), r_{13}, -1) - w(y, r_{23}, -1) = w(z(y), r_{13}, -1) - (ay \mp bz(y))/c$.

Since $z(y)$ is increasing and $z(-r_{23}) = -r_{13}$, $z((a - c^{+}r_{23})/b) = 1$ (UPC), respectively $z(y)$ is decreasing and $z(-r_{23}) = -r_{13}$, $z((a - c^{-}r_{23})/b) = -1$ (LPC), its inverse takes the "-" sign for both (UPC) and (LPC), that is $y = y(z) = \pm(a - c^{\pm}r_{23})/b \cdot z - c^{\pm} \cdot \sqrt{1-z^2}$. Insert it into the function $\Delta^*(z) = \Delta^*(y(z), z) = w(z, r_{23}, -1) - (ay(z) \mp bz)/c^{\pm}$ to see that $\Delta(x,y) \ge \Delta^*(y, z(y)) = \Delta^*(z) = 0$ as desired.

Case 2: Similarly to Case 1, one shows that $z(y)$ is decreasing (UPC) (respectively increasing (LPC)) in the interval $(a - c^{\pm}r_{23})/b \le y \le 1$, and in particular $x \ge z(y) \le z(1) = (a - c^{+}r_{23})/b \ge r_{13}$ (UPC), $x \le z(y) \le z(1) = (c^{-}r_{23} - a)/b \le r_{13}$ (LPC), where the last inequalities are true because $a - br_{13} - c^{+}r_{23} \ge c^{+} \cdot (1 - r_{23}) \ge 0$ (UPC), $br_{13} + a - c^{-}r_{23} \ge c^{-} \cdot (1 - r_{23}) \ge 0$

(LPC). Now, we show that $\Delta(x,y) = (ay \mp bx)/c^{\pm} - w(x, r_{13}, 1) \ge 0$ under the constraints $z(y) \le x \le 1$ (UPC), $-1 \le x \le z(y)$ (LPC), $(a - c^{\pm}r_{23})/b \le y \le 1$. The function $h(x) = w(x, r_{13}, 1) \pm bx/c$ is decreasing in the interval $(a - c^{+}r_{23})/b \le x \le 1$ (UPC), respectively increasing in the interval $-1 \le x \le (c^{-}r_{23} - a)/b$ (LPC):

$$h'(x) = \pm(b \pm cr_{13})/c^{\pm} - a \cdot x/\sqrt{1-x^2} \le (\ge) 0 \quad \Leftrightarrow \quad x \ge (a - c^{+}r_{23})/b, \quad x \le (c^{-}r_{23} - a)/b$$

It follows that $\Delta(x,y) \ge \Delta^*(y, z(y)) = (ay \mp bz(y))/c^{\pm} - w(z(y), r_{13}, 1)$. Since $z(y)$ is decreasing and $z((a - c^{+}r_{23})/b) = 1$, $z(1) = (a - c^{+}r_{23})/b$ (UPC), respectively $z(y)$ is increasing and $z((a - c^{-}r_{23})/b) = -1$, $z(1) = (c^{-}r_{23} - a)/b$ (LPC), its inverse takes the "+" sign for both (UPC) and (LPC), that is $y = y(z) = \pm(a - c^{\pm}r_{23})/b \cdot z + c^{\pm} \cdot \sqrt{1-z^2}$. Insert it into $\Delta^*(z) = \Delta^*(y(z), z) = (ay(z) \mp bz)/c^{\pm} - w(z, r_{13}, 1)$ to see that $\Delta(x,y) \ge \Delta^*(y, z(y)) = \Delta^*(z) = 0$.

(3) $(x, t) \in E(r_{13})$ and $(y, t) \in E(r_{23})$

The validity of the inequality $\frac{\partial}{\partial t} C_{13}(x,t) \le \frac{\partial}{\partial t} C_{23}(y,t)$ in the Cases (I.a) and (I.b) follows from (A2.1). In Case (I.c) we have $(ay - bx)/c^{+} \le -1 \le t$, hence $\frac{\partial}{\partial t} C_{13}(x,t) \ge \frac{\partial}{\partial t} C_{23}(y,t)$. Similarly, the validity of the inequality $\frac{\partial}{\partial t} C_{13}(x,t) + \frac{\partial}{\partial t} C_{23}(y,t) - \frac{1}{2} \ge 0$ in the Cases (I.a) and (I.b) follows from (A2.2). In Case (I.c) we have $(ay + bx)/c^{-} \le -1 \le t$, hence $\frac{\partial}{\partial t} C_{13}(x,t) + \frac{\partial}{\partial t} C_{23}(y,t) - \frac{1}{2} \le 0$.

(4) $(x, t) \in E^{+}(r_{13})$, (y, t) arbitrary

(4.1) $x - r_{13} < 0$ (UPC), $x - r_{13} \ge 0$ (LPC)

We first note that $\frac{\partial}{\partial t} C_{13}(x,t) = 0$ (UPC), $\frac{\partial}{\partial t} C_{13}(x,t) = \frac{1}{2}$ (LPC):

(UPC): (i) using $x - r_{13} < 0$: $t \le 1 < 1 + r_{13} - x \Rightarrow (x, t) \notin L_1(r_{13})$
(ii) using (P5) of Lemma A1.2: $t \ge w(x, r_{13}, 1) > x + r_{13} - 1 \Rightarrow (x, t) \notin L_4(r_{13})$
(LPC): (P2) of Lemma A1.2: $t \ge w(x, r_{13}, 1) \ge 1 + r_{13} - x \Rightarrow (x, t) \in L_1(r_{13}) \Rightarrow$

We must have $y - r_{23} < 0$ in Cases (I.b) and (I.c). We argue as follows:

(UPC) Case (I.b): $x < r_{13} \Rightarrow y < (c + bx)/a \le (c + br_{13})/a = r_{23}$
(UPC) Case (I.c): if $y \ge r_{23}$, $-x > -r_{13}$ then $ay - bx \ge ar_{23} - br_{13} = c$ in contradiction to the assumption $ay - bx \le -c$ defining Case (I.c).
(LPC) Case (I.b): $-x \le -r_{13} \Rightarrow y < (c^{-} - bx)/a \le (c^{-} - br_{13})/a = r_{23}$
(LPC) Case (I.c): if $y \ge r_{23}$, $x \ge r_{13}$ then $ay + bx \ge ar_{23} + br_{13} = c^{-}$ in contradiction to the assumption $ay + bx \le -c^{-}$ defining Case (I.c).

In this situation, if one assumes $t \ge w(y, r_{23}, 1)$, then $\frac{\partial}{\partial t} C_{23}(y,t) = 0$ in virtue of (3.11):

(UPC): (i) using $y - r_{23} < 0$: $t \le 1 < 1 + r_{23} - y \Rightarrow (y, t) \notin L_1(r_{23})$
(ii) using (P5) of Lemma 3.2: $t \ge w(y, r_{23}, 1) > y + r_{23} - 1 \Rightarrow (y, t) \notin L_4(r_{23})$
(LPC): (i) using $y - r_{23} < 0$: $t \le 1 < 1 + r_{23} - y \Rightarrow (y, t) \notin L_1(r_{23})$
(ii) using (P5) of Lemma 3.2: $t \ge w(y, r_{23}, 1) > y + r_{23} - 1 \Rightarrow (y, t) \notin L_4(r_{23})$

We have three main cases:

(a) $c^{\pm}t \le ay \mp bx$ (Case (I.a) and possibly Case (I.b))

We have necessarily $\frac{\partial}{\partial t}C_{13}(x,t)=0\le\frac{\partial}{\partial t}C_{23}(y,t)$ (UPC), $\frac{\partial}{\partial t}C_{13}(x,t)+\frac{\partial}{\partial t}C_{23}(y,t)-\frac{1}{2}\ge 0$ (LPC).

(b) $c^{\pm}t \ge ay \mp bx > c^{\pm}\cdot w(y, r_{23}, 1)$ (possible in Cases (I.b) and (I.c))

By the above we have $\frac{\partial}{\partial t}C_{13}(x,t)=\frac{\partial}{\partial t}C_{23}(y,t)=0$ (UPC), $\frac{\partial}{\partial t}C_{13}(x,t)+\frac{\partial}{\partial t}C_{23}(y,t)-\frac{1}{2}=0$ (LPC), which is consistent with the desired inequalities

(c) $c^{\pm}t \ge ay \mp bx$ and $ay \mp bx \le c^{\pm}\cdot w(y, r_{23}, 1)$ (possible in Cases (I.b) and (I.c))

We derive the inequality (C4.1) $t \ge w(x, r_{13}, 1) \ge w(y, r_{23}, 1)$, which again implies that $\frac{\partial}{\partial t}C_{13}(x,t)=\frac{\partial}{\partial t}C_{23}(y,t)=0$ (UPC), $\frac{\partial}{\partial t}C_{13}(x,t)+\frac{\partial}{\partial t}C_{23}(y,t)-\frac{1}{2}=0$ (LPC). We use the transformation $z = z(y) = \pm(ay - c^{\pm}\cdot w(y, r_{23}, 1))/b$ and its inverse $y = y(z) = \pm(a - c^{\pm} r_{23})/b\cdot z \pm c^{\pm}\cdot\sqrt{1-z^2}$, where the second "±" signs hold for both (UPC) and (LPC). We show that $\Delta(x, y) = w(x, r_{13}, 1) - w(y, r_{23}, 1) \ge 0$ under the constraints $z(y) \le x < r_{13}$ (UPC), $r_{13} \le x \le z(y)$ (LPC), $-1 \le y < r_{23}$. By (P4) of Lemma A1.2 the function $w(x, r_{13}, 1)$ is increasing in x (UPC), respectively decreasing in x (LPC). It follows that $\Delta(x, y) \ge \Delta^*(y, z(y)) = w(z(y), r_{13}, 1) - w(y, r_{23}, 1) = w(z(y), r_{13}, 1) - (ay \mp bz(y))/c^{\pm}$.

Insert the inverse $y = y(z)$ into $\Delta^*(z) = \Delta^*(y(z), z) = w(z, r_{13}, 1) - (ay(z) \mp bz)/c^{\pm}$ to see that $\Delta(x,y)\ge\Delta^*(y,z(y))=\Delta^*(z)=a\sqrt{1-z^2}(1\mp 1)\ge 0$ as desired.

(4.2) $x - r_{13} \ge 0$ (UPC), $x - r_{13} < 0$ (LPC)

We first note that $\frac{\partial}{\partial t}C_{13}(x,t)=\frac{1}{2}$ (UPC), $\frac{\partial}{\partial t}C_{13}(x,t)=0$ (LPC):

(UPC): Property (P2) implies $t \ge w(x, r_{13}, 1) \ge 1 + r_{13} - x \Rightarrow (x, t) \in L_1(r_{13})$

(LPC): (i) using $x - r_{13} < 0$: $t \le 1 < 1 + r_{13} - x \Rightarrow (x, t) \notin L_1(r_{13})$

(ii) using (P5) of Lemma 3.2: $t \ge w(x, r_{13}, 1) > x + r_{13} - 1 \Rightarrow (x, t) \notin L_4(r_{13})$

We distinguish between two main cases:

(a) $c^{\pm}t \ge ay \mp bx$ (Case (I.c) and possibly Case (I.b))

We have necessarily $\frac{\partial}{\partial t}C_{13}(x,t)=\frac{1}{2}\ge\frac{\partial}{\partial t}C_{23}(y,t)$ (UPC), $\frac{\partial}{\partial t}C_{13}(x,t)+\frac{\partial}{\partial t}C_{23}(y,t)-\frac{1}{2}\le 0$ (LPC)

(b) $ay \mp bx \ge c^{\pm}t \ge c^{\pm}\cdot w(x, r_{13}, 1)$ (Case (I.a) and possibly Case (I.b))

We show that , $\frac{\partial}{\partial t}C_{23}(y,t)=\frac{1}{2}$ which is consistent with the desired inequalities.

We show Case 1: $x \le (b + c^{+}r_{13})/a$ (UPC), $x \ge (c^{-}r_{13} - b)/a$ (LPC) (C4.2.1) $t \ge w(x, r_{13}, 1) \ge w(y, r_{23}, 1)$, and $y - r_{23} \ge 0$ (UPC), $y - r_{23} > 0$ (LPC)

Case 2: $x \ge (b + c^{+}r_{13})/a$ (UPC), $x \le (c^{-}r_{13} - b)/a$ (LPC) (C4.2.2) $c^{\pm}\cdot w(x, r_{13}, 1) \le c^{\pm}t \le ay \mp bx < c^{\pm}\cdot w(y, r_{23}, -1)$, and $y + r_{23} \ge 0$

Indeed, if these conditions hold, then we have $\frac{\partial}{\partial t}C_{23}(y,t)=\frac{1}{2}$ in virtue of (3.11) as follows:

Case 1: (P2) $\Rightarrow$ $t \ge w(y, r_{23}, 1) \ge 1 + r_{23} - y \Rightarrow (y, t) \in L_1(r_{23})$

Case 2: (P1) $\Rightarrow$ $t < w(y, r_{23}, -1) \le y + r_{23} - 1 \Rightarrow (y, t) \in L_4(r_{23})$

To show the above inequalities we use the transformation $z = z(x) = (c^{\pm}\cdot w(x, r_{13}, 1) \pm bx)/a$.

Case 1: The function $z(x)$ is increasing in the interval $r_{13} \le x \le (b + c^{+}r_{13})/a$ (UPC), respectively decreasing in the interval $(c^{-}r_{13} - b)/a \le x < r_{13}$ (UPC): $z'(x)=\pm(b\pm c^{\pm}r_{13})/a - c^{\pm}\cdot x/\sqrt{1-x^2}\ge(\le)0 \Leftrightarrow x\le(b+c^{+}r_{13})/a,\ x\ge(c^{-}r_{13}-b)/a$.

In particular $y \ge z(x) > z(r_{13}) = (c^{+} + br_{13})/a = r_{23}$ (UPC), $y \ge z(x) > z(r_{13}) = (c^{-} - br_{13})/a = r_{23}$ (LPC). Now, we show that $\Delta(x, y) = w(x, r_{13}, 1) - w(y, r_{23}, 1) \ge 0$ under the constraints $z(x) \le y \le 1$, $r_{13} \le x \le (b + c^{+}r_{13})/a$ (UPC), $(c^{-}r_{13} - b)/a \le x < r_{13}$ (LPC). Since $w(y, r_{23}, 1)$ is decreasing in y by (P4) of Lemma 3.2, we have $\Delta(x,y) \geq \Delta^*(x, z(x)) = w(x, r_{13}, 1) - w(z(x), r_{23}, 1) = (az(x) \mp bx)/c^{\pm} - w(z(x), r_{23}, 1)$.

Since $z(x)$ is increasing and $z(r_{13}) = r_{23}$, $z((b + c^+r_{13})/a) = 1$ (UPC), respectively $z(x)$ is decreasing, $z(r_{13}) = r_{23}$, $z((c^-r_{13} - b)/a) = 1$ its inverse in (A1.3) takes the "-" sign (UPC), respectively the "+" sign (LPC), that is $x = x(z) = \pm(b \pm c^{\pm}r_{13})/a \cdot z \mp c \cdot \sqrt{1-z^2}$. Insert it into $\Delta^*(z) = \Delta^*(x(z), z) = (az \mp bx(z))/c^{\pm} - w(z, r_{23}, 1)$ to see that $\Delta(x,y) \geq \Delta^*(x, z(x)) = \Delta^*(z) = 0$ as desired.

Case 2: Similarly to Case 1, one shows that $z(x)$ is decreasing in the interval $(b + c^+r_{13})/a \leq x \leq 1$ (UPC), respectively $z(x)$ is increasing in the interval $-1 \leq x \leq (c^-r_{13} - b)/a$ (LPC). In particular $y \geq z(x) \geq z(1) = (b + c^+r_{13})/a \geq -r_{23}$ (UPC), $y \geq z(x) \geq z(-1) = (b - c^-r_{13})/a \geq -r_{23}$ (LPC), where the last inequalities are true because $ar_{23} + b + c^+r_{13} = (b + c^+) \cdot (1 + r_{13}) \geq 0$ (UPC), $ar_{23} + b - c^-r_{13} \geq c^- \cdot (1 - r_{13}) \geq 0$ (LPC).

Now, we show that $\Delta(x,y) = w(y, r_{23}, -1) - (ay \mp bx)/c^{\pm} \geq 0$ under the constraints $z(x) \leq y \leq 1$, $(b + c^+r_{13})/a \leq x \leq 1$ (UPC), $-1 \leq x \leq (c^-r_{13} - a)/b$ (LPC). The function $h(y) = w(y, r_{23}, -1) - ay/c^{\pm}$ is increasing in the interval $(b \pm c^{\pm}r_{13})/a \leq y \leq 1$:

$$h'(y) = \left(c^{\pm}r_{23} - a\right)/c^{\pm} + b \cdot y/\sqrt{1-y^2} \geq 0 \quad \Leftrightarrow \quad y \geq \left(a - c^{\pm}r_{23}\right)/b = \left(b \pm c^{\pm}r_{13}\right)/a$$

It follows that $\Delta(x,y) \geq \Delta^*(x, z(x)) = w(z(x), r_{23}, -1) - (az(x) \mp bx)/c^{\pm}$. Since $z(x)$ is decreasing and $z((b + c^+r_{13})/a) = 1$, $z(1) = (b + c^+r_{13})/a$ (UPC), respectively $z(x)$ is increasing and $z((c^-r_{13} - b)/a) = 1$, $z(-1) = (a - c^-r_{23})/b$ (LPC), its inverse takes the "+" sign (UPC), respectively the "-" sign (LPC), that is $x = x(z) = \pm(b \pm c^{\pm}r_{13})/a \cdot z \pm c^{\pm} \cdot \sqrt{1-z^2}$. Insert it into $\Delta^*(z) = \Delta^*(x(z), z) = w(z, r_{23}, 1) - (az \mp bx(z))/c^{\pm}$ to see that $\Delta(x,y) \geq \Delta^*(x, z(x)) = \Delta^*(z) = 0$.

(5) $(y, t) \in E^+(r_{23})$, (x, t) arbitrary

(5.1) $y - r_{23} \geq 0$

Property (P2) implies $t \geq w(y, r_{23}, 1) \geq 1 + r_{23} - y \Rightarrow (y, t) \in L_1(r_{23}) \rightarrow \frac{\partial}{\partial t} C_{23}(y, t) = \frac{1}{2}$

We must have $x - r_{13} \geq 0$ (UPC), $x - r_{13} \geq 0$ (LPC), in Cases (I.b) and (I.c):

(UPC) Case (I.b): $x > (ay - c^+)/b \leq (ar_{23} - c^+)/b = r_{13}$

(UPC) Case (I.c): if $-x > -r_{13}$, $y \geq r_{23}$ then $ay - bx > ar_{23} - br_{13} = c^+$ in contradiction to the assumption $ay - bx \leq -c^+$ defining Case (I.c).

(LPC) Case (I.b): $-y \leq -r_{23} \Rightarrow x < (c^- - ay)/b \leq (c^- - ar_{23})/b = r_{13}$

(LPC) Case (I.c): if $x \geq r_{13}$, $y \geq r_{23}$ then $ay + bx \geq ar_{23} + br_{13} = c^-$ in contradiction to the assumption $ay + bx \leq -c^-$ defining Case (I.c).

In this situation, if $t \geq w(x, r_{13}, 1)$, then $\frac{\partial}{\partial t} C_{13}(x, t) = \frac{1}{2}$ (UPC), $\frac{\partial}{\partial t} C_{13}(x, t) = 0$ (LPC):

(UPC): using (P2) of Lemma A1.2: $t \geq w(x, r_{13}, 1) \geq 1 + r_{13} - x \Rightarrow (x, t) \in L_1(r_{13})$

(LPC):

(i) using $x - r_{13} < 0$: $t \leq 1 < 1 + r_{13} - x \Rightarrow (x, t) \notin L_1(r_{13})$

(ii) using (P5) of Lemma A1.2: $t \geq w(x, r_{13}, 1) > x + r_{13} - 1 \Rightarrow (x, t) \notin L_4(r_{13})$

We distinguish between three main cases:

(a) $c^{\pm}t \leq ay \mp bx$ (Case (I.a) and possibly Case (I.b))

We have necessarily $\frac{\partial}{\partial t}C_{13}(x,t) \le \frac{\partial}{\partial t}C_{23}(y,t) = \frac{1}{2}$ (UPC), $\frac{\partial}{\partial t}C_{13}(x,t) + \frac{\partial}{\partial t}C_{23}(y,t) - \frac{1}{2} \ge 0$ (LPC)

(b) $c^{\pm}t \ge ay \mp bx > c^{\pm} \cdot w(x, r_{13}, 1)$ (possible in Cases (I.b) and (I.c))

By the above we have $\frac{\partial}{\partial t}C_{13}(x,t) = \frac{\partial}{\partial t}C_{23}(y,t) = \frac{1}{2}$ (UPC), $\frac{\partial}{\partial t}C_{13}(x,t) + \frac{\partial}{\partial t}C_{23}(y,t) - \frac{1}{2} = 0$ (LPC), which is consistent with the desired inequalities

(c) $c^{\pm}t \ge ay \mp bx$ and $ay \mp bx \le c^{\pm} \cdot w(x, r_{13}, 1)$ (possible in Cases (I.b) and (I.c))

We derive the inequality (C5.1) $t \ge w(y, r_{23}, 1) \ge w(x, r_{13}, 1)$, which again implies $\frac{\partial}{\partial t}C_{13}(x,t) = \frac{\partial}{\partial t}C_{23}(y,t) = \frac{1}{2}$ (UPC), $\frac{\partial}{\partial t}C_{13}(x,t) + \frac{\partial}{\partial t}C_{23}(y,t) - \frac{1}{2} = 0$ (LPC). We use the transformation $z = z(x) = (c \cdot w(x, r_{13}, 1) \pm bx)/a$ and its inverse in (A1.3) to show that $\Delta(x,y) = w(y, r_{23}, 1) - w(x, r_{13}, 1) \ge 0$ under the constraints $r_{23} \le y \le z(x)$, $r_{13} \le x \le 1$ (UPC), $-1 \le x < r_{13}$ (LPC). By (P4) of Lemma A1.2 the function $w(y, r_{23}, 1)$ is decreasing in y, hence $\Delta(x,y) \ge \Delta^*(x, z(x)) = w(z(x), r_{23}, 1) - w(x, r_{13}, 1) = w(z(x), r_{23}, 1) - (az(x) \div bx)/c^{\pm}$.

Insert the inverse $x = x(z)$ into $\Delta^*(z) = \Delta^*(x(z), z) = w(z, r_{23}, 1) - (az \mp bx(z))/c^{\pm}$ to see that $\Delta(x,y) \ge \Delta^*(x, z(x)) = \Delta^*(z) = b\sqrt{1-z^2}(1 \pm 1) \ge 0$ as desired.

(5.2) $y - r_{23} < 0$

We first note that $\frac{\partial}{\partial t}C_{23}(y,t) = 0$:

(i) using $y - r_{23} < 0$: $t \le 1 < 1 + r_{23} - y \Rightarrow (y,t) \notin L_1(r_{23})$

(ii) using (P5) of Lemma 3.2: $t \ge w(y, r_{23}, 1) > y + r_{23} - 1 \Rightarrow (y,t) \notin L_4(r_{23})$

We distinguish between two main cases:

(a) $c^{\pm}t \ge ay \mp bx$ (Case (I.c) and possibly Case (I.b))

We have necessarily $\frac{\partial}{\partial t}C_{13}(x,t) \ge \frac{\partial}{\partial t}C_{23}(y,t) = 0$ (UPC), $\frac{\partial}{\partial t}C_{13}(x,t) + \frac{\partial}{\partial t}C_{23}(y,t) - \frac{1}{2} \le 0$ (LPC)

(b) $c^{\pm} \cdot w(y, r_{23}, 1) \le c^{\pm}t \le ay \mp bx$ (Case (I.a) and possibly Case (I.b))

We show that $\frac{\partial}{\partial t}C_{13}(x,t) = 0$ (UPC), $\frac{\partial}{\partial t}C_{13}(x,t) = \frac{1}{2}$ (LPC), which is consistent with the desired inequalities $\frac{\partial}{\partial t}C_{13}(x,t) \le \frac{\partial}{\partial t}C_{23}(y,t) = 0$ (UPC), $\frac{\partial}{\partial t}C_{13}(x,t) + \frac{\partial}{\partial t}C_{23}(y,t) - \frac{1}{2} \ge 0$ (LPC)

Case 1 $y \ge (c^{\pm}r_{23} - a)/b$

(C5.2.1) $t \ge w(y, r_{23}, 1) \ge w(x, r_{13}, 1)$, and $x - r_{13} < 0$ (UPC), $x - r_{13} > 0$ (LPC)

Case 2 $y \le (c^{\pm}r_{23} - a)/b$

(C5.2.2) $c^{\pm} \cdot w(y, r_{23}, 1) \le c^{\pm}t \le ay \mp bx < c^{\pm} \cdot w(x, r_{13}, -1)$, $x + r_{13} < 0$ (UPC), $x + r_{13} \ge 0$ (LPC)

Indeed, if these conditions hold, then $\frac{\partial}{\partial t}C_{13}(x,t) = 0$ in virtue of (3.11) as follows:

UPC) Case 1 (i) using $x - r_{13} < 0$: $t \le 1 < 1 + r_{13} - x \Rightarrow (x,t) \notin L_1(r_{13})$

(ii) using (P5) of Lemma 3.2: $t \ge w(x, r_{13}, 1) > x + r_{13} - 1 \Rightarrow (x,t) \notin L_4(r_{13})$

(LPC) Case 1: (P2)$\Rightarrow$ $t \ge w(x, r_{13}, 1) \ge 1 + r_{13} - x \Rightarrow (x,t) \in L_1(r_{13})$

(UPC) Case 2: (i) using (P6) of Lemma 3.2: $t \le w(x, r_{13}, -1) < 1 + r_{13} - x \Rightarrow (x,t) \notin L_1(r_{13})$

(ii) using $x + r_{13} < 0$ $t \ge -1 > x + r_{13} - 1 \Rightarrow (x,t) \notin L_4(r_{13})$

(LPC) Case 2: (P1) $\Rightarrow$ $t < w(x, r_{13}, -1) \le x + r_{13} - 1 \Rightarrow (x,t) \in L_4(r_{13})$

To show the above inequalities we use the transformation $z = z(y) = \pm(ay - c^{\pm} \cdot w(y, r_{23}, 1))/b$.

Case 1: Since the function $z(y)$ is increasing (UPC) (respectively decreasing (LPC)) in the interval $(c^{\pm}r_{23} - a)/b \le y < r_{23}$, we have in particular $x \le z(y) < z(r_{23}) = (ar_{23} - c^{+})/b = r_{13}$ (UPC), $x \ge z(y) > z(r_{23}) = (c^{-} - ar_{23})/b = r_{13}$ (LPC). We show that $\Delta(x,y) = w(y, r_{23}, 1) - w(x, r_{13}, 1) \ge 0$ under the constraints $-1 \le x \le z(y)$ (UPC), $z(y) \le x \le 1$ (LPC), $(c^{\pm}r_{23} - a)/b \le y < r_{23}$. With (P4) of Lemma A1.2, $w(x, r_{13}, 1)$ is increasing in x (UPC), decreasing in x (LPC), hence $\Delta(x,y) \ge \Delta^*(y, z(y)) = w(y, r_{23}, 1) - w(z(y), r_{13}, 1) = (ay \mp bz(y))/c^{\pm} - w(z(y), r_{13}, 1)$.

Since $z(y)$ is increasing and $z(r_{23}) = r_{13}$, $z((c^{+}r_{23} - a)/b) = -1$ (UPC), respectively $z(y)$ is decreasing and $z(r_{23}) = r_{13}$, $z((c^{-}r_{23} - a)/b) = 1$ (LPC), its inverse in (A1.3) takes the "+" sign: $y(z) = \pm(a - cr_{23})/b \cdot z + c\sqrt{1-z^2}$. Insert into $\Delta^*(z) = \Delta^*(y(z), z) = (ay(z) \mp bz)/c^{\pm} - w(z, r_{23}, 1)$ to see that $\Delta(x,y) \ge \Delta^*(y, z(y)) = \Delta^*(z) = 0$ as desired.

Case 2: One shows that $z(y)$ is decreasing (UPC) (respectively increasing (LPC)) in the interval $-1 \le y \le (c^{\pm}r_{23} - a)/b$. In particular $x \le z(y) \le z(-1) = (c^{+}r_{23} - a)/b \le -r_{13}$ (UPC), $x \ge z(y) \ge z(-1) = (a - c^{-}r_{23})/b \ge -r_{13}$ (LPC), where the last inequalities are true because $a - br_{13} - c^{+}r_{23} \ge c^{+} \cdot (1 - r_{23}) \ge 0$ (UPC), $a + br_{13} - c^{-}r_{23} \ge c^{-} \cdot (1 - r_{23}) \ge 0$ (LPC).

Now, we show that $\Delta(x,y) = w(x, r_{13}, -1) - (ay \mp bx)/c^{\pm} \ge 0$ under the constraints $-1 \le x \le z(y)$ (UPC), $z(y) \le x \le 1$ (LPC), $-1 \le y \le (c^{\pm}r_{23} - a)/b$. Since the function $h(x) = w(x, r_{13}, -1) \pm bx/c$ is decreasing (UPC), respectively increasing (LPC), we have $\Delta(x, y) \ge \Delta^*(y, z(y)) = w(z(y), r_{13}, -1) - (ay \mp bz(y))/c^{\pm}$. Since $z(y)$ is decreasing and $z((c^{+}r_{23} - a)/b) = -1$, $z(-1) = (c^{+}r_{23} - a)/b$ (UPC), respectively $z(y)$ is increasing and $z((c^{-}r_{23} - a)/b) = 1$, $z(-1) = (a - c^{-}r_{23})/b$ (LPC), its inverse in (A1.3) takes the "-" sign, that is $y(z) = \pm(a - c^{\pm}r_{23})/b \cdot z - c^{\pm} \cdot \sqrt{1-z^2}$. Insert it into $\Delta^*(z) = \Delta^*(y(z), z) = w(z, r_{13}, -1) - (ay(z) - bz)/c$ to see that $\Delta(x,y) \ge \Delta^*(y, z(y)) = \Delta^*(z) = 0$ as desired.

Authors information
URL: http://sites.google.com/site/whurlimann/

Acknowledgement
The author is grateful to the referees for their useful comments, which led to some improvement in content.

References

Allen, DE, Ashraf, MA, McAleer, M, Powell, RJ, Singh, AK: Financial dependence analysis: applications of vine copulae. (2013). http://eprints.ucm.es/17819/1/1305.pdf Accessed 31 May 2013

Bedford, T, Cooke, R: Probability density decomposition for conditionally dependent random variables modelled by vines. Ann. Math. Artif. Intell. **32**, 245–268 (2001)

Bedford, T, Cooke, R: Vines – a new graphical model for dependent random variables. Ann. Statist. **30**(4), 1031–1068 (2002)

Bernard, C, Czado, C: Multivariate Option Pricing Using Copulae. Appl. Stoch. Models Bus. Ind. (2012). doi:10.1002/asmb.1934

Brechmann, EC, Czado, C: Risk management with high-dimensional vine copulas: An analysis of the Euro Stoxx 50. Statist. Risk Modeling (2013) in press

Carathéodory, C: Über den Variabilitätsbereich der Fourierschen Konstanten von positiven harmonischen Funktionen. Rend. del Circolo Matem. di Palermo **32**, 193–217 (1911)

Cherubini, U, Luciano, E, Vecchiato, W: Copula Methods in Finance. J. Wiley Finance Series, New York (2004)

Czado, C: Pair-copula construction of multivariate copulas. In: Jaworski, P., Durante, F., Härdle, W., Rychlik, T. (eds.) Copula Theory and its Applications. Proc. Workshop held in Warsaw, 25–26 Sep. 2009. Lecture Notes in Statistics, vol. 198, pp. 93–109. Springer, Heidelberg (2010)

Czado, C, Schepsmeier, U, Min, A: Maximum likelihood estimation of mixed C-vines with application to exchange rates. Statist. Modelling **12**(3), 229–255 (2012)

Devroye, L, Letac, G: Copulas in three dimensions with prescribed correlations. (2010). http://arxiv.org/abs/1004.3146v1 [math.ST] Accessed 31 May 2013

Durante, F, Klement, EP, Quesada-Molina, JJ: Copulas: compatibility and Fréchet classes. (2007a). http://arxiv.org/abs/0711.2409v1 [math.ST] (2007a). Accessed 31 May 2013

Durante, F, Klement, EP, Quesada-Molina, JJ, Sarkoci, P: Remarks on two product-like constructions for copulas. Kybernetika **43**(2), 235–244 (2007b)

Ghosh, S: Dependence in stochastic simulation models. Cornell University, Dissertation (2004). http://www.research.ibm.com/people/g/ghoshs/pubs/thesis.pdf. Accessed 31 May 2013

Ghosh, S, Henderson, SG: Properties of the NORTA method in higher dimensions. In: Proceedings of the 2002 Winter Simulation Conference, Sch. of Operations Res. & Ind. Eng.,Cornell Univ., Ithaca, NY, USA, pp. 263–269. (2002)

Hürlimann, W: Random loss development factor curves and stochastic claims reserving. JP J Fund Appl Stat **1**(1), 49–62 (2011)

Hürlimann, W: Compatibility conditions for the multivariate normal copula with given rank correlation matrix. Pioneer J Theor Appl Stat **3**(2), 71–86 (2012a)

Hürlimann, W: Positive semi-definite correlation matrices: recursive algorithmic generation and volume measure. Pure Math Sci **1**(3), 137–149 (2012b)

Hürlimann, W: On trivariate copulas with bivariate linear Spearman marginal copulas. J. Math. Syst Sci **2**, 368–383 (2012c)

Joe, H: Families of m-variate distributions with given margins and m(m-1)/2 bivariate dependence parameters. In: Rüschendorf, L., Schweizer, B, Taylor, MD (eds.) Distributions with Fixed Marginals and Related Topics. IMS Lecture Notes Monograph Series 28, pp. 120–141. Hayward, CA (1996)

Joe, H: Multivariate models and dependence concepts. In: Monographs on Statistics and Applied Probability, vol. 73. Chapman & Hall, London (1997)

Joe, H, Kurowicka, D (eds.): Dependence Modeling – Vine Copula Handbook. World Scientific Publishing Co., Singapore (2011)

Kurowicka, D, Misiewicz, J, Cooke, R, Elliptical copulae: Proceedings of the International Conference on Monte Carlo Simulation, Monte Carlo. In: GI Schuëller & PD Spanos (Eds.) Proceedings Monte Carlo Simulation, Lisse: Balkema, pp. 209–214. (2000)

Kurowicka, D, Cooke, R: Conditional, partial and rank correlation for the elliptical copula; dependence modelling in uncertainty analysis. TR, Delft Univ Technol. In: E Zio, M Demichela & N Piccinini (Eds.), Proceedings ESREL 2001, Torino: Politecnico di Torino, pp. 1795–1802.

Kurowicka, D, Cooke, R: Uncertainty Analysis with High Dimensional Dependence Modelling. J. Wiley, Chichester (2006)

Letac, G: Jacobi polynomials and joint distributions in Rn with beta margins and prescribed correlation matrices. Warwick Workshop, Centre Res Stat Methodol (2010). http://www2.warwick.ac.uk/fac/sci/statistics/crism/workshops/orthogonal-polynomials/speakers/gerard_letac.pdf. Accessed 31 May 2013

Mauguis, P-A, Guégan, D: An econometric study of vine copulas. Int. J. Economics and Finance **2**(5), 2–14 (2010)

Perlman, MD, Wellner, JA: Squaring the circle and cubing the sphere: circular and spherical copulas. Symmetry **3**, 574–599 (2011)

Sklar, A: Fonctions de répartition et leurs marges. Publications de l'Institut Statistique de l'Université de Paris **8**, 229–31 (1959)

Steinitz, E: Bedingt konvergente Reihen und konvexe Systeme. J Reine Angew. Math. **143**, 128–175 (1914)

Vaz de Melo Mendes, B, Mendes Semeraro, M, Câmara Leal, RP: Pair-copula modeling in finance. Financial Markets and Portfolio Management **24**(2), 193–213 (2010)

Ycart, B: Extreme points in convex sets of symmetric matrices. Proc. Amer. Math. Soc. **95**(4), 607–612 (1985)

4

The Kumaraswamy-geometric distribution

Alfred Akinsete[1*†], Felix Famoye[2†] and Carl Lee[2†]

*Correspondence:
akinsete@marshall.edu
†Equal contributors
[1] Department of Mathematics, Marshall University, Huntington, West Virginia 25755, USA
Full list of author information is available at the end of the article

Abstract

In this paper, the Kumaraswamy-geometric distribution, which is a member of the *T*-geometric family of discrete distributions is defined and studied. Some properties of the distribution such as moments, probability generating function, hazard and quantile functions are studied. The method of maximum likelihood estimation is proposed for estimating the model parameters. Two real data sets are used to illustrate the applications of the Kumaraswamy-geometric distribution.

AMS 2010 Subject Classification: 60E05; 62E15; 62F10; 62P20

Keywords: Transformation; Moments; Entropy; Estimation

1 Introduction

Eugene *et al.* (2002) introduced the beta-generated family of univariate continuous distributions. Suppose X is a random variable with cumulative distribution function (CDF) $F(x)$, the CDF for the beta-generated family is obtained by applying the inverse probability transformation to the beta density function. The CDF for the beta-generated family of distributions is given by

$$G(x) = \frac{1}{B(\alpha,\beta)} \int_0^{F(x)} t^{\alpha-1}(1-t)^{\beta-1}\, dt,\ 0 < \alpha, \beta < \infty, \tag{1}$$

where $B(\alpha,\beta) = \Gamma(\alpha)\Gamma(\beta)/\Gamma(\alpha+\beta)$. The corresponding probability density function (PDF) is given by

$$g(x) = \frac{1}{B(\alpha,\beta)} [F(x)]^{\alpha-1} [1-F(x)]^{\beta-1} \left[\frac{d}{dx}F(x)\right]. \tag{2}$$

Eugene *et al.* (2002) used a normal random variable X to define and study the beta-normal distribution. Following the paper by Eugene *et al.* (2002), many other authors have defined and studied a number of the beta-generated distributions, using various forms of known $F(x)$. See for example, beta-Gumbel distribution by Nadarajah and Kotz (2004), beta-Weibull distribution by Famoye *et al.* (2005), beta-exponential distribution by Nadarajah and Kotz (2006), beta-gamma distribution by Kong *et al.* (2007), beta-Pareto distribution by Akinsete *et al.* (2008), beta-Laplace distribution by Cordeiro and Lemonte (2011), beta-generalized Weibull distribution by Singla *et al.* (2012), and beta-Cauchy distribution by Alshawarbeh *et al.* (2013), amongst others. After the paper by Jones (2009), on the tractability properties of the Kumaraswamy's distribution (Kumaraswamy 1980), Cordeiro and de Castro (2011) replaced the classical beta generator distribution with the Kumaraswamy's distribution and introduced the Kumaraswamy generated family.

Detailed statistical properties on some Kumaraswamy generated distributions include the Kumaraswamy generalized gamma distribution by de Pascoa *et al.* (2011), Kumaraswamy log-logistic distribution by de Santana *et al.* (2012) and Kumaraswamy Gumbel distribution by Cordeiro *et al.* (2012). Alexander *et al.* (2012) replaced the beta generator distribution with the generalized beta type I distribution. The authors referred to this form as the generalized beta-generated distributions (GBGD) and the generator has three shape parameters.

The above technique of generating distributions is possible, only when the generator distributions are continuous and the random variable of the generator lies between 0 and 1. In a recent work by Alzaatreh *et al.* (2013b), the authors proposed a new method for generating family of distributions, referred to by the authors as the T-X family, where a continuous random variable T is the *transformed*, and any random variable X is the *transformer*. See also Alzaatreh *et al.* (2012a, 2013a). These works opened a wide range of techniques for generating distributions of random variables with supports on $\mathbb{R}$. The T-X family enables one to easily generate, not only the continuous distributions, but the discrete distributions as well. As a result, Alzaatreh *et al.* (2012b) defined and studied the T-geometric family, which are the discrete analogues of the distribution of the random variable T.

Suppose $F(x)$ denotes the CDF of any random variable X and $r(t)$ denotes the PDF of a continuous random variable T with support $[a, b]$. Alzaatreh *et al.* (2013b) gave the CDF of the T-X family of distributions as

$$G(x) = \int_a^{W(F(x))} r(t)dt = R\{W(F(x))\}, \tag{3}$$

where $R(t)$ is the CDF of the random variable T, $W(F(x)) \in [a, b]$ is a non-decreasing and absolutely continuous function. Common support $[a, b]$ are $[0, 1]$, $(0, \infty)$, and $(-\infty, \infty)$. Alzaatreh *et al.* (2013b) studied in some details the case of a non-negative continuous random variable T with support $(0, \infty)$. With this technique, it is much easier to generate any discrete distribution. If X is a discrete random variable, the T-X family, is a family of discrete distributions, transformed from the non-negative continuous random variable T. The probability mass function (PMF) of the T-X family of discrete distributions may now be written as

$$g(x) = G(x) - G(x-1) = R\{W(F(x))\} - R\{W(F(x-1))\}. \tag{4}$$

The T-geometric family studied in Alzaatreh *et al.* (2012b) is a special case of (4) by defining $W(F(x)) = -\ln(1 - F(x))$. The rest of the paper is outlined as follows: Section 2 defines the Kumaraswamy geometric distribution (KGD). In Section 3, we discuss some properties of the distribution. In Section 4, the moments of KGD are provided, while Section 5 contains the hazard function and the Shannon entropy. In Section 6, we discuss the maximum likelihood method for estimating the parameters of the distribution. A simulation study is also discussed. Section 7 details the results of applications of the distribution to two real data sets with comparison to other distributions, and Section 8 contains some concluding remarks.

2 The Kumaraswamy-geometric distribution

Following the T-X generalization technique by Alzaatreh *et al.* (2013b), we allow the *transformed* random variable T to have the Kumaraswamy's distribution, the *transformer* random variable X to have the geometric distribution, and $W(F(x)) = F(x)$.

Kumaraswamy (1980) proposed and discussed a probability distribution for handling double-bounded random processes with varied hydrological applications. Let T be a random variable with the Kumaraswamy's distribution. The PDF and CDF are defined, respectively, as

$$r(t) = \alpha\beta t^{\alpha-1}\left(1 - t^{\alpha}\right)^{\beta-1}, \quad 0 < t < 1, \quad \text{and} \tag{5}$$

$$R(t) = 1 - \left(1 - t^{\alpha}\right)^{\beta}, \quad 0 < t < 1, \tag{6}$$

where both $\alpha > 0$ and $\beta > 0$ are the shape parameters. The beta and Kumaraswamy distributions share similar properties. For example, the Kumaraswamy's distribution, also referred to as the minimax distribution, is unimodal, uniantimodal, increasing, decreasing or constant depending on the values of its parameters. A more detailed description, background and genesis, and properties of Kumaraswamy's distribution are outlined in Jones (2009). The author highlighted several advantages of the Kumaraswamy's distribution over the beta distribution, namely; its simple normalizing constant, simple explicit formulas for the distribution and quantile functions, and simple random variate generation procedure.

The geometric distribution, also referred to as the Pascal distribution, is a special case of the negative binomial distribution. It is thought of as the discrete analogue of the continuous exponential distribution (Johnson *et al.* 2005). Many characterizations of the geometric distribution are analogous to the characterization of the exponential distribution. The geometric distribution has been used extensively in the literature in modeling the distribution of the lengths of waiting times. If X is a random variable having the geometric distribution with parameter p, the PMF of X may be written as

$$P(X = x) = pq^{x}, \quad x = 0, 1, 2, \ldots, \quad p + q = 1, \tag{7}$$

where p is the probability of success in a single Bernoulli trial. The CDF of the geometric distribution is given by

$$P(X \leq x) = 1 - q^{x+1}, \quad x = 0, 1, 2, \ldots \tag{8}$$

The Kumaraswamy-geometric distribution (KGD) is defined by using Equation (3) with $a = 0$, where the random variable T has the Kumaraswamy's distribution with the CDF (6) and the random variable X has the geometric distribution with the CDF (8). Since the random variable T is defined on $(0, 1)$, we use the function $W(F(x)) = F(x)$ in (3) to obtain the CDF of KGD as

$$G(x) = \int_0^{F(x)} r(t)dt = R(F(x)) = 1 - \left[1 - \left(1 - q^{x+1}\right)^{\alpha}\right]^{\beta}, \quad x = 0, 1, 2, \ldots \tag{9}$$

The corresponding PMF for the KGD now becomes

$$g(x) = \left[1 - \left(1 - q^{x}\right)^{\alpha}\right]^{\beta} - \left[1 - \left(1 - q^{x+1}\right)^{\alpha}\right]^{\beta}, \quad x = 0, 1, 2, \ldots, \ \alpha > 0, \ \beta > 0, \tag{10}$$

by using Equation (4). Thus, a random variable X having the PMF expressed in Equation (10) is said to follow the Kumaraswamy-geometric distribution with parameters α, β and q, or simply $X \sim \text{KGD}(\alpha, \beta, q)$. One can show that the PMF in Equation (10) satisfies $\sum_0^\infty g(x) = 1$ by telescopic cancellation.

It is interesting to note that the KGD can be generated from a different random variable T and a different $W(F(x))$ function. Suppose a random variable Y follows the Kumaraswamy's distribution in (5), then its PDF is

$$f(y) = \alpha\beta y^{\alpha-1}\left(1-y^{\alpha}\right)^{\beta-1}, \quad 0 < y < 1.$$

Suppose we define a new random variable as $T = -\ln(1 - Y)$. By using the transformation technique, the PDF of T is given by

$$f(t) = \alpha\beta e^{-t}\left(1-e^{-t}\right)^{\alpha-1}\left[1-\left(1-e^{-t}\right)^{\alpha}\right]^{\beta-1}, \quad t > 0. \tag{11}$$

The corresponding CDF is given by

$$F(t) = 1-\left[1-\left(1-e^{-t}\right)^{\alpha}\right]^{\beta}, \quad t > 0. \tag{12}$$

A random variable T with the CDF in (12) will be called the log-Kumaraswamy's distribution (LKD). We are unable to find any reference to this distribution in the literature. However, it is a special case of the log-exponentiated Kumaraswamy distribution studied by Lemonte *et al.* (2013). By using the LKD and the *T-X* distribution by Alzaatreh *et al.* (2013b), we can define the log-Kumaraswamy-geometric distribution (LKGD) by using Equation (3), where T follows the LKD, X follows the geometric distribution and $W(F(x)) = -\ln(1-F(x))$. By using $1-F(x) = q^{x+1}$ and $-\ln(1-F(x)) = -\ln q^{x+1}$, the probability mass function of LKGD can be obtained as

$$g(x) = G(x) - G(x-1) = R\left[-\ln q^{x+1}\right] - R\left[-\ln q^{x}\right] = \left[1-\left(1-q^{x}\right)^{\alpha}\right]^{\beta} - \left[1-\left(1-q^{x+1}\right)^{\alpha}\right]^{\beta}, \tag{13}$$

which is the same as the KGD in (10) defined by using Kumaraswamy's and geometric distributions. The LKGD, and hence the KGD, is the discrete analogue of log-Kumaraswamy's distribution.

Special cases of KGD

The following are special cases of KGD:

(a) When $\alpha = \beta = 1$, the KGD in (10) reduces to the geometric distribution in (7) with parameter p.

(b) When $\alpha = 1$, the KGD with parameters α, β and q reduces to the geometric distribution with parameter p_*, where $p_* = 1 - q^{\beta}$.

(c) When $\beta = 1$, the KGD reduces to the exponentiated-exponential-geometric distribution (EEGD) discussed in Alzaatreh *et al.* (2012b).

It is easy to verify that $\lim_{x\to\infty} G(x) = 1$. The plots of the PMF of the KGD for various values of α, β and q are given in Figure 1.

3 Some properties of Kumaraswamy-geometric distribution

Suppose X follows the KGD with CDF $G(x)$ in (9). The quantile function $X_*(= Q(U), 0 < U < 1)$ of KGD is the inverse of the cumulative distribution. That is,

$$X_* = Q(U) = (\log q)^{-1}\log\left\{1-\left[1-(1-U)^{1/\beta}\right]^{1/\alpha}\right\}, \tag{14}$$

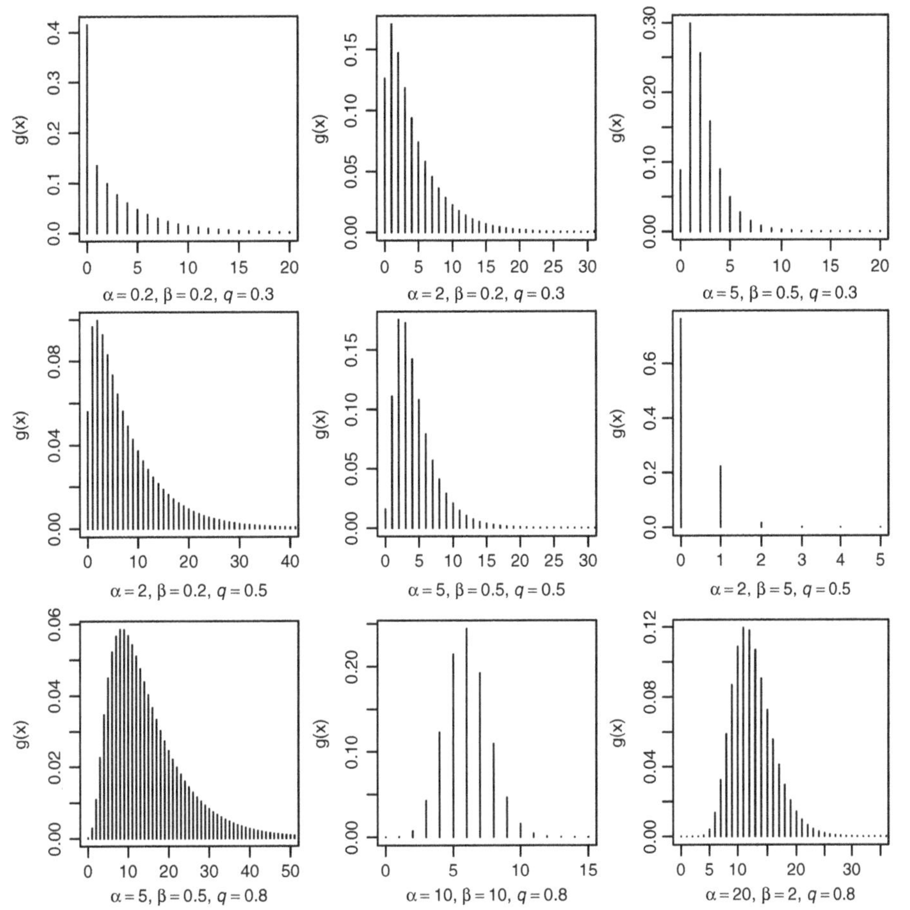

Figure 1 Probability mass function for values of α, β and q.

where U has a uniform distribution with support on $(0, 1)$. Equation (14) can be used to simulate the Kumaraswamy-geometric random variable. First, simulate a random variable U and compute the value of X_* in (14), which is not necessarily an integer. The Kumaraswamy-geometric random variate X is the largest integer $\leq X_*$, which can be denoted by $[X_*]$.

Transformation: The relationship between the KGD and the Kumaraswamy's, exponential, exponentiated-exponential, Pareto, Weibull, Rayleigh, and the logistic distributions are given in the following lemma.

Lemma 1. *Suppose $[v]$ denotes the largest integer less than or equal to the quantity v.*

(a) *If Y has Kumaraswamy's distribution with parameters α and β, then the distribution of $X = \left[\log_q(1-Y)\right]$ is KGD.*

(b) *If Y is standard exponential, then $X = \left[\log_q\left\{1-\left(1-e^{-Y/\beta}\right)^{1/\alpha}\right\}\right]$ has KGD.*

(c) *If Y follows an exponentiated-exponential distribution with scale parameter λ and index parameter c, then $X = \left[\log_q\left\{1-\left[1-(1-e^{-\lambda Y})^{c/\beta}\right]^{1/\alpha}\right\}\right]$ has KGD.*

(d) *If the random variable Y has a Pareto distribution with parameters θ, k and CDF* $F(y) = 1 - \left(\frac{\theta}{\theta+y}\right)^k$, *then* $X = \left[\log_q \left\{1 - \left(1 - \left(\frac{\theta}{\theta+Y}\right)^{k/\beta}\right)^{1/\alpha}\right\}\right]$ *has KGD.*

(e) *If the random variable Y has a Weibull distribution with* $F(y) = 1 - \exp\{-(y/\gamma)^c\}$ *as CDF, then* $X = \left[\log_q \left\{1 - (1 - \exp[-(Y/\gamma)^c/\beta])^{1/\alpha}\right\}\right]$ *has KGD.*

(f) *If the random variable Y has a Rayleigh distribution with* $F(y) = 1 - \exp\left[\frac{-y^2}{2b^2}\right]$ *as CDF, then* $X = \left[\log_q \left\{1 - \left(1 - \exp\left[\frac{-Y^2}{2b^2\beta}\right]\right)^{1/\alpha}\right\}\right]$ *has KGD.*

(g) *If Y is a logistic random variable with* $F(y) = \left[1 + \exp(\{-(y-a)/b\})\right]^{-1}$ *as CDF, then* $X = \left[\log_q \left\{1 - \left(1 - [1 + \exp\{-(Y-a)/b\}]^{-1/\beta}\right)^{1/\alpha}\right\}\right]$ *has KGD.*

Proof. By using the transformation technique, it is easy to show that the random variable X has KGD as given in (10). We will show the result for part (a). Let R be the CDF of the Kumaraswamy's distribution.

$$\begin{aligned} P(X = x) &= P\left(\left[\log_q(1-Y)\right] = x\right) = P\left(x \le \log_q(1-Y) < x+1\right) \\ &= P\left(1 - q^x \le Y < 1 - q^{x+1}\right) = R\left(1 - q^{x+1}\right) - R\left(1 - q^x\right) \\ &= \{1 - (1 - q^x)^\alpha\}^\beta - \left\{1 - \left(1 - q^{x+1}\right)^\alpha\right\}^\beta, \end{aligned}$$

which is the PMF of the KGD in (10). □

In general, if we have a continuous random variable Y and its CDF is $F(y)$, then $X = \left[\log_q \left\{1 - \left(1 - F^{1/\beta}(Y)\right)^{1/\alpha}\right\}\right]$ has KGD.

Limiting behavior: As $x \to \infty$, $\lim_{x\to\infty} g(x) = 0$. Also, as $x \to 0$, $\lim_{x\to 0} g(x) = 1 - [1 - (1-q)^\alpha]^\beta$. This limit becomes 0 if $q \to 1$ and/or $\alpha \to \infty$. Thus, the distribution starts with probability zero or a constant probability as evident from Figure 1.

Mode of the KGD: Since the KGD is also LKGD, a T-geometric distribution, we use Lemma 2 in Alzaatreh *et al.* (2012b), which states that a T-geometric distribution has a reversed J-shape if the distribution of the random variable T has a reversed J-shape. We only need to show when the distribution of log-Kumaraswamy distribution has a reversed J-shape.

On taking the first derivative of (11) with respect to t, we obtain

$$f'(t) = \left[\alpha e^{-t} - 1 + \left(1 - e^{-t}\right)^\alpha - \alpha\beta e^{-t}\left(1 - e^{-t}\right)^\alpha\right] V(t) = Q(t)V(t), \quad (15)$$

where $V(t) = \alpha\beta e^{-t}\left(1 - e^{-t}\right)^{\alpha-2}\left[1 - \left(1 - e^{-t}\right)^\alpha\right]^{\beta-2}$ is positive. For $\beta \ge 1$ and $\alpha \le 1$, it is straight forward to show that $Q(t) \le 0$. For $\beta < 1$ and $\alpha \le 1$, the function $Q(t)$ is an increasing function of t. It is not difficult to show that $\lim_{t\to 0} Q(t) = \alpha - 1 \le 0$ and $\lim_{t\to\infty} Q(t) = 0$. Thus, for $\alpha \le 1$ and any value of β and q, $Q(t) \le 0$ and so the PDF of the log-Kumaraswamy distribution is monotonically decreasing or has a reversed J-shape. Hence, the KGD has a reversed J-shape and a unique mode at $x = 0$ when $\alpha \le 1$.

When $\alpha > 1$, it is not easy to show that the KGD is unimodal. However, through numerical analysis of the behavior of the PMF, and its plots in Figure 1 for various values of β and q, we observe, that for values of $\alpha > 1$, the KGD is concave down or has a reversed J-shape with a unique mode.

4 Moments

Using Equation (10), the rth raw moment is given by

$$E\left(X^r\right) = \mu_r' = \sum_{x=1}^{\infty} x^r \left\{1 - (1 - q^x)^\alpha\right\}^\beta - \sum_{x=1}^{\infty} x^r \left\{1 - \left(1 - q^{x+1}\right)^\alpha\right\}^\beta$$

$$= \sum_{x=1}^{\infty} x^r \left[\left(\sum_{i=1}^{\infty} (-1)^{i-1} \binom{\alpha}{i} q^{xi} \right)^\beta - \left(\sum_{i=1}^{\infty} (-1)^{i-1} \binom{\alpha}{i} q^{(x+1)i} \right)^\beta \right].$$

The two inner summations terminate at α, if α is a positive integer. When $\beta = 1$ in the above, we have,

$$\mu_r' = \sum_{x=1}^{\infty} x^r \sum_{i=1}^{\alpha} (-1)^{i-1} \binom{\alpha}{i} q^{xi} (1 - q^i), \quad \text{for } \alpha \in \mathcal{Z}^+.$$

In particular, let $r = 1$ and $\alpha = 1$, the expression for the first moment, or the mean of the KGD may be written as,

$$E(X) = \mu_1' = q(1-q) \sum_{x=1}^{\infty} x q^{x-1} = \frac{q}{p},$$

which is the mean of the geometric distribution, a special case of KGD.

We discuss in what follows, an alternative approach of expressing the PMF of the KGD in Equation (10).

$$\begin{aligned} g(x) &= \left\{1 - (1 - q^x)^\alpha\right\}^\beta - \left\{1 - \left(1 - q^{x+1}\right)^\alpha\right\}^\beta \\ &= \sum_{i=0}^{\infty} (-1)^i \binom{\beta}{i} (1 - q^x)^{\alpha i} - \sum_{i=0}^{\infty} (-1)^i \binom{\beta}{i} \left(1 - q^{x+1}\right)^{\alpha i} \\ &= \sum_{i=0}^{\infty} (-1)^i \binom{\beta}{i} \sum_{j=0}^{\infty} (-1)^j \binom{\alpha i}{j} q^{xj} - \sum_{i=0}^{\infty} (-1)^i \binom{\beta}{i} \sum_{j=0}^{\infty} (-1)^j \binom{\alpha i}{j} q^{(x+1)j} \\ &= \sum_{i=1}^{\infty} \sum_{j=1}^{\infty} (-1)^{i+j} \binom{\beta}{i} \binom{\alpha i}{j} (1 - q^j) q^{xj}. \end{aligned} \tag{16}$$

Using Equation (16), it is now easy to write the expressions for the moment, moment generating function, and probability generating function for the KGD respectively as follows:

$$\mu_r' = \sum_{x=0}^{\infty} \sum_{i=1}^{\infty} \sum_{j=1}^{\infty} (-1)^{i+j} \binom{\beta}{i} \binom{\alpha i}{j} \left(1 - q^j\right) x^r \left(q^j\right)^x; \quad \left|q^j\right| < 1, \tag{17}$$

$$M(t) = \sum_{x=0}^{\infty} \sum_{i=1}^{\infty} \sum_{j=1}^{\infty} (-1)^{i+j} \binom{\beta}{i} \binom{\alpha i}{j} \left(1 - q^j\right) \left(e^t q^j\right)^x; \quad \left|e^t q^j\right| < 1, \tag{18}$$

$$\varphi(t) = \sum_{x=0}^{\infty} \sum_{i=1}^{\infty} \sum_{j=1}^{\infty} (-1)^{i+j} \binom{\beta}{i} \binom{\alpha i}{j} \left(1 - q^j\right) \left(tq^j\right)^x; \quad \left|tq^j\right| < 1. \tag{19}$$

Equation (17) is equivalent to the series $\sum_{x=0}^{\infty} x^r g(x)$, where $g(x)$ is given by (10). Observe that the series is absolutely convergent by using the ratio test and hence the series in (17) is absolutely convergent. Thus, interchanging the order of summation has no effect. Using Equation (17), the rth moment may be written as

$$E(X^r) = \mu'_r = \sum_{x=0}^{\infty}\sum_{i=1}^{\infty}\sum_{j=1}^{\infty}(-1)^{i+j}\binom{\beta}{i}\binom{\alpha i}{j}\left(1-q^j\right)\left(q^j\right)^x x^r$$

$$= \sum_{i=1}^{\infty}\sum_{j=1}^{\infty}(-1)^{i+j}\binom{\beta}{i}\binom{\alpha i}{j}\left(1-q^j\right)\sum_{x=1}^{\infty}\left(q^j\right)^x x^r$$

$$= \sum_{i=1}^{\infty}\sum_{j=1}^{\infty}(-1)^{i+j}\frac{\beta^{(i)}}{i!}\frac{(\alpha i)^{(j)}}{j!}\left(1-q^j\right)L_{-r}\left(q^j\right),$$

where $\beta^{(i)} = \beta(\beta-1)(\beta-2)\cdots(\beta-i+1)$, and similarly for $(\alpha i)^{(j)}$. Also,

$$L_{-r}(u) = \sum_{k=1}^{\infty}\frac{u^k}{k^{-r}}, \quad u = q^j,$$

is the polylogarithm function, (http://mathworld.wolfram.com/Polylogarithm.html).

Expressions for the first few moments are thus:

$$\mu'_1 = \sum_{i=1}^{\infty}\sum_{j=1}^{\infty}(-1)^{i+j}\frac{\beta^{(i)}}{i!}\frac{(\alpha i)^{(j)}}{j!}\frac{q^j}{1-q^j}, \tag{20}$$

$$\mu'_2 = \sum_{i=1}^{\infty}\sum_{j=1}^{\infty}(-1)^{i+j}\frac{\beta^{(i)}}{i!}\frac{(\alpha i)^{(j)}}{j!}\frac{q^j\left(1+q^j\right)}{\left(1-q^j\right)^2}, \tag{21}$$

$$\mu'_3 = \sum_{i=1}^{\infty}\sum_{j=1}^{\infty}(-1)^{i+j}\frac{\beta^{(i)}}{i!}\frac{(\alpha i)^{(j)}}{j!}\frac{q^j\left(1+4q^j+q^{2j}\right)}{(1-q^j)^3}, \tag{22}$$

$$\mu'_4 = \sum_{i=1}^{\infty}\sum_{j=1}^{\infty}(-1)^{i+j}\frac{\beta^{(i)}}{i!}\frac{(\alpha i)^{(j)}}{j!}\frac{q^j\left(1+q^j\right)\left(1+10q^j+q^{2j}\right)}{\left(1-q^j\right)^4}. \tag{23}$$

The expression for the variance may be written as

$$\sigma^2 = \sum_{i=1}^{\infty}\sum_{j=1}^{\infty}(-1)^{i+j}\frac{\beta^{(i)}}{i!}\frac{(\alpha i)^{(j)}}{j!}\frac{q^j\left(1+q^j\right)}{\left(1-q^j\right)^2} - \left(\sum_{i=1}^{\infty}\sum_{j=1}^{\infty}(-1)^{i+j}\frac{\beta^{(i)}}{i!}\frac{(\alpha i)^{(j)}}{j!}\frac{q^j}{1-q^j}\right)^2.$$

Expressions for the skewness and kurtosis for the KGD may be obtained by combining appropriate expressions in Equations (20), (21), (22), and (23). In the particular case for which $\alpha = 1 = \beta$, the expressions for the central moments of the geometric distribution are as follows:

$$\mu_1 = \mu'_1 = \frac{q}{p}, \quad \mu_2 = \sigma^2 = \frac{q}{p^2}, \quad \mu_3 = \frac{q(1+q)}{p^3}, \quad \mu_4 = \frac{q\left(p^2+9q\right)}{p^4}.$$

The results for this special case may be found in standard textbooks on probability. See for example, Zwillinger and Kokoska (2000).

Both the moment generating function $(M(t))$ and the probability generating function $(\varphi(t))$ can be simplified further. In the case of $\varphi(t)$, we have

$$\varphi(t) = \sum_{i=1}^{\infty}\sum_{j=1}^{\infty}(-1)^{i+j}\binom{\beta}{i}\binom{\alpha i}{j}\left(1-q^j\right)\sum_{x=0}^{\infty}\left(q^j\right)^x t^x, \quad |tq^j| < 1\ \forall j.$$

After further simplification, the above reduces to,

$$\varphi(t) = \sum_{i=1}^{\infty}\sum_{j=1}^{\infty}(-1)^{i+j}\binom{\beta}{i}\binom{\alpha i}{j}\frac{(1-q^j)}{(1-tq^j)}.$$

By letting

$$A(\alpha,\beta|i,j) = \sum_{i=1}^{\infty}\sum_{j=1}^{\infty}(-1)^{i+j}\binom{\beta}{i}\binom{\alpha i}{j},$$

the first two factorial moments may be expressed as

$$\varphi'(t=1) = \mu_{[1]} = E(X) = A(\alpha,\beta|i,j)\frac{q^j}{1-q^j}$$
$$\varphi''(t=1) = \mu_{[2]} = E(X(X-1)) = A(\alpha,\beta|i,j)\frac{2q^{2j}}{(1-q^j)^2}.$$

In general,

$$\varphi^{(m)}(t) = A(\alpha,\beta|i,j)\frac{m!\,(1-q^j)q^{mj}}{(1-tq^j)^{m+1}},$$

which reduces to the result in Alzaatreh *et al.* (2012b) when $\beta = 1$.

Through numerical computation, we obtain the mode, the mean, the standard deviation (SD), the skewness and the kurtosis of the KGD. The values of α and β for the numerical computation are from 0.2 to 10 at an increment of 0.1, while the values of q are from 0.2 to 0.9 at an increment of 0.1. For brevity, we report the mode, the mean and the standard deviation in Table 1 and the skewness and kurtosis in Table 2 for some values of q, β and α. From the numerical computation, the mean, mode and standard deviation are increasing functions of q. From Table 1, the mean, mode and standard deviation are decreasing functions of β but increasing functions of α. For $\alpha \le 1$, the skewness and kurtosis are decreasing functions of q but increasing functions of β. For $\alpha > 1$, the skewness and kurtosis first decrease and then increase as both q and β increase. The skewness and kurtosis are decreasing functions of α. Some of these observations can be seen in Table 2 while others are from the numerical computation. Instead of Tables 1 and 2, contour plots may be used to present the results in the tables. However, it becomes difficult to see the patterns described above.

5 Hazard rate and Shannon entropy

The hazard rate function is defined as

$$h(x) = \frac{g(x)}{1-G(x)},$$

where $G(x) = \sum_{y=0}^{x} g(y)$. For the KGD, we have, after substituting expressions for the PMF and CDF (Equations (10) and (9)),

$$h(x) = \left(\frac{1-(1-q^x)^\alpha}{1-\left(1-q^{x+1}\right)^\alpha}\right)^\beta - 1. \tag{24}$$

The asymptotic behaviors of the hazard function are such that,

$$\lim_{x\to 0} h(x) = (1-p^\alpha)^{-\beta} - 1 = L_1,$$

and in particular, $\lim_{x\to 0} h(x; \alpha = 1 = \beta) = p/q = 1/\mathrm{E}(X)$. Also, $\lim_{x\to\infty} h(x) = q^{-\beta} - 1 = L_2$, after using the L'Hôspital's rule. This result generalizes the limiting behavior of

Table 1 Mode, mean and standard deviation (SD) of KGD for some values of α, β and q

		$q = 0.4$			$q = 0.6$			$q = 0.8$		
α	β	Mode	Mean	SD	Mode	Mean	SD	Mode	Mean	SD
0.4	0.4	0	1.60	2.48	0	3.16	4.51	0	7.74	10.40
	0.6	0	0.83	1.52	0	1.72	2.81	0	4.39	6.54
	0.8	0	0.48	1.04	0	1.06	1.96	0	2.82	4.60
	1.5	0	0.10	0.39	0	0.29	0.79	0	0.92	1.96
	2.0	0	0.04	0.22	0	0.13	0.49	0	0.49	1.26
	4.0	0	0.001	0.04	0	0.01	0.11	0	0.07	0.35
0.6	0.4	0	1.87	2.59	0	3.67	4.69	0	8.97	10.80
	0.6	0	1.04	1.65	0	2.14	3.02	0	5.43	6.99
	0.8	0	0.64	1.17	0	1.41	2.18	0	3.71	5.08
	1.5	0	0.18	0.51	0	0.48	1.00	0	1.48	2.42
	2.0	0	0.08	0.32	0	0.26	0.67	0	0.91	1.67
	4.0	0	0.005	0.07	0	0.04	0.20	0	0.21	0.62
0.8	0.4	0	2.08	2.66	0	4.08	4.81	0	9.93	11.04
	0.6	0	1.21	1.74	0	2.49	3.16	0	6.27	7.27
	0.8	0	0.79	1.27	0	1.71	2.33	0	4.47	5.38
	1.5	0	0.26	0.60	0	0.68	1.16	0	2.01	2.74
	2.0	0	0.13	0.40	0	0.41	0.82	0	1.35	1.99
	4.0	0	0.01	0.12	0	0.08	0.31	0	0.43	0.87
1.5	0.4	1	2.61	2.79	1	5.06	5.00	3	12.22	11.44
	0.6	0	1.68	1.89	1	3.39	3.39	2	8.38	7.76
	0.8	0	1.21	1.44	1	2.54	2.59	2	6.43	5.92
	1.5	0	0.54	0.81	0	1.31	1.47	1	3.61	3.35
	2.0	0	0.35	0.62	0	0.95	1.14	1	2.77	2.61
	4.0	0	0.08	0.29	0	0.38	0.63	0	1.42	1.46
2.0	0.4	1	2.87	2.83	2	5.55	5.06	4	13.35	11.57
	0.6	1	1.93	1.94	1	3.85	3.47	4	9.45	7.92
	0.8	1	1.44	1.50	1	2.97	2.68	3	7.45	6.10
	1.5	0	0.73	0.89	1	1.69	1.58	2	4.51	3.57
	2.0	0	0.51	0.71	1	1.30	1.26	2	3.61	2.83
	4.0	0	0.18	0.40	0	0.65	0.77	1	2.11	1.69
4.0	0.4	2	3.56	2.89	3	6.79	5.16	8	16.19	11.79
	0.6	1	2.59	2.02	3	5.05	3.59	7	12.21	8.19
	0.8	1	2.09	1.59	2	4.14	2.81	6	10.13	6.41
	1.5	1	1.32	1.01	2	2.77	1.74	5	7.00	3.94
	2.0	1	1.08	0.84	2	2.34	1.43	4	5.99	3.23
	4.0	1	0.64	0.61	1	1.57	0.96	3	4.24	2.10
6.0	0.4	2	3.99	2.90	4	7.55	5.19	10	17.92	11.87
	0.6	2	3.01	2.04	4	5.79	3.63	9	13.90	8.29
	0.8	2	2.50	1.61	3	4.87	2.86	8	11.80	6.53
	1.5	1	1.72	1.03	3	3.47	1.80	7	8.59	4.08
	2.0	1	1.46	0.87	2	3.02	1.50	6	7.55	3.38
	4.0	1	1.02	0.62	2	2.21	1.02	5	5.71	2.27

the hazard rate function for the EEGD discussed in Alzaatreh *et al.* (2012b). Observe that $L_1 > L_2$ when $\alpha < 1$. For $\alpha < 1$, we check the behavior of $h(x)$. The function $h(x)$ is monotonically decreasing when $\alpha < 1$ if $h(x) \geq h(x+1)$ for all x. When $\alpha = 1$, observe that $h(x)$ is a constant. For values of $\alpha < 1$, we numerically evaluate $d(x) = h(x) - h(x+1)$ for α and q from 0.1 to 0.9 at an increment of 0.1. All the values are

Table 2 Skewness and kurtosis of KGD for some values of α, β and q

		$q = 0.4$		$q = 0.6$		$q = 0.8$	
α	β	Skewness	Kurtosis	Skewness	Kurtosis	Skewness	Kurtosis
0.4	0.4	2.4529	8.4883	2.3789	8.0612	2.3360	7.8215
	0.6	2.8022	10.920	2.6534	9.9332	2.5656	9.3787
	0.8	3.1918	14.035	2.9430	12.172	2.7954	11.133
	1.5	4.9662	32.736	4.0792	23.245	3.5894	18.553
	2.0	6.8722	60.013	5.0708	35.519	4.1724	25.214
	4.0	30.524	990.63	12.671	196.82	7.1320	72.567
0.6	0.4	2.2480	7.2663	2.1956	7.0004	2.1684	6.8666
	0.6	2.4465	8.4649	2.3387	7.8659	2.2823	7.5649
	0.8	2.6694	9.9363	2.4883	8.8459	2.3932	8.3020
	1.5	3.6723	17.839	3.0582	13.119	2.7529	10.999
	2.0	4.6998	27.856	3.5337	17.209	2.9979	13.043
	4.0	14.964	238.38	6.6621	54.450	4.0916	23.691
0.8	0.4	2.1208	6.5743	2.0847	6.4108	2.0685	6.3383
	0.6	2.2269	7.1361	2.1498	6.7675	2.1151	6.6063
	0.8	2.3522	7.8283	2.2198	7.1642	2.1603	6.8772
	1.5	2.9553	11.535	2.5010	8.8497	2.3055	7.8006
	2.0	3.5861	16.072	2.7466	10.413	2.4048	8.4499
	4.0	9.2202	90.141	4.3332	22.881	2.8543	11.487
1.5	0.4	1.9024	5.5182	1.9015	5.5273	1.9038	5.5384
	0.6	1.8485	5.2051	1.8400	5.2063	1.8433	5.2269
	0.8	1.8105	4.9322	1.7862	4.9087	1.7894	4.9374
	1.5	1.8191	4.4671	1.6645	4.1552	1.6529	4.1894
	2.0	1.9494	4.6197	1.6281	3.8307	1.5890	3.8375
	4.0	3.3608	11.0492	1.7629	3.5961	1.4574	3.0767
2.0	0.4	1.8322	5.2176	1.8440	5.2733	1.8503	5.2983
	0.6	1.7260	4.6855	1.7442	4.7835	1.7563	4.8321
	0.8	1.6352	4.2020	1.6551	4.3386	1.6733	4.4121
	1.5	1.4620	3.0364	1.4309	3.2239	1.4664	3.3814
	2.0	1.4564	2.6068	1.3301	2.7152	1.3700	2.9254
	4.0	2.0958	3.6392	1.1838	1.6496	1.1546	1.9922
4.0	0.4	1.7344	4.8228	1.7559	4.9001	1.7627	4.9251
	0.6	1.5605	4.0589	1.6019	4.1951	1.6151	4.2403
	0.8	1.4043	3.3984	1.4675	3.5897	1.4874	3.6530
	1.5	1.0049	1.9031	1.1406	2.2381	1.1833	2.3432
	2.0	0.8155	1.2870	0.9925	1.7112	1.0500	1.8310
	4.0	0.4669	−0.1633	0.6666	0.8077	0.7728	0.9349
6.0	0.4	1.7069	4.7081	1.7246	4.7731	1.7310	4.7964
	0.6	1.5197	3.8941	1.5522	4.0017	1.5641	4.0428
	0.8	1.3558	3.2107	1.4034	3.3521	1.4207	3.4078
	1.5	0.9572	1.7788	1.0513	1.9565	1.0843	2.0406
	2.0	0.7743	1.2870	0.8984	1.4383	0.9394	1.5299
	4.0	0.3102	0.6548	0.5863	0.5921	0.6444	0.6837

positive, which indicates that the function $h(x)$ is monotonically decreasing. Similarly, we analytically evaluate $d(x)$ for small values of $\alpha > 1$ and the difference $d(x)$ is always negative. Numerically, we use the values of q from 0.1 to 0.9 at an increment of 0.1 with values of α from 1.5 to 10.0 at an increment of 0.5. All the $d(x)$ values are negative which indicates that the function $h(x)$ is monotonically increasing. Thus, we have a decreasing hazard rate when $\alpha < 1$ and an increasing hazard rate when $\alpha > 1$. For $\alpha = 1$, $L_1 = L_2$ and we have a constant hazard rate. The graphs of the hazard rate function defined in Equation (24) are shown in Figure 2 for various values of the parameters. We see in Figure 2 that the hazard rate decreases for values of $\alpha < 1$ and increases for $\alpha > 1$.

The entropy of a random variable is a measure of variation of uncertainty. For a discrete random variable X with probability mass function $g(x)$, the Shannon entropy is defined as,

$$\mathcal{S}(x) = -\sum_x g(x)\log_2 g(x) \geq 0. \tag{25}$$

In probabilistic context, $\mathcal{S}(x)$ is a measure of the information carried by $g(x)$, with higher entropy corresponding to less information. Substituting Equation (10) in Equation (25), we have

$$\begin{aligned}\mathcal{S}(x) &= -\sum_x \left\{[1-(1-q^x)^\alpha]^\beta - \left[1-\left(1-q^{x+1}\right)^\alpha\right]^\beta\right\}\\ &\quad \times \log_2\left\{[1-(1-q^x)^\alpha]^\beta - \left[1-\left(1-q^{x+1}\right)^\alpha\right]^\beta\right\}.\end{aligned}$$

Suppose we write the PMF as

$$[1-(1-q^x)^\alpha]^\beta \left\{1-\left(\frac{1-\left(1-q^{x+1}\right)^\alpha}{1-(1-q^x)^\alpha}\right)^\beta\right\}.$$

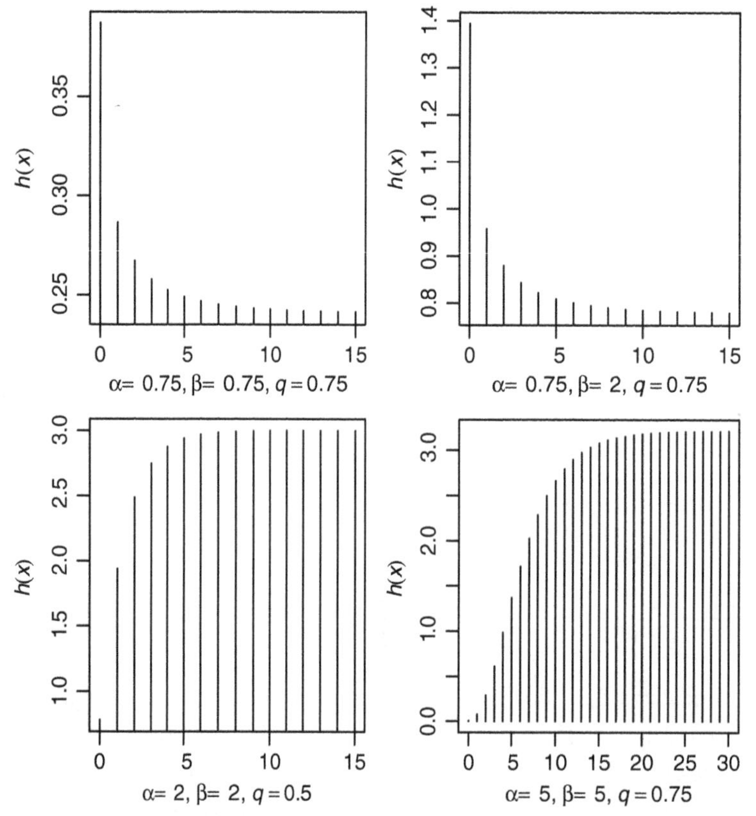

Figure 2 Hazard functions for values of α, β and q.

Let $\alpha = 1$ for simplicity. We may now write the entropy as

$$\mathcal{S}(x) = -\sum_{x=0}^{\infty}\left(1-q^{\beta}\right)q^{x\beta}\log_2\left\{\left(1-q^{\beta}\right)q^{x\beta}\right\}. \tag{26}$$

After some algebra, Equation (26) becomes,

$$\mathcal{S}(x) = \frac{-\left(1-q^{\beta}\right)\log_2\left(1-q^{\beta}\right)-q^{\beta}\log_2 q^{\beta}}{1-q^{\beta}} > 0. \tag{27}$$

On setting $\beta = 1$ in Equation (27), we have, for the geometric distribution,

$$\mathcal{S}(x) = -\frac{q}{1-q}\log_2 q - \log_2(1-q) = \frac{-(1-p)\log_2(1-p) - p\log_2 p}{p}.$$

Note that when $\beta = 1$, $p = q = 1/2$, $\mathcal{S}(x) = 2$. It is not difficult to show that $\mathcal{S}(x)$ is an increasing function of q for any given β. This is consistent with the pattern of the standard deviation. We also note that $\lim_{\beta\to\infty}\mathcal{S}(x) = 0$, with the proviso that $0\log 0 = 0$. This indicates that smaller values of β increase the uncertainty in the distribution, while higher values of β increase the amount of information measured in terms of the probability. Actually, a zero entropy indicates that all information needed is measured solely in terms of the probability. In a way, the KGD has smaller entropy (more probabilistic information) than the geometric distribution for values of $\beta > 1$.

6 Maximum likelihood estimation

We discuss the maximum likelihood estimation of the parameters of the KGD in subsection 6.1. Subsection 6.2 contains the results of a simulation that is conducted to evaluate the performance of the maximum likelihood estimation method.

6.1 Estimation

Let a random sample of size n be taken from KGD, with observed frequencies n_x, $x = 0, 1, 2, \ldots, k$, where $\sum_{x=0}^{k} n_x = n$. From Equation (10), the likelihood function for a random sample of size n may be expressed as

$$L(x|\alpha,\beta,q) = \prod_{x=0}^{k}\left\{\left[1-\left(1-q^{x}\right)^{\alpha}\right]^{\beta} - \left[1-\left(1-q^{x+1}\right)^{\alpha}\right]^{\beta}\right\}^{n_x}. \tag{28}$$

The log-likelihood function is

$$\begin{aligned} l(\alpha,\beta,q) = \ln L(x|\alpha,\beta,q) &= n_0 \ln\left\{1-\left[1-(1-q)^{\alpha}\right]^{\beta}\right\} \\ &+ \sum_{x=1}^{k} n_x \ln\left\{\left[1-\left(1-q^{x}\right)^{\alpha}\right]^{\beta} - \left[1-\left(1-q^{x+1}\right)^{\alpha}\right]^{\beta}\right\}. \end{aligned} \tag{29}$$

Differentiating the log-likelihood function with respect to the parameters, we obtain

$$\frac{\partial l(\alpha,\beta,q)}{\partial \alpha} = \frac{n_0\beta(1-q)^{\alpha}\left[1-(1-q)^{\alpha}\right]^{\beta-1}\ln(1-q)}{1-\left[1-(1-q)^{\alpha}\right]^{\beta}} + \sum_{x=1}^{k}\frac{n_x\left[A_x+B_x\right]}{\left\{1-(1-q^x)^{\alpha}\right\}^{\beta}-\left\{1-\left(1-q^{x+1}\right)^{\alpha}\right\}^{\beta}}, \tag{30}$$

$$\frac{\partial l(\alpha,\beta,q)}{\partial \beta} = \frac{-n_0\left[1-(1-q)^{\alpha}\right]^{\beta}\ln\left[1-(1-q)^{\alpha}\right]}{1-\left[1-(1-q)^{\alpha}\right]^{\beta}} + \sum_{x=1}^{k}\frac{n_x\left[C_x-D_x\right]}{\left\{1-(1-q^x)^{\alpha}\right\}^{\beta}-\left\{1-\left(1-q^{x+1}\right)^{\alpha}\right\}^{\beta}}, \tag{31}$$

$$\frac{\partial l(\alpha,\beta,q)}{\partial q} = \frac{-n_0\alpha\beta\left[1-(1-q)^{\alpha}\right]^{\beta-1}(1-q)^{\alpha-1}}{1-[1-(1-q)^{\alpha}]^{\beta}} + \sum_{x=1}^{k}\frac{n_x\left[E_x-F_x\right]}{\left\{1-(1-q^x)^{\alpha}\right\}^{\beta}-\left\{1-\left(1-q^{x+1}\right)^{\alpha}\right\}^{\beta}}, \tag{32}$$

where,

$$\begin{aligned}
A_x &= -\beta\left[1-\left(1-q^x\right)^{\alpha}\right]^{\beta-1}\left(1-q^x\right)^{\alpha}\ln\left(1-q^x\right),\\
B_x &= \beta\left[1-\left(1-q^{x+1}\right)^{\alpha}\right]^{\beta-1}\left(1-q^{x+1}\right)^{\alpha}\ln\left(1-q^{x+1}\right),\\
C_x &= \left[1-\left(1-q^x\right)^{\alpha}\right]^{\beta}\ln\left[1-\left(1-q^x\right)^{\alpha}\right],\\
D_x &= \left[1-\left(1-q^{x+1}\right)^{\alpha}\right]^{\beta}\ln\left[1-\left(1-q^{x+1}\right)^{\alpha}\right],\\
E_x &= \alpha\beta x q^{x-1}\left[1-\left(1-q^x\right)^{\alpha}\right]^{\beta-1}\left(1-q^x\right)^{\alpha-1},\\
F_x &= \alpha\beta\left(x+1\right)q^x\left[1-\left(1-q^{x+1}\right)^{\alpha}\right]^{\beta-1}\left(1-q^{x+1}\right)^{\alpha-1}.
\end{aligned}$$

Setting the non-linear Equations (30), (31) and (32) to zero and solving them iteratively, we get the estimates $\hat{\theta} = (\hat{\alpha},\hat{\beta},\hat{q})^T$ for the parameter vector $\theta = (\alpha,\beta,q)^T$. The initial values of parameters α and β can be set to 1 and that of parameter q can be set to 0.5.

For interval estimation and hypothesis tests on the parameters, we require the information matrix $\mathcal{I}(\theta)$, with elements $-\partial^2 l/(\partial i\partial j) = -l_{ij}$, where, $i,j \in \{\alpha,\beta,q\}$. Under conditions that are fulfilled for parameters in the interior of the parameter space but not on the boundary, the asymptotic distribution of $\sqrt{n}(\hat{\theta}-\theta)$ is $N_3(0,\mathcal{I}^{-1}(\theta))$. The asymptotic multivariate normal distribution $N_3(0,\mathcal{I}^{-1}(\hat{\theta}))$ of $\hat{\theta}$ can be used to construct approximate confidence intervals for the parameters. For example, the $100(1-\xi)\%$ asymptotic confidence interval for the ith parameter θ_i is given by

$$\left(\hat{\theta}_i - z_{\xi/2}\sqrt{\mathcal{I}_{i,i}},\ \hat{\theta}_i + z_{\xi/2}\sqrt{\mathcal{I}_{i,i}}\right),$$

where $\mathcal{I}_{i,i}$ is the ith diagonal element of $\mathcal{I}^{-1}(\theta)$ for $i = 1,2,3$, and $z_{\xi/2}$ is the upper $\xi/2$ point of standard normal distribution. See for example, Mahmoudi (2011).

The information matrix $\mathcal{I}(\theta)$ is of the form,

$$\mathcal{I}(\alpha,\beta,q) = \begin{pmatrix} l_{\alpha\alpha} & l_{\alpha\beta} & l_{\alpha q} \\ l_{\beta\alpha} & l_{\beta\beta} & l_{\beta q} \\ l_{q\alpha} & l_{q\beta} & l_{qq} \end{pmatrix}.$$

The expressions for the elements l_{ij} are given in the Appendix.

6.2 Simulation

A simulation study is conducted to evaluate the performance of the maximum likelihood estimation method. Equation (14) is used to generate a random sample from the KGD with parameters α, β and q. The different sample sizes considered in the simulation are $n = 250,\ 500$ and 750. The parameter combinations for the simulation study are shown in Table 3. The combinations were chosen to reflect the following cases of the distribution: under-dispersion ($\alpha = 6,\ \beta = 4,\ q = 0.8$), over-dispersion (all other cases), monotonically decreasing ($\alpha = 1.6,\ \beta = 2.0,\ q = 0.6$), and unimodal with mode greater than 0 (all other cases). For each parameter combination and each sample size, the simulation process is repeated 100 times. The average bias (actual − estimate) and standard deviation of the parameter estimates are reported in Table 3. The biases are relatively small when compared to the standard deviations. In most cases, as the sample size increases, the standard deviations of the estimators decrease.

7 Applications of KGD

We apply the KGD to two data sets. The first data set is the observed frequencies of the distribution of purchases of a brand X breakfast cereals purchased by consumers over a period of time (Consul 1989). The other data set is the number of absences among shift-workers in a steel industry (Gupta and Ong 2004). Comparisons are made with the generalized negative binomial distribution (GNBD) defined by Jain and Consul (1971) and

Table 3 Bias and standard deviation for maximum likelihood estimates

	Actual values			Bias with standard deviation in parentheses			
n	α	β	q	$\hat{\alpha}$	$\hat{\beta}$	$\hat{q}$	Mode
250	1.6	2.0	0.6	−0.1295(0.2099)	0.1128(0.4697)	0.0409(0.0733)	0
	4.0	2.0	0.4	−0.3136(0.7844)	0.1047(0.4617)	0.0237(0.0698)	1
	4.0	2.0	0.6	−0.1977(0.5714)	0.1021(0.4571)	0.0198(0.0571)	2
	4.0	2.0	0.8	−0.2467(0.5521)	0.0852(0.4020)	0.0126(0.0293)	4
	6.0	4.0	0.8	−0.3931(0.8570)	0.2084(0.9507)	0.0121(0.0278)	5
500	1.6	2.0	0.6	−0.0437(0.1436)	0.0121(0.4496)	0.0153(0.0665)	0
	4.0	2.0	0.4	−0.3142(0.5876)	0.1383(0.4421)	0.0290(0.0639)	1
	4.0	2.0	0.6	−0.2015(0.4728)	0.0808(0.4504)	0.0194(0.0527)	2
	4.0	2.0	0.8	−0.2747(0.4630)	0.1526(0.3781)	0.0157(0.0278)	4
	6.0	4.0	0.8	−0.3283(0.7771)	0.2187(0.9375)	0.0116(0.0267)	5
750	1.6	2.0	0.6	−0.0533(0.1305)	0.1184(0.4250)	0.0278(0.0669)	0
	4.0	2.0	0.4	−0.3181(0.5939)	0.1707(0.4195)	0.0328(0.0622)	1
	4.0	2.0	0.6	−0.1787(0.4775)	0.0562(0.4469)	0.0159(0.0522)	2
	4.0	2.0	0.8	−0.1717(0.4739)	0.0813(0.4117)	0.0101(0.0299)	4
	6.0	4.0	0.8	−0.2472(0.5940)	0.1947(0.8090)	0.0087(0.0227)	5

the exponentiated-exponential geometric distribution (EEGD) defined by Alzaatreh *et al.* (2012b).

7.1 Purchases by consumers

Consul (1989), p. 128 stated that, "The number of units of different commodities purchased by consumers over a period of time", appears to follow the generalized Poisson distribution (GPD). In support of this assertion, the author analyzed some relevant data sets and observed that the GPD model provided adequate fits. One of the data sets considered by Consul (1989) consists of observed frequencies of the distribution of purchases of a brand X breakfast cereals. The original data in Table 4 was taken from Chatfield (1975).

The data contains the frequency of consumers who bought r units of brand X over a number of weeks. The data is fitted to the KGD, EEGD, and the GNBD. We are not sure how the frequencies for the (10-11), (12-15) and (16-35) were handled in previous applications of the data set. In our analysis, the probabilities for the classes (10-11) and (12-15) were obtained by adding the corresponding individual probabilities in each class. When finding the maximum likelihood estimates, the probability for the last class was obtained by subtracting the sum of all previous probabilities from 1. The results in Table 4 show that all the three distributions provide adequate fit to the data. Since the parameter β in KGD is not significantly different from 1, it may be more appropriate to apply the two-parameter EEGD to fit the data. Also, the likelihood ratio test to compare the EEGD with the KGD is not significant at 5% level.

Table 4 The number of units r purchased by observed number (f_r) of consumers

		Expected		
r-value	Observed	EEGD	GNBD	KGD
0	299	300.11	298.98	299.33
1	69	63.10	68.03	65.59
2	37	41.13	41.12	42.02
3	34	30.68	29.89	30.92
4	23	24.36	23.51	24.28
5	20	20.06	19.32	19.81
6	12	16.92	16.32	16.58
7	18	14.52	14.04	14.14
8	14	12.61	12.25	12.21
9	9	11.06	10.80	10.67
10-11	17	18.47	18.16	17.72
12-15	27	26.69	26.63	25.50
16-35	63	62.29	62.95	63.23
Total	642	642	642	642
Parameter		$\hat{\alpha} = 0.2910(0.0212)$	$\hat{\theta} = 0.9658(0.0148)$	$\hat{\alpha} = 0.3373(0.0924)$
Estimates		$\hat{\theta} = 0.9267(0.0076)$	$\hat{m} = 0.2264(0.0339)$	$\hat{\beta} = 1.5114(1.3648)$
			$\hat{\beta} = 0.9882(0.0126)$	$\hat{q} = 0.9592(0.0544)$
χ^2		4.34	3.92	4.11
df		10	9	9
p-value		0.9307	0.9166	0.9040
AIC		2471.3	2472.8	2473.0
LL*		−1233.64	−1233.38	−1233.50

*LL = log-likelihood value.

7.2 Number of absences by shift-workers

The KGD is also applied to a data set from Gupta and Ong (2004), which represents the observed frequencies of the number of absences among shift-workers in a steel industry. The data in Table 5 was originally studied by Arbous and Sichel (1954) in an attempt to create a model that can describe the distribution of absences to a group of people in single- and double-exposure periods. The original data contains the number of absences, x-value, of 248 shift workers in the years 1947 and 1948. Arbous and Sichel (1954) used the negative binomial distribution (NBD) to fit the data. Gupta and Ong (2004) proposed a four-parameter generalized negative binomial distribution to fit the data and compared it to the NBD and the GPD. The chi-square value for their distribution was 8.27 with 15

Table 5 The number of absences among shift-workers in a steel industry

	Observed	Expected		
x-value	Frequency	EEGD	GNBD	KGD
0	7	10.18	9.86	6.52
1	16	16.41	16.44	18.21
2	23	18.30	19.32	21.40
3	20	18.57	19.84	20.92
4	23	18.02	19.09	19.31
5	24	17.03	17.73	17.45
6	12	15.82	16.13	15.64
7	13	14.52	14.50	13.97
8	9	13.22	12.93	12.46
9	9	11.95	11.49	11.11
10	8	10.74	10.17	9.90
11	10	9.62	9.00	8.83
12	8	8.59	7.95	7.87
13	7	7.65	7.02	7.01
14	2^a	6.79	6.21	6.25
15	12^a	6.02	5.49	5.57
16	3^b	5.33	4.86	4.96
17	5^b	4.71	4.30	4.42
18	4^c	4.16	3.81	3.94
19	2^c	3.67	3.38	3.51
20	2^d	3.23	3.00	3.13
21	5^d	2.85	2.67	2.79
22	5^e	2.51	2.38	2.49
23	2^e	2.21	2.11	2.21
24	1^e	1.94	1.88	1.97
25-48	16	13.96	16.44	16.16
Total	248	248	248	248
Parameter Estimates		$\hat{\alpha} = 1.5255(0.1471)$ $\hat{\theta} = 0.8767(0.0093)$	$\hat{\theta} = 0.0026(0.0001)$ $\hat{m} = 1242.96(4.259)$ $\hat{\beta} = 254.77(20.785)$	$\hat{\alpha} = 3.1859(1.6702)$ $\hat{\beta} = 0.1292(0.0882)$ $\hat{q} = 0.4099(0.2374)$
χ^2		12.18	9.62	7.78
df		17	16	16
p-value		0.7891	0.8857	0.9551
AIC		1517.8	1517.3	1515.2
Log-likelihood		−756.88	−755.66	−754.62

Observed frequencies with the same letters are pooled together.

degrees of freedom. The chi-square value obtained by Gupta and Ong for the GPD was 27.79 with 17 degrees of freedom (DF). The DF $= k - s - 1$, where k is the number of classes and s is the number of estimated parameters.

We re-analyzed the data for the EEGD, the GNBD, and the GPD. We obtained a chi-square of 9.62 for the GPD, which is much smaller than the 27.79 provided by Gupta and Ong. Thus, our estimates from the GPD (not reported in Table 5) differ significantly from the results in Gupta and Ong (2004). When finding the maximum likelihood estimates, the probability for the last class was obtained by subtracting the sum of all previous probabilities from 1. In view of this, the results obtained from the EEGD are slightly different from those of Alzaatreh *et al.* (2012b) who applied the EEGD to fit the data. We apply the KGD to model the data in Table 5, and the results from the table indicate that the KGD, EEGD and GNBD provide good fit to the data.

If $m \to \infty$, the GNBD with parameters θ, m and β goes to the GPD with parameters α and λ, where $\alpha = m\theta$ and $\lambda = \theta\beta$ on page 218 of Consul and Famoye (2006). We fitted the GPD to the data and we got the same log-likelihood with $\hat{\alpha} = 3.2250$ and $\hat{\lambda} = 0.6597$. From the GNBD, we obtained $\hat{m}\hat{\theta} = 1242.96 \times 0.002591 = 3.22 \approx 3.2250 = \hat{\alpha}$ and $\hat{\beta}\hat{\theta} = 254.77 \times 0.002591 = 0.66 \approx 0.6597 = \hat{\lambda}$.

We observe that the parameter β in the KGD is significantly different from 1. This makes the KGD a more appropriate distribution over the EEGD. The likelihood ratio statistic for testing the EEGD against the KGD is $\chi_1^2 = -2(754.62 - 756.88) = 4.52$ with a p-value of 0.0335. Thus, we reject the null hypothesis that the data follows the EEGD at the 5% level. The likelihood ratio test supports the claim that the parameter β is significantly different from 1, and hence the KGD appears to be superior to the EEGD.

8 Conclusion

Discrete distributions are often derived by using the Lagrange expansions framework (see for example Consul and Famoye 2006) or using difference equations (see for example Johnson *et al.* 2005). Recently, Alzaatreh *et al.* (2012b, 2013b) developed a general method for generating distributions and these distributions are members of the *T-X* family. The method can be applied to derive both the discrete and continuous distributions. This article used the *T-X* family framework to define a new discrete distribution named the Kumaraswamy-geometric distribution (KGD).

Some special cases, and properties of the KGD are discussed, which include moments, hazard rate and entropy. The method of maximum likelihood estimation is used in estimating the parameters of the KGD. The distribution is applied to model two real life data sets; one consisting of the observed frequencies of the distribution of purchases of a brand X breakfast cereals, and the other, the observed frequencies of the number of absences among shift-workers in a steel industry. Two other distributions, the EEGD and the GNBD are compared with KGD. It is found that the KGD performed as well as the EEGD in modeling the observed numbers of consumers. The results also show that the KGD outperformed the EEGD in modeling the number of absences among shift-workers. It is expected that the additional parameter offered by the Kumaraswamy's distribution will enable the use of the KGD in modeling events where the EEGD or the geometric distribution may not provide adequate fits.

Appendix

Elements of the information matrix

$$
\begin{aligned}
l_{\alpha\alpha} &= \frac{A_0\left[1-\beta(1-q)^{\alpha}\right]\ln(1-q)}{P_0\left[1-(1-q)^{\alpha}\right]} - \frac{A_0 B_0}{P_0^2} + \sum_{x=1}^{k}\frac{n_x}{P_x}\left[\frac{\partial\left(A_x+B_x\right)}{\partial\alpha} - \frac{(A_x+B_x)^2}{P_x}\right],\\
l_{\alpha\beta} &= \frac{A_0}{P_0\beta} + \frac{A_0\ln\left[1-(1-q)^{\alpha}\right]}{P_0^2} + \sum_{x=1}^{k}\frac{n_x}{P_x}\left[\frac{\partial\left(A_x+B_x\right)}{\partial\beta} - \frac{(A_x+B_x)\left(C_x-D_x\right)}{P_x}\right],\\
l_{\alpha q} &= \frac{A_0}{P_0}\left[\frac{\alpha\{\beta(1-q)^{\alpha}-1+\left[1-(1-q)^{\alpha}\right]^{\beta}\}}{(1-q)\left[1-(1-q)^{\alpha}\right]P_0} - \frac{1}{(1-q)\ln(1-q)}\right]\\
&\quad + \sum_{x=1}^{k}\frac{n_x}{P_x}\left[\frac{\partial\left(A_x+B_x\right)}{\partial q} - \frac{(A_x+B_x)\left(E_x-F_x\right)}{P_x}\right],\\
l_{\beta\beta} &= \frac{C_0\ln\left[1-(1-q)^{\alpha}\right]}{P_0^2} + \sum_{x=1}^{k}\frac{n_x}{P_x}\left[\frac{\partial(C_x-D_x)}{\partial\beta} - \frac{(C_x-D_x)^2}{P_x}\right],\\
l_{\beta q} &= \frac{C_0}{P_0}\left[\frac{\alpha\beta(1-q)^{\alpha-1}}{\left[1-(1-q)^{\alpha}\right]P_0} + \frac{\alpha(1-q)^{\alpha-1}}{\left[1-(1-q)^{\alpha}\right]\ln\left[1-(1-q)^{\alpha}\right]}\right]\\
&\quad + \sum_{x=1}^{k}\frac{n_x}{P_x}\left[\frac{\partial\left(C_x-D_x\right)}{\partial q} - \frac{(C_x-D_x)\left(E_x-F_x\right)}{P_x}\right],\\
l_{qq} &= \frac{E_0}{P_0}\left[\frac{\alpha\beta(1-q)^{\alpha-1}\left[1-(1-q)^{\alpha}\right]^{\beta-1}}{P_0} + \frac{(\alpha\beta-1)(1-q)^{\alpha}-\alpha+1}{(1-q)\left[1-(1-q)^{\alpha}\right]}\right]\\
&\quad + \sum_{x=1}^{k}\frac{n_x}{P_x}\left[\frac{\partial\left(E_x-F_x\right)}{\partial q} - \frac{(E_x-F_x)^2}{P_x}\right],
\end{aligned}
$$

where,

$$
\begin{aligned}
P_x &= \left[1-\left(1-q^{x}\right)^{\alpha}\right]^{\beta} - \left[1-\left(1-q^{x+1}\right)^{\alpha}\right]^{\beta},\\
P_0 &= 1-\left[1-(1-q)^{\alpha}\right]^{\beta},\\
A_0 &= n_0\beta(1-q)^{\alpha}\left[1-(1-q)^{\alpha}\right]^{\beta-1}\ln(1-q),\\
B_0 &= \beta(1-q)^{\alpha}\left[1-(1-q)^{\alpha}\right]^{\beta-1}\ln(1-q),\\
C_0 &= -n_0\left[1-(1-q)^{\alpha}\right]^{\beta}\ln\left[1-(1-q)^{\alpha}\right],\\
E_0 &= -n_0\alpha\beta\left[1-(1-q)^{\alpha}\right]^{\beta-1}(1-q)^{\alpha-1},\\
\frac{\partial A_x}{\partial\alpha} &= \frac{A_x\left[1-\beta\left(1-q^{x}\right)^{\alpha}\right]\ln\left(1-q^{x}\right)}{1-\left(1-q^{x}\right)^{\alpha}},\\
\frac{\partial A_x}{\partial\beta} &= A_x\left(1/\beta+\ln\left[1-\left(1-q^{x}\right)^{\alpha}\right]\right),\\
\frac{\partial A_x}{\partial q} &= \frac{A_x\alpha x q^{x-1}\left[\beta\left(1-q^{x}\right)^{\alpha}-1\right]}{\left(1-q^{x}\right)\left[1-\left(1-q^{x}\right)^{\alpha}\right]} - \frac{A_x x q^{x-1}}{\left(1-q^{x}\right)\ln\left(1-q^{x}\right)},\\
\frac{\partial B_x}{\partial\alpha} &= \frac{B_x\left[1-\beta\left(1-q^{x+1}\right)^{\alpha}\right]\ln\left(1-q^{x+1}\right)}{1-\left(1-q^{x+1}\right)^{\alpha}},\\
\frac{\partial B_x}{\partial\beta} &= B_x\left(1/\beta+\ln\left[1-\left(1-q^{x+1}\right)^{\alpha}\right]\right),
\end{aligned}
$$

$$
\begin{aligned}
\frac{\partial B_x}{\partial q} &= \frac{B_x\alpha(x+1)q^x[\beta(1-q^{x+1})^\alpha - 1]}{(1-q^{x+1})\left[1-(1-q^{x+1})^\alpha\right]} - \frac{B_x(x+1)q^x}{(1-q^{x+1})\ln(1-q^{x+1})},\\
\frac{\partial C_x}{\partial \alpha} &= \frac{-C_x(1-q^x)^\alpha \ln(1-q^x)}{1-(1-q^x)^\alpha}\left(\beta + \frac{1}{\ln[1-(1-q^x)^\alpha]}\right),\\
\frac{\partial C_x}{\partial \beta} &= C_x \ln[1-(1-q^x)^\alpha],\\
\frac{\partial C_x}{\partial q} &= \frac{C_x\alpha x q^{x-1}(1-q^x)^{\alpha-1}}{1-(1-q^x)^\alpha}\left(\beta + \frac{1}{\ln[1-(1-q^x)^\alpha]}\right),\\
\frac{\partial D_x}{\partial \alpha} &= \frac{-D_x(1-q^{x+1})^\alpha \ln(1-q^{x+1})}{1-(1-q^{x+1})^\alpha}\left(\beta + \frac{1}{\ln\left[1-(1-q^{x+1})^\alpha\right]}\right),\\
\frac{\partial D_x}{\partial \beta} &= D_x \ln\left[1-(1-q^{x+1})^\alpha\right],\\
\frac{\partial D_x}{\partial q} &= \frac{D_x\alpha(x+1)q^x(1-q^{x+1})^{\alpha-1}}{1-(1-q^{x+1})^\alpha}\left(\beta + \frac{1}{\ln\left[1-(1-q^{x+1})^\alpha\right]}\right),\\
\frac{\partial E_x}{\partial \alpha} &= E_x\left(1/\alpha + \frac{[1-\beta(1-q^x)^\alpha]\ln(1-q^x)}{1-(1-q^x)^\alpha}\right),\\
\frac{\partial E_x}{\partial \beta} &= E_x\left(1/\beta + \ln[1-(1-q^x)^\alpha]\right),\\
\frac{\partial E_x}{\partial q} &= \frac{E_x(x-1)}{q} + \frac{E_x[(\alpha\beta-1)(1-q^x)^\alpha - \alpha + 1]\,xq^{x-1}}{(1-q^x)[1-(1-q^x)^\alpha]},\\
\frac{\partial F_x}{\partial \alpha} &= F_x\left(1/\alpha + \frac{\left[1-\beta(1-q^{x+1})^\alpha\right]\ln(1-q^{x+1})}{1-(1-q^{x+1})^\alpha}\right),\\
\frac{\partial F_x}{\partial \beta} &= F_x\left(1/\beta + \ln\left[1-(1-q^{x+1})^\alpha\right]\right), \text{and}\\
\frac{\partial F_x}{\partial q} &= \frac{F_x x}{q} + \frac{F_x\left[(\alpha\beta-1)(1-q^{x+1})^\alpha - \alpha + 1\right](x+1)q^x}{(1-q^{x+1})\left[1-(1-q^{x+1})^\alpha\right]}.
\end{aligned}
$$

The values of A_x, B_x, C_x, D_x, E_x, and F_x are given in Section 6.

Competing interests

The authors declare that they have no competing interests.

Authors' contributions

The authors, viz AA, FF and CL with the consultation of each other carried out this work and drafted the manuscript together. All authors read and approved the final manuscript.

Acknowledgment

The authors are grateful to the Associate Editor and the anonymous referees for many constructive comments and suggestions that have greatly improved the paper.

Author details

[1]Department of Mathematics, Marshall University, Huntington, West Virginia 25755, USA. [2]Department of Mathematics, Central Michigan University, Mount Pleasant, Michigan 48859, USA.

References

Akinsete, A, Famoye, F, Lee, C: The beta-Pareto distributions. Statistics. **42**(6), 547–563 (2008)

Alexander, C, Cordeiro, GM, Ortega, EMM, Sarabia, JM: Generalized beta-generated distributions. Comput. Stat. Data Anal. **56**, 1880–1897 (2012)

Alshawarbeh, E, Famoye, F, Lee, C: Beta-Cauchy distribution: some properties and its applications. J. Stat. Theory Appl. **12**, 378–391 (2013)

Alzaatreh, A, Famoye, F, Lee, C: Gamma-Pareto distribution and its applications. J. Mod. Appl. Stat. Meth. **11**(1), 78–94 (2012a)

Alzaatreh, A, Famoye, F, Lee, C: Weibull-Pareto distribution and its applications. Comm. Stat. Theor. Meth. **42**(7), 1673–1691 (2013a)

Alzaatreh, A, Lee, C, Famoye, F: On the discrete analogues of continuous distributions. Stat. Meth. **9**, 589–603 (2012b)

Alzaatreh, A, Lee, C, Famoye, F: A new method for generating families of continuous distributions. Metron. **71**, 63–79 (2013b)

Arbous, AG, Sichel, HS: New techniques for the analysis of asenteeism data. Biometrika. **41**, 77–90 (1954)

Chatfield, C: A marketing application of characterization theorem. In: Patil, GP, Kotz, S, Ord, JK (eds.) Statistical Distributions in Scientific Work, volume 2, pp. 175–185. D. Reidel Publishing Company, Boston, (1975)

Consul, PC: Generalized Poisson Distributions: Properties and Applications. Marcel Dekker, Inc., New York (1989)

Consul, PC, Famoye, F: Lagrangian Probability Distributions. Birkhäuser, Boston (2006)

Cordeiro, GM, de Castro, M: A new family of generalized distributions. J. Stat. Comput. Simulat. **81**(7), 883–898 (2011)

Cordeiro, GM, Lemonte, AJ: The beta Laplace distribution. Stat. Probability Lett. **81**, 973–982 (2011)

Cordeiro, GM, Nadarajah, S, Ortega, EMM: The Kumaraswamy Gumbel distribution. Stat. Methods Appl. **21**, 139–168 (2012)

de Pascoa, MAR, Ortega, EMM, Cordeiro, GM: The Kumaraswamy generalized gamma distribution with application in survival analysis. Stat. Meth. **8**, 411–433 (2011)

de Santana, TV, Ortega, EMM, Cordeiro, GM, Silva, GO: The Kumaraswamy-log-logistic distribution. Stat. Theory Appl. **3**, 265–291 (2012)

Eugene, N, Lee, C, Famoye, F: The beta-normal distribution and its applications. Comm. Stat. Theor. Meth. **31**(4), 497–512 (2002)

Famoye, F, Lee, C, Olumolade, O: The beta-Weibull distribution. J. Stat. Theory Appl. **4**(2), 121–136 (2005)

Gupta, RC, Ong, SD: A new generalization of the negative binomial distribution. Comput. Stat. Data Anal. **45**, 287–300 (2004)

Jain, GC, Consul, PC: A generalized negative binomial distribution. SIAM J. Appl. Math. **21**, 501–513 (1971)

Johnson, NL, Kemp, AW, Kotz, S: Univariate Discrete Distributiuons. third edition. John Wiley & Sons, New York (2005)

Jones, MC: Kumaraswamy's distribution: a beta-type distribution with some tractability advantages. Stat. Methodologies. **6**, 70–81 (2009)

Kong, L, Lee, C, Sepanski, JH: On the properties of of beta-gamma distribution. J. Mod. Appl. Stat. Meth. **6**(1), 187–211 (2007)

Kumaraswamy, P: A generalized probability density function for double-bounded random processes. Hydrology **46**, 79–88 (1980)

Lemonte, AJ, Barreto-Souza, W, Cordeiro, GM: The exponentiated Kumaraswamy distribution and its log-transform. Braz. J. Probability Stat. **27**, 31–53 (2013)

Mahmoudi, E: The beta generalized Pareto distribution with application to lifetime data. Math. Comput. Simulations. **81**, 2414–2430 (2011)

Nadarajah, S, Kotz, S: The beta Gumbel distribution. Math. Probl. Eng. **2004**(4), 323–332 (2004)

Nadarajah, S, Kotz, S: The beta exponential distribution. Reliability Eng. Syst. Saf. **91**, 689–697 (2006)

Singla, N, Jain, K, Sharma, SK: The beta generalized Weibull distribution: properties and applications. Reliability Eng. Syst. Saf. **102**, 5–15 (2012)

Zwillinger, D, Kokoska, S: Standard Probability and Statistics Tables and Formulae. Chapman and Hall/CRC, Boca Raton (2000)

Joint distribution of rank statistics considering the location and scale parameters and its power study

Wan-Chen Lee

Correspondence:
umlee223@cc.umanitoba.ca
Department of Statistics, University of Manitoba, Winnipeg, Canada

Abstract
The ranking method used for testing the equivalence of two distributions has been studied for decades and is widely adopted for its simplicity. However, due to the complexity of calculations, the power of the test is either estimated by a normal approximation or found when an appropriate alternative is given. Here, via the Finite Markov chain imbedding technique, we are able to establish the marginal and joint distributions of the rank statistics considering the shift and scale parameters, respectively and simultaneously, under two different continuous distribution functions. Furthermore, the procedures of distribution equivalence tests and their power functions are discussed. Numerical results of a joint distribution of rank statistics under the standard normal distribution and the powers for a sequence of alternative normal distributions with means from −20 to 20 and standard deviations from 1 to 9 and their reciprocal are presented. In addition, we discuss the powers of the rank statistics under the Lehmann alternatives.

2010 Mathematics Subject Classification: Primary 62G07; Secondary 62G10

Keywords: FMCI; Lehmann alternative; Rank statistic; Rank-sum test; Power

1 Introduction

Suppose that on the basis of observations $X_1, \ldots, X_m; Y_1, \ldots, Y_n$ from the cumulative distribution functions F and G, two major topics in the hypothesis testing are to test the equivalence of either the center or the dispersion of the two populations of interest. The hypotheses are stated, for some $\theta \neq 0$,

$$H_o : F(x) = G(x) \quad \text{versus} \quad H_a : F(x) = G(x - \theta), \quad \text{for all } x,$$

which is known as the shift alternative and, for some $\sigma \neq 1$,

$$H_o : F(x) = G(x) \quad \text{versus} \quad H_a : F(x) = G\left(x\sigma^{-1}\right), \quad \text{for all } x.$$

Wilcoxon (1945) proposed the ranking method for testing the significance of the difference of the two populations means, also known as the Wilcoxon rank-sum test, and defined a statistic W_Y, as the sum of the ranks of the $y's$ in the combined and ordered sequence of $x's$ and $y's$, equivalent to

$$\sum_{j=1}^{n} \{ \# \text{ of } x_i's < y_j \} + \frac{n(n+1)}{2}.$$

Mann and Whitney (1947) introduced an elaboration of the ranking test, proposed the statistic $U_X = mn - W_Y + \frac{n(n+1)}{2}$, and proved that the limiting distribution of the test statistic U_X is

$$\frac{U_X - E(U_X)}{\sqrt{Var(U_X)}} \xrightarrow{L} N(0,1)$$

as m and n go to infinity in any arbitrary manner where

$$E(U_X) = mnp_1$$

and

$$Var(U_X) = mnp_1(1-p_1) + mn(n-1)\left(p_2 - p_1^2\right) + mn(m-1)\left(p_3 - p_1^2\right),$$

with

$$\begin{aligned} p_1 &= P(X > Y), \\ p_2 &= P(X > Y \text{ and } X > Y'), \\ p_3 &= P(X > Y \text{ and } X' > Y), \end{aligned} \tag{1}$$

where X, X' and Y, Y' are independently distributed, X, X' with the distribution F, and Y, Y' with the distribution G. Intuitively, the power for the right-sided test can be found as

$$P\left(\frac{U_X - E(U_X)}{\sqrt{Var(U_X)}} > \frac{c - E(U_X)}{\sqrt{Var(U_X)}} \,\middle|\, H_a\right), \tag{2}$$

where c is the value such that

$$\Phi\left(\frac{c - \frac{1}{2}mn}{\sqrt{\frac{1}{12}mn(m+n+1)}} \,\middle|\, H_o\right) \geq 1 - \alpha.$$

Over the years, there have been studies on finding the exact or approximate power for the rank-sum test. By choosing an appropriate alternative distribution function, Shieh et al. (2006) derived the exact power for the uniform, normal, double exponential and exponential shift models. Rosner and Glynn (2009) discussed power against the family of alternatives of the form

$$\Phi^{-1}(F_Y(y)) = \Phi^{-1}(F_X(y)) + \mu \quad \text{for some } \mu \neq 0,$$

where the underlying distributions F_X and F_Y are normal. Collings and Hamilton (1988) presented a bootstrap method to find the empirical distribution functions in order to approximate the power against the shift alternative. Lehmann (1953) derived the power function as

$$P(S_1 = s_1, S_2 = s_2, \cdots, S_n = s_n) = \frac{k^n}{\binom{m+n}{m}} \prod_{j=1}^{n} \frac{\Gamma(s_j + jk - j)}{\Gamma(s_j)} \frac{\Gamma(s_{j+1})}{\Gamma(s_{j+1} + jk - j)},$$

where s_j is the rank of y_j in the combined samples for the alternative hypothesis of

$$G_Y(x) = F_X(x)^k, \quad \text{for all } x,$$

where k is a positive integer. However, Lehmann (1998) pointed out that the power function of the rank-sum test, Equation (2), was only qualitative. Since the numerical values for assessing the probabilities in Equation (1) are considerably complicated in computation when F and G are continuous distributions with $F \neq G$.

As the rank-sum test is widely adopted for testing the center differences of two distributions, it is natural to study the efficiency of a rank-sum test for variability (Ansari and Bradley 1960). For decades, studies have focused on proposing new definitions of the rank statistic and using the methods of Chernoff and Savage to show the relative efficiency of the proposed statistic to the F-test, see for example Mood (1954), Siegel and Tukey (1960), Ansari and Bradley (1960), and Klotz (1962). Ansari and Bradley (1960) mentioned that if the means of the X and Ysamples cannot be considered equal, differences in location have a severe impact on all the tests of dispersion. Klotz (1962) showed the power of a rank test can be found by integrating the joint density of X and Y samples over that part of the $m + n$ dimensional space defined by the alternative orderings which lie in the critical region of the test, for which conditions are very strict.

Our approach aims at releasing some of the conditions for finding the distribution of the proposed rank statistic. We systematically imbed the random vector $\boldsymbol{U}_n$ into a Markov chain to induce the marginal and joint distributions of the rank statistics considering the shift and scale parameter, respectively, under any form of two distribution functions. A joint distribution of rank statistics, to the best of our knowledge, has not been studied in the literature. The main strength of using the finite Markov chain imbedding approach (FMCI) is to derive the distribution of the rank statistic without giving any conditions. Therefore, under the null hypothesis of $F = G$, we are able to identify a proper critical region and, under the alternative assumption, the power of the test can be determined naturally. The distribution of the random vector $\boldsymbol{U}_n$, independent of the form of the distribution function F, is also demonstrated under the null hypothesis of the distribution equivalence.

The main contributions of this paper are as follows. In Section 2.1, we introduce the procedures of deriving the distribution of the rank statistic considering the shift parameter and its power function by using FMCI. The procedures are general and can be applied to either two identical distribution functions of interest or two different continuous density functions. In Section 2.2, we address the steps for finding the distribution of the rank statistic considering the scale parameter and its power function. In Section 2.3, we retrieve the joint distribution of the rank statistics considering the location and scale parameters simultaneously as well as its power function. Numerical results of a joint distribution and some powers of the rank statistics against shift parameter and scale parameter, individually and simultaneously, are presented in Section 3. We also discuss the powers of the rank statistics under the Lehmann alternatives. We end this paper with a short conclusion in Section 4.

2 Methods

2.1 Distributions of the rank statistic in the shift case

Let $\{X_1, \ldots, X_m\}$ and $\{Y_1, \ldots, Y_n\}$ be two independent samples from the continuous cumulative density distributions $F(x)$ and $G(x - \theta)$, respectively. Given $\boldsymbol{x} = \{x_1, \ldots, x_m\}$ and $x_{[i]}$ is the ith smallest number in the sample, we have

$$p_i = P\left(x_{[i-1]} < Y < x_{[i]}\right) = \int_{x_{[i-1]}}^{x_{[i]}} g(y)dy = G\left(x_{[i]}\right) - G\left(x_{[i-1]}\right),$$

for $i = 1, 2, \ldots, m+1$ where $x_{[0]} = -\infty$ and $x_{[m+1]} = \infty$. Therefore, we define the sampling distribution of Y in the $(m+1)$ intervals as

$$\begin{aligned} \boldsymbol{p} &= \left(G\left(x_{[1]}\right) - G\left(x_{[0]}\right), \ldots, G\left(x_{[m+1]}\right) - G\left(x_{[m]}\right)\right) \\ &= \left(p_1, p_2, \ldots, p_{m+1}\right). \end{aligned} \tag{3}$$

Given m, for $t = 1, 2, \ldots, n,$ let

$$\Omega_t = \left\{ \boldsymbol{u}_t = (u_1(t), \cdots, u_{m+1}(t)) : \sum_{i=1}^{m+1} u_i(t) = t \text{ and } u_i(t) \geq 0, \ i = 1, \ldots, m+1 \right\},$$

where $u_i(t)$ is the number of y's in the interval $[x_{[i-1]}, x_{[i]})$ among $y_1, \ldots, y_t$. For each $\boldsymbol{u}_n = (u_1(n), \cdots, u_{m+1}(n))$, we have a corresponding rank-sum of y's in the combined sample

$$R_l(\boldsymbol{U}_n = \boldsymbol{u}_n | X) = \frac{\sum_{i=1}^{m+1} u_i^2(n) + \sum_{i=1}^{m+1} u_i(n)}{2} + \sum_{i=1}^{m} (u_i(n) + 1) \left(\sum_{j=i+1}^{m+1} u_j(n) \right). \tag{4}$$

Theorem 1. *The statistic R_l is equivalent to the statistic W_Y, which is addressed by Wilcoxon in 1945.*

Proof. Let

$$I(x_i, y_j) = \begin{cases} 1 & \text{if } x_i < y_j \\ 0 & \text{otherwise.} \end{cases}$$

The rank statistic W_Y, sum of the ranks of y's observations, can be determined by

$$\begin{aligned} \sum_{j=1}^{n} \left(\sum_{i=1}^{m} I(x_i, y_j) + j \right) &= \sum_{j=1}^{n} \sum_{i=1}^{m} I(x_i, y_j) + \sum_{j=1}^{n} j \\ &= \sum_{i=1}^{m} \sum_{j=1}^{n} I(x_i, y_j) + \frac{n(n+1)}{2}. \end{aligned} \tag{5}$$

The first summation of the first term in Equation (5) can be interpreted as the number of y observations larger than $x_{[i]}$ which is $\sum_{j=i+1}^{m+1} u_j(n)$ in our expression. It is not difficult to see that $\sum_{i=1}^{m+1} u_i(n)$ equals n, the size of y sample. Therefore, the equation can be rewritten as

$$\sum_{i=1}^{m} \left(\sum_{j=i+1}^{m+1} u_j(n) \right) + \frac{\sum_{i=1}^{m+1} u_i(n)^2 + 2 \sum_{i=1}^{m} u_i(n) \left(\sum_{j=i+1}^{m+1} u_j(n) \right) + \sum_{i=1}^{m+1} u_i(n)}{2}.$$

It is then easy to see that

$$\sum_{i=1}^{m} (u_i(n) + 1) \left(\sum_{j=i+1}^{m+1} u_j(n) \right) + \frac{\sum_{i=1}^{m+1} u_i(n)^2 + \sum_{i=1}^{m+1} u_i(n)}{2} = R_l.$$

□

Next, we demonstrate that for two random samples from the same population, the distribution of the random vector $\boldsymbol{U}_n$ is independent of the form of the distribution function.

Theorem 2. *Distribution-free property of $\boldsymbol{U}_n$.*

$$P(\boldsymbol{U}_n = \boldsymbol{u}_n | H_o) = \frac{1}{Card(\Omega_n)} = \frac{1}{\binom{m+n}{n}}. \tag{6}$$

Proof. We know the joint PDF of the ordered sample of $x's$ is given by

$$f\left(x_{[1]}, \ldots, x_{[m]}\right) = m! \prod_{i=1}^{m} f(x_i)$$

and, when $F = G$, the conditional probability of the random vector $\boldsymbol{U}_n$ given $\boldsymbol{X} = (x_1, x_2, \ldots, x_m)$ is

$$P(\boldsymbol{U}_n = \boldsymbol{u}_n | \, x_1, x_2, \ldots, x_m \,) = \frac{n!}{\prod_{i=1}^{m+1} u_i(n)!} \prod_{i=1}^{m+1} \left(\int_{x_{[i-1]}}^{x_{[i]}} f(y) dy \right)^{u_i(n)}, \tag{7}$$

where $x_{[0]} = -\infty$ and $x_{[m+1]} = \infty$. By taking the expected value of the conditional probability, we have

$$\begin{aligned} & P(\boldsymbol{U}_n = \boldsymbol{u}_n | H_o) \\ &= \int_{-\infty \le x_{[1]} \le \cdots \le x_{[m]} \le \infty} \cdots \int P(\boldsymbol{u}_n | \, x_1, \ldots, x_m \,) f\left(x_{[1]}, \ldots, x_{[m]}\right) \, dx_{[1]} \cdots dx_{[m]} \\ &= \int_{-\infty}^{\infty} \int_{x_{[1]}}^{\infty} \cdots \int_{x_{[m-1]}}^{\infty} \frac{n!}{\prod_{i=1}^{m+1} u_i(n)!} \left(F\left(x_{[1]}\right)\right)^{u_1(n)} \left(F\left(x_{[2]}\right) - F\left(x_{[1]}\right)\right)^{u_2(n)} \\ & \quad \cdots \left(1 - F\left(x_{[m]}\right)\right)^{u_{m+1}(n)} m! \, dF\left(x_{[1]}\right) \cdots dF\left(x_{[m]}\right). \end{aligned} \tag{8}$$

Using variable transformation, it is clear to see that the random variables $F\left(x_{[1]}\right), \ldots, F\left(x_{[m]}\right)$ have a Dirichlet distribution with parameters $u_1(n) + 1, u_2(n) + 1, \ldots, u_{m+1}(n) + 1$. Therefore, we have

$$P(\boldsymbol{U}_n = \boldsymbol{u}_n | H_o) = \frac{n! \, m!}{(n+m)!} = \frac{1}{Card(\Omega_n)}$$

which is independent of the distribution function. □

This is the reason that the distribution of the rank statistic $\boldsymbol{U}_n$ is distribution-free under the null hypothesis. However, the distribution of the random vector $\boldsymbol{U}_n$ is discrete uniform with the mass function one over the number of possible outcomes of the random vector $\boldsymbol{U}_n$ only when assuming $F = G$. In other words, the distribution of the random vector $\boldsymbol{U}_n$ can be found by the traditional combinatorial analysis when $F = G$. Unfortunately, when $F \neq G$, we will not be able to establish the distribution of $\boldsymbol{U}_n$ through Equation (7) as solving the multiple integral in Equation (8) is either tedious given some appropriate alternative distribution function or difficult. Our understanding is that finding the power of the test has not been solved in most cases. To overcome this situation, we bring in the finite Markov chain imbedding approach.

Let Ω_t, $t = 0, 1, \ldots, n$, be the state space which has

$$\binom{m+t}{t}$$

possible states, $\Gamma_n = \{0, 1, \ldots, n\}$ be an index set, and $\{Z_t : t \in \Gamma_n\}$ be a non-homogeneous Markov chain on the state space Ω_t. As a transition probability matrix $\boldsymbol{M}_t$ for this chain, $t = 1, \ldots, n$, consider

$$\boldsymbol{M}_t = \begin{array}{c} \\ \Omega_{t-1} \end{array} \!\!\begin{array}{c} \Omega_t \\ \left[\begin{array}{c} p_{\boldsymbol{u}_{t-1}, \boldsymbol{u}_t} \end{array} \right]_{\binom{m+t-1}{t-1} \times \binom{m+t}{t}} \end{array},$$

where

$$\begin{aligned} p_{\boldsymbol{u}_{t-1}, \boldsymbol{u}_t} &= P(Z_t = \boldsymbol{u}_t | Z_{t-1} = \boldsymbol{u}_{t-1}) \\ &= \begin{cases} p_i & \text{if } u_i(t-1) + 1 = u_i(t) \text{ and } u_j(t-1) = u_j(t) \ \forall j \neq i \\ 0 & \text{otherwise} \end{cases}, \end{aligned}$$

and p_i is defined in Equation (3).

Theorem 3. *$R_l(\boldsymbol{U}_n|\boldsymbol{X})$ is finite Markov chain imbeddable, and*

$$P(R_l(\boldsymbol{U}_n) = r|\boldsymbol{X}) = \boldsymbol{\xi} \left(\prod_{t=1}^{n} \boldsymbol{M}_t \right) \boldsymbol{B}'(C_r),$$

where $\boldsymbol{B}(C_r) = \sum_{k: R_l(\boldsymbol{U}_n)=r} e_k$, e_k is a $1 \times \binom{m+n}{n}$ unit row vector corresponding to state $\boldsymbol{u}_n$, $\boldsymbol{\xi} (= P(Z_0 = 1) = 1)$ is the initial probability and $\boldsymbol{M}_t$, $t = 1, \ldots, n$, are the transition probability matrices of the imbedded Markov chain defined on the state space Ω_t.

Proof. For each $\boldsymbol{u}_n = (u_1(n), \cdots, u_{m+1}(n))$ in the state space Ω_n, we have a corresponding rank R_l as shown in Equation (4). Intuitively, the minimum rank r_{ls} is $n(n+1)/2$ and the maximum rank r_{lb} is $n(2m + n + 1)/2$. In accordance with the possible values of the rank R_l, we define a finite partition $\{C_r : r = r_{ls}, \ldots, r_{lb}\}$ such that

$$P(Z_n \in C_r|\boldsymbol{p}) = \boldsymbol{\xi} \left(\prod_{t=1}^{n} \boldsymbol{M}_t \right) \boldsymbol{B}'(C_r) \tag{9}$$

where $\boldsymbol{B}(C_r) = \sum_{k: R_l(\boldsymbol{U}_n)=r} e_k$, e_k is a $1 \times \binom{m+n}{n}$ unit row vector corresponding to state $\boldsymbol{U}_n$, we then obtain the conditional probability of the rank R_l. □

Then, the Law of Large Numbers is used to determine the probability of $\boldsymbol{U}_n$ for any continuous F and G

$$\frac{1}{N} \sum_{i=1}^{N} P(\boldsymbol{U}_n = \boldsymbol{u}_n | \, \mathbf{X}_i) \xrightarrow{p} P(\boldsymbol{U}_n = \boldsymbol{u}_n)$$

where $\boldsymbol{X}_i$ is the ith sample of size m from the distribution function F. It is easy to see that

$$P(R_l(\boldsymbol{U}_n) = r) = \sum_{\boldsymbol{u}_n : R(\boldsymbol{u}_n) = r} P(\boldsymbol{U}_n = \boldsymbol{u}_n). \tag{10}$$

To test

$$H_o : F(x) = G(x) \quad \text{versus} \quad H_a : F(x) = G(x - \theta),$$

for some $\theta \neq 0$, the power function is approximated by

$$\begin{aligned}
&P(R_l(\boldsymbol{U}_n) \leq r_{1\alpha}|H_a) + P(R_l(\boldsymbol{U}_n) \geq r_{2\alpha}|H_a) \\
&= \sum_{r=r_{ls}}^{r_{1\alpha}} P(R_l(\boldsymbol{U}_n) = r|H_a) + \sum_{r=r_{2\alpha}}^{r_{lb}} P(R_l(\boldsymbol{U}_n) = r|H_a) \\
&= \sum_{r=r_{ls}}^{r_{1\alpha}} \sum_{\boldsymbol{u}_n:R(\boldsymbol{u}_n)=r} P(\boldsymbol{U}_n = \boldsymbol{u}_n|H_a) + \sum_{r=r_{2\alpha}}^{r_{lb}} \sum_{\boldsymbol{u}_n:R(\boldsymbol{u}_n)=r} P(\boldsymbol{U}_n = \boldsymbol{u}_n|H_a) \\
&\approx \sum_{r=r_{ls}}^{r_{1\alpha}} \sum_{\boldsymbol{u}_n:R(\boldsymbol{u}_n)=r} \frac{1}{N} \sum_{i=1}^{N} P(\boldsymbol{U}_n|H_a;\, \mathbf{X}_i) + \sum_{r=r_{2\alpha}}^{r_{lb}} \sum_{\boldsymbol{u}_n:R(\boldsymbol{u}_n)=r} \frac{1}{N} \sum_{i=1}^{N} P(\boldsymbol{U}_n|H_a;\, \mathbf{X}_i) \\
&= \frac{1}{N} \left(\sum_{r=r_{ls}}^{r_{1\alpha}} \sum_{i=1}^{N} \sum_{\boldsymbol{u}_n:R(\boldsymbol{u}_n)=r} P(\boldsymbol{U}_n|H_a;\, \mathbf{X}_i) + \sum_{r=r_{2\alpha}}^{r_{lb}} \sum_{i=1}^{N} \sum_{\boldsymbol{u}_n:R(\boldsymbol{u}_n)=r} P(\boldsymbol{U}_n|H_a;\, \mathbf{X}_i) \right) \\
&= \frac{1}{N} \sum_{i=1}^{N} \left(\sum_{r=r_{ls}}^{r_{1\alpha}} P(R_l(\boldsymbol{U}_n) = r|H_a;\, \mathbf{X}_i) + \sum_{r=r_{2\alpha}}^{r_{lb}} P(R_l(\boldsymbol{U}_n) = r|H_a;\, \mathbf{X}_i) \right),
\end{aligned}$$

where

$$P(R_l(\boldsymbol{U}_n) \leq r_{1\alpha}|H_o) + P(R_l(\boldsymbol{U}_n) \geq r_{2\alpha}|H_o) \leq \alpha.$$

Note that the alternative hypothesis is subject to the purpose of the test. This simply needs to be slightly modified if a one-sided test is adopted.

2.2 Distributions of the rank statistic in the scale case

We studied the distribution and the power function of the rank statistic R_l considering a shift in location. Now, the distribution and the power function of the rank statistic considering the scale parameter will be addressed. For this purpose, we consider $F(x) = G\left(x\sigma^{-1}\right)$ and state the null and alternative hypotheses as

$$H_o : \sigma = 1 \quad \text{versus} \quad H_a : \sigma \neq 1.$$

To do so, we begin with the procedure of finding the distribution of the rank statistic, denoted R_s, considering the scale parameter through the random vector $\mathbf{U}_n$. The array of ranks are given by

$$(m+n)/2, \ldots, 3, 2, 1, \qquad 1, 2, 3, \ldots, (m+n)/2;$$

if $m+n$ is even, and

$$(m+n-1)/2, \ldots, 3, 2, 1, \qquad 0 \qquad 1, 2, 3, \ldots, (m+n-1)/2$$

if $m+n$ is odd. We first introduce how to determine the rank-sum of $y's$ observations in the combined samples, R_s, with respect to

$$\Omega_n = \left\{ \mathbf{u}_n = (u_1(n), \ldots, u_{m+1}(n)) : \sum_{i=1}^{m+1} u_i(n) = n \right\}$$

where $u_i(n)$ means the number of y observations belonging to $[x_{[i-1]}, x_{[i]})$. Let $med(x, y)$ be the median among $x's$ and $y's$ and belongs to $[x_{[i]}, x_{[i+1]})$ which will then break $\mathbf{U}_n$ into two parts $\boldsymbol{U}_n^-$ and $\boldsymbol{U}_n^+$. If $m+n$ is odd and $med(x,y) = x_{[i]}$, then

$$\boldsymbol{U}_n^- = (u_1^- = u_i(n)\ ,\ u_2^- = u_{i-1}(n)\ ,\ \cdots\ ,\ u_i^- = u_1(n))$$

is a $1 \times i$ vector and

$$\boldsymbol{U}_n^+ = \left(u_1^+ = u_{i+1}(n)\ ,\ u_2^+ = u_{i+2}(n)\ ,\ \cdots\ ,\ u_{m+1-i}^+ = u_{m+1}(n)\right)$$

is a $1 \times (m+1-i)$ vector. The second possible case is, if $m+n$ is odd and $med(x,y) = y_{\left[\sum_{k=1}^{i} u_k(n)+j\right]}$, then $\boldsymbol{U}_n^-$, a row vector with length $i+1$, has the form

$$\left(u_1^- = j-1\ ,\ u_2^- = u_i(n)\ ,\ \cdots\ ,\ u_{i+1}^- = u_1(n)\right)$$

and $\boldsymbol{U}_n^+$, a row vector with length $m+1-i$, is given by

$$\left(u_1^+ = u_{i+1}(n) - j\ ,\ u_2^+ = u_{i+2}(n)\ ,\ \cdots\ ,\ u_{m+1-i}^+ = u_{m+1}(n)\right).$$

The third possible case is, if $m+n$ is even and $x_{[i]}$ is the smallest number larger than $med(x,y)$, the vectors are now defined as

$$\boldsymbol{U}_n^- = (u_1^- = u_i(n)\ ,\ u_2^- = u_{i-1}(n)\ ,\ \cdots\ ,\ u_i^- = u_1(n))$$

and

$$\boldsymbol{U}_n^+ = \left(u_1^+ = 0\ ,\ u_2^+ = u_{i+1}(n)\ ,\ \cdots\ ,\ u_{m+2-i}^+ = u_{m+1}(n)\right).$$

The last possibility is, if $m+n$ is even, $y_{\left[\sum\limits_{k=1}^{i} u_k(n)+j\right]}$ is the smallest number larger than $med(x,y)$. The vectors are now defined as

$$\boldsymbol{U}_n^- = (u_1^- = j-1\ ,\ u_2^- = u_i(n)\ ,\ \cdots\ ,\ u_{i+1}^- = u_1(n))$$

and

$$\boldsymbol{U}_n^+ = \left(u_1^+ = u_{i+1}(n) - j + 1\ ,\ u_2^+ = u_{i+2}(n)\ ,\ \cdots\ ,\ u_{m+1-i}^+ = u_{m+1}(n)\right).$$

Let n^- be the length of the vector $\boldsymbol{U}_n^-$ and n^+ be the length of the vector $\boldsymbol{U}_n^+$.

Theorem 4. *$R_s(\mathbf{U}_n|\boldsymbol{X})$ is finite Markov chain imbeddable, and*

$$P(R_s(\mathbf{U}_n) = r|\boldsymbol{X}) = \boldsymbol{\xi}\left(\prod_{t=1}^{n} M_t\right)\boldsymbol{B}'(C_r),$$

where $\boldsymbol{B}(C_r) = \sum_{k:R_s(\mathbf{U}_n)=r} e_k$, e_k is a $1 \times \binom{m+n}{n}$ unit row vector corresponding to state $\mathbf{U}_n$, $\boldsymbol{\xi}(= P(Z_0 = 1) = 1)$ is the initial probability and M_t, $t = 1, \ldots, n$ are the transition probability matrices of the imbedded Markov chain defined on the state space Ω_t.

Proof. For each $\mathbf{U}_n$ in the state space Ω_n, we have a corresponding

$$\begin{aligned} R_s(\mathbf{U}_n|\boldsymbol{X}) &= R_s(\boldsymbol{U}_n^-|\boldsymbol{X}) + R_s(\boldsymbol{U}_n^+|\boldsymbol{X}) \\ &= \frac{\sum_{k=1}^{n^-}\left(u_k^-\right)^2 + \sum_{k=1}^{n^-} u_k^-}{2} + \sum_{k=1}^{n^- -1}\left(u_k^- + 1\right)\left(\sum_{j=k+1}^{n^-} u_j^-\right) \\ &\quad + \frac{\sum_{k=1}^{n^+}\left(u_k^+\right)^2 + \sum_{k=1}^{n^+} u_k^+}{2} + \sum_{k=1}^{n^+ -1}\left(u_k^+ + 1\right)\left(\sum_{j=k+1}^{n^+} u_j^+\right). \end{aligned} \tag{11}$$

The smallest possible value of $R_s(\mathbf{U}_n)$ is

$$r_{ss} = \begin{cases} \frac{n(n+2)}{4} & \text{if } m+n \text{ is even and } n \text{ is even} \\ \frac{(n+1)(n+3)}{4} & \text{if } m+n \text{ is even and } n \text{ is odd} \\ \frac{n^2}{4} & \text{if } m+n \text{ is odd and } n \text{ is even} \\ \frac{(n+1)(n-1)}{4} & \text{if } m+n \text{ is odd and } n \text{ is odd} \end{cases} \tag{12}$$

and the largest possible value is

$$r_{sb} = \begin{cases} \frac{n(2m+n+2)}{4} & \text{if } m+n \text{ is even and } n \text{ is even} \\ \frac{n(2m+n+2)-1}{4} & \text{if } m+n \text{ is even and } n \text{ is odd} \\ \frac{n(2m+n-1)}{4} & \text{if } m+n \text{ is odd and } n \text{ is even} \\ \frac{n(2m+n)-1}{4} & \text{if } m+n \text{ is odd and } n \text{ is odd} \end{cases} \tag{13}$$

In accordance with Equation (11), we use the possible value of R_s as a rule of the partition. The rest of the proof follows along the same line as that of Theorem 3, and here, is omitted.

□

Similarly, we apply the LLN to conclude that

$$\frac{1}{N}\sum_{i=1}^{N} P(R_s|\,\mathbf{X}_i\,) \xrightarrow{p} P(R_s)$$

which establishes the distribution of R_s.

Through FMCI we, again, successfully retrieved the distribution of R_s under selected alternative distributions, for which the procedures are similar to those in the previous section. In addition, it is quite intuitive to approximate the power function by

$$\frac{1}{N}\sum_{i=1}^{N}\left(\sum_{s=r_{ss}}^{s_{1\alpha}} P(R_s(\mathbf{U}_n)=s|\,\mathbf{X}_i) + \sum_{s=s_{2\alpha}}^{r_{sb}} P(R_s(\mathbf{U}_n)=s|\,\mathbf{X}_i)\right),$$

where

$$P(R_s(\mathbf{U}_n) \le s_{1\alpha}|H_o) + P(R_s(\mathbf{U}_n) \ge s_{2\alpha}|H_o) \le \alpha.$$

2.3 Joint distributions of the rank statistics in the shift and scale case

We have derived the marginal distributions of R_l and R_s in terms of $\boldsymbol{U}_n$, respectively, which yield the following theorem.

Theorem 5. *$(R_l(\boldsymbol{U}_n|\boldsymbol{X}), R_s(\boldsymbol{U}_n|\boldsymbol{X}))$ is finite Markov chain imbeddable, and*

$$P(R_l(\boldsymbol{U}_n)=r_1; R_s(\boldsymbol{U}_n)=r_2|\boldsymbol{X}) = \boldsymbol{\xi}\left(\prod_{t=1}^{n}\boldsymbol{M}_t\right)\boldsymbol{B}'(C_r)$$

where $\boldsymbol{B}(C_r) = \sum_{k:R_l(\boldsymbol{U}_n)=r_1 \,\&\, R_s(\boldsymbol{U}_n)=r_2} e_k$, e_k is a $1\times\binom{m+n}{n}$ unit row vector corresponding to state $\boldsymbol{u}_n$, $\boldsymbol{\xi}(=P(Z_0=1)=1)$ is the initial probability and $\boldsymbol{M}_t$, $t=1,\ldots,n$ are the transition probability matrices of the imbedded Markov chain defined on the state space Ω_t.

Proof. By Equations (4) and (11), we know each $\boldsymbol{u}_n$ in the state space Ω_n has corresponding values of R_l and R_s. The combinations of the values R_l and R_s are used to be the standard of the partition. The rest of the proof follows along the same line as that of Theorem 3. □

The joint distribution of the ranks considering both the location and scale parameters which can be determined through our algorithm is yet to be studied in the literature. Our result allows us to test the homogeneity of the distribution functions $F(x) = G\left((x-\theta)\sigma^{-1}\right)$. We state the hypotheses as follows

$$H_o : \theta = 0 \text{ and } \sigma = 1 \text{ v.s. } H_a : \theta \neq 0 \text{ or } \sigma \neq 1. \tag{14}$$

Also we are able to identify a proper critical region under the null hypothesis and discuss its power when $F \neq G$. For example, a rectangular critical region can be

$$C_\alpha = \{R_l \leq r_{1l},\ R_l \geq r_{2l},\ R_s \leq r_{1s} \quad \text{or} \quad R_s \geq r_{2s}\}$$

where r_{1l}, r_{2l}, r_{1s} and r_{2s} are the critical values such that

$$P(R_l \leq r_{1l}|H_o) + P(R_l \geq r_{2l}|H_o) + P(r_{1l} < R_l < r_{2l}, R_s \leq r_{1s}|H_o) \\ + P(r_{1l} < R_l < r_{2l}, R_s \geq r_{2s}|H_o) \leq \alpha$$

or an elliptic critical region

$$C'_\alpha = \left\{\frac{R_l^2}{a} + \frac{R_s^2}{b} > C\right\}$$

for some positive constants a and b such that

$$P\left(\frac{R_l^2}{a} + \frac{R_s^2}{b} > C|H_o\right) \leq \alpha.$$

According to the above defined rejection region, the power of the test can be found as

$$P(R_l \leq r_{1l}|H_a) + P(R_l \geq r_{2l}|H_a) + P(r_{1l} < R_l < r_{2l}, R_s \leq r_{1s}|H_a) \\ + P(r_{1l} < R_l < r_{2l}, R_s \geq r_{2s}|H_a) \tag{15}$$

or

$$P\left(\frac{R_l^2}{a} + \frac{R_s^2}{b} > C|H_a\right). \tag{16}$$

Note that unless having a conjecture about the values of θ and σ, we tend to use a two-sided test. However, with the knowledge of the center and shape of the distribution of interest, deciding a sectorial critical region is a better choice, for which an example is demonstrated in the numerical studies.

3 Numerical results and discussion

3.1 A joint distribution of R_l and R_s

Let $\{X_1, \ldots, X_5\} \sim N(0, 1)$ and $\{Y_1, \ldots, Y_7\} \sim N(\theta, \sigma)$. Figure 1 gives the joint distribution of the random variables R_l and R_s under the null hypothesis of $\theta = 0$ and $\sigma = 1$. The marginal distributions of R_l and R_s can be easily established from their joint distribution. Figure 1 also shows that the two random variables R_l and R_s are dependent. We construct two critical regions as shown in Figure 2, according to their joint distribution. Outside the yellow area in Figure 2 is the selected rectangular critical region $C_{0.1738}$ and outside the red shadow is the elliptic one $C'_{0.1738}$.

3.2 Powers for a joint test using R_l and R_s

The alternative of interest is stated in the preceding section (see Equation (14)). The power functions of the test statistics R_l and R_s for a sequence of normally distributed populations with θ from -20 to 20 with an increment of 0.5 and σ from 1 to 10 with an increment of 1, and its reciprocal under two types of critical regions are provided in Figures 3 and 4. We adopt a two-sided test because of the selected values of the parameters. It should be slightly modified the critical region in the previous step in order to calculate the powers if a one-sided test is adopted. Both critical regions roughly perform equally well as shown in Figures 3 and 4. Figure 5 presents the performance of the two critical regions for given various parameter settings. Figures 5(a) and (b) show that given a standard deviation of 1 or a mean of 0, the powers of the two critical regions, rectangular and elliptic, are high and similar. However, when the variation of the alternative population reduces ($\sigma = 1/10$) or increases ($\sigma = 10$), the elliptic critical region performs better than the rectangular one as shown in Figures 5(c) and (d). Therefore, we suggest that when conducting a test for the equivalence of two distributions, an elliptic rejection area should be used.

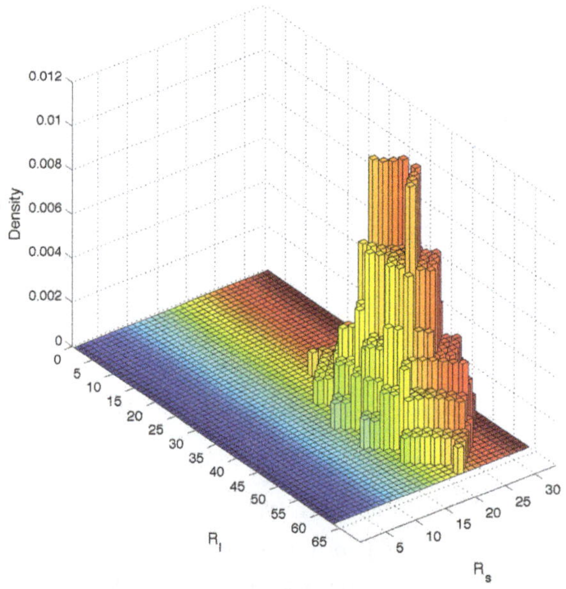

Figure 1 Joint distribution of R_l and R_s in the case where $m = 5, n = 7$ and $F = G \sim N(0, 1)$.

0	0	0	0	0	0	0.00108	0	0	0	0	0	0	0	0	0	0	0
0	0	0	0	0	0	0	0.0012	0	0	0	0	0	0	0	0	0	0
0	0	0	0	0	0	0	0.00123	0.00125	0	0	0	0	0	0	0	0	0
0	0	0	0	0	0	0.00125	0	0.00128	0.00115	0	0	0	0	0	0	0	0
0	0	0	0	0	0.00126	0	0.00125	0	0.00251	0.00116	0	0	0	0	0	0	0
0	0	0	0	0.00127	0	0.00125	0	0.00253	0	0.00252	0.00118	0	0	0	0	0	0
0	0	0	0.00122	0	0.00126	0	0.00251	0.00126	0.00254	0	0.0037	0	0	0	0	0	0
0	0	0.00111	0	0.00126	0	0.00251	0.00127	0.00252	0.00129	0.0038	0	0.00252	0	0	0	0	0
0	0	0	0.0012	0	0.00249	0.00248	0.00253	0.00127	0.0038	0.00252	0.00258	0	0.00244	0	0	0	0
0	0	0	0	0.00241	0.00247	0.00253	0.00251	0.00377	0.00253	0.00256	0.00386	0.00251	0	0.00121	0	0	0
0	0	0	0	0.00367	0.00249	0.0025	0.00371	0.00502	0.00254	0.00386	0.00253	0.00378	0.00127	0	0.00123	0	0
0	0	0	0.00244	0	0.00375	0.00356	0.00505	0.00253	0.00765	0.00252	0.00381	0.00129	0.00381	0.00129	0	0	0
0	0	0.00244	0	0.00248	0	0.00754	0.00249	0.00763	0.00378	0.00759	0.00128	0.00385	0.00132	0.00384	0	0	0
0	0.00122	0	0.00247	0	0.00505	0	0.0114	0.00378	0.00755	0.00249	0.00767	0.00129	0.00386	0	0.00251	0	0
0.00118	0	0.0012	0	0.005	0	0.00768	0.0026	0.01132	0.00248	0.00761	0.0038	0.00773	0	0.00256	0	0.00125	0
0	0.00121	0	0.00247	0	0.00759	0.00391	0.00759	0.00254	0.01136	0.00369	0.00761	0.00257	0.00511	0	0.0013	0	0.00127
0	0	0.00244	0	0.00377	0.00395	0.00754	0.00383	0.0076	0.0051	0.01146	0.00253	0.00508	0.00388	0.00262	0	0.00132	0
0	0	0	0.00372	0.00398	0.00377	0.00384	0.00753	0.00767	0.00762	0.00506	0.00755	0.00378	0.00258	0.0025	0.00266	0	0
0	0	0	0.00391	0.00373	0.0039	0.00375	0.00763	0.00762	0.00764	0.00508	0.00754	0.0038	0.00253	0.0026	0.00255	0	0
0	0	0.00264	0	0.00384	0.00368	0.00765	0.0038	0.00753	0.00509	0.01146	0.00255	0.00503	0.00379	0.00258	0	0.00127	0
0	0.00132	0	0.00261	0	0.00763	0.00376	0.00756	0.00256	0.01135	0.00387	0.0076	0.00257	0.00512	0	0.0013	0	0.00131
0.00135	0	0.00131	0	0.00509	0	0.00757	0.00245	0.01137	0.00246	0.00756	0.00381	0.00762	0	0.00261	0	0.00133	0
0	0.00128	0	0.00257	0	0.00504	0	0.01132	0.00375	0.00757	0.00252	0.00764	0.00122	0.00381	0	0.00261	0	0
0	0	0.00257	0	0.00254	0	0.00745	0.00254	0.00753	0.00375	0.00759	0.00128	0.00383	0.00124	0.00382	0	0	0
0	0	0	0.0025	0	0.00372	0.00376	0.00501	0.00252	0.00753	0.00256	0.00379	0.00129	0.00379	0.00126	0	0	0
0	0	0	0	0.00369	0.0025	0.00249	0.00378	0.00496	0.00252	0.00374	0.00258	0.00379	0.00128	0	0.00133	0	0
0	0	0	0	0.0024	0.00251	0.00256	0.00244	0.00379	0.00248	0.00252	0.00375	0.00257	0	0.00131	0	0	0
0	0	0	0.00118	0	0.0025	0.00249	0.00259	0.00122	0.00378	0.00248	0.0025	0	0.00251	0	0	0	0
0	0	0.00118	0	0.00125	0	0.00257	0.00126	0.00257	0.00121	0.00373	0	0.00251	0	0	0	0	0
0	0	0	0.00125	0	0.00128	0	0.00257	0.00121	0.00252	0	0.00374	0	0	0	0	0	0
0	0	0	0	0.00128	0	0.00129	0	0.00252	0	0.00249	0.00131	0	0	0	0	0	0
0	0	0	0	0	0.00128	0	0.00128	0	0.00245	0.00127	0	0	0	0	0	0	0
0	0	0	0	0	0	0.00125	0	0.00126	0.00119	0	0	0	0	0	0	0	0
0	0	0	0	0	0	0	0.00124	0.00124	0	0	0	0	0	0	0	0	0
0	0	0	0	0	0	0	0.00123	0	0	0	0	0	0	0	0	0	0
0	0	0	0	0	0	0.00122	0	0	0	0	0	0	0	0	0	0	0

Figure 2 Critical Regions at size 17.38% for R_l and R_s for $m = 5$ and $n = 7$.

Next, we consider the problem of determining an optimum rank test. To conduct a test of distributions equivalency, we can use either R_l or R_s as the test statistic. As mentioned earlier, the marginal distribution R_l or R_s can be easily established from their joint distribution. Figures 6 and 7 provide the power functions for the test statistics R_l and R_s at the level of significance 17.38%, respectively. Figure 7 shows that the rank test against scale parameter is badly effected by the centre of the alternative population. This was seen before by Ansari and Bradley (1960). By comparing Figures 6 and 7 with Figure 4, it seems that the joint test would be much more reliable than either R_l or R_s alone for distributions equivalence tests. A joint test for distributions equivalency would like a better option under most circumstances.

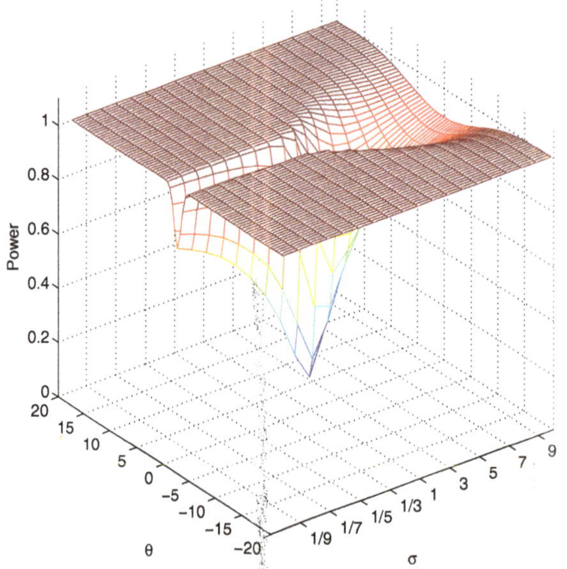

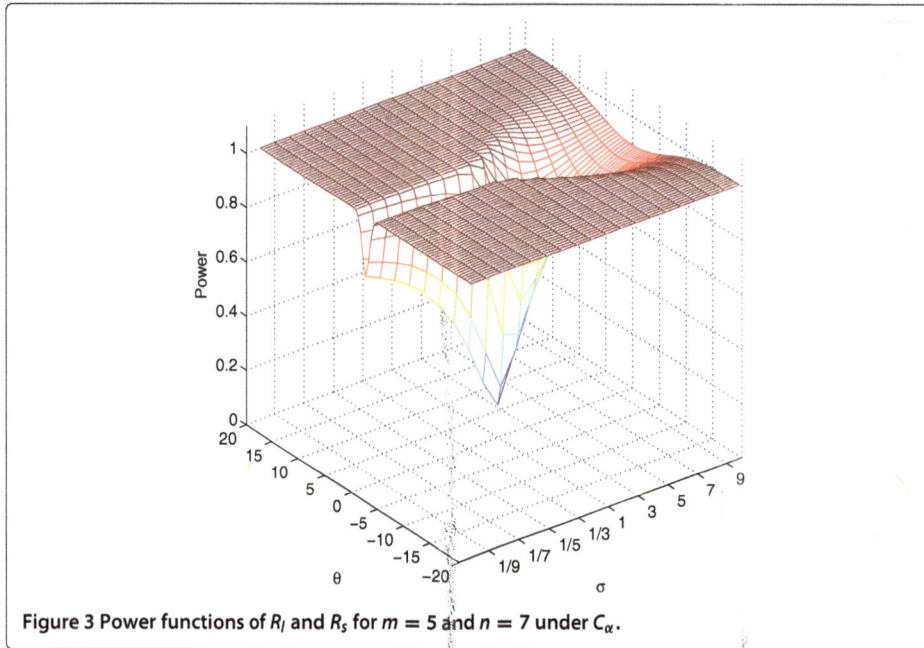

Figure 3 Power functions of R_l and R_s for $m = 5$ and $n = 7$ under C_α.

3.3 Lehmann alternatives

Consider the one-sided alternative $F(x;\theta,\sigma) > G(x;\theta,\sigma)$, Lehmann (1953) proposed a test of $H_o : F(x;\theta,\sigma) = G(x;\theta,\sigma)$ against $H_a : F(x;\theta,\sigma)^k = G(x;\theta,\sigma)$ which is known as the family of Lehmann alternative. Note $F(x;\theta,\sigma)^k$ is the cumulative distribution of $\max_{1\leq i\leq k}(x_i)$ when $X_i \sim F$ and, under the alternative hypothesis, $G(x;\theta,\sigma)$ is stochastically larger than $F(x;\theta,\sigma)$. First of all, we know

$$\begin{aligned} E_k(X) &= \int_{-\infty}^{0} -G(x)dx + \int_{0}^{\infty} 1 - G(x)dx \\ &> \int_{-\infty}^{0} -F(x)dx + \int_{0}^{\infty} 1 - F(x)dx = E(X). \end{aligned} \tag{17}$$

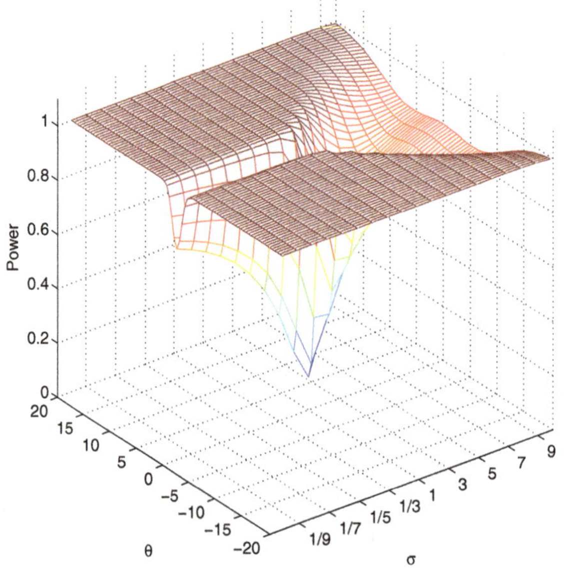

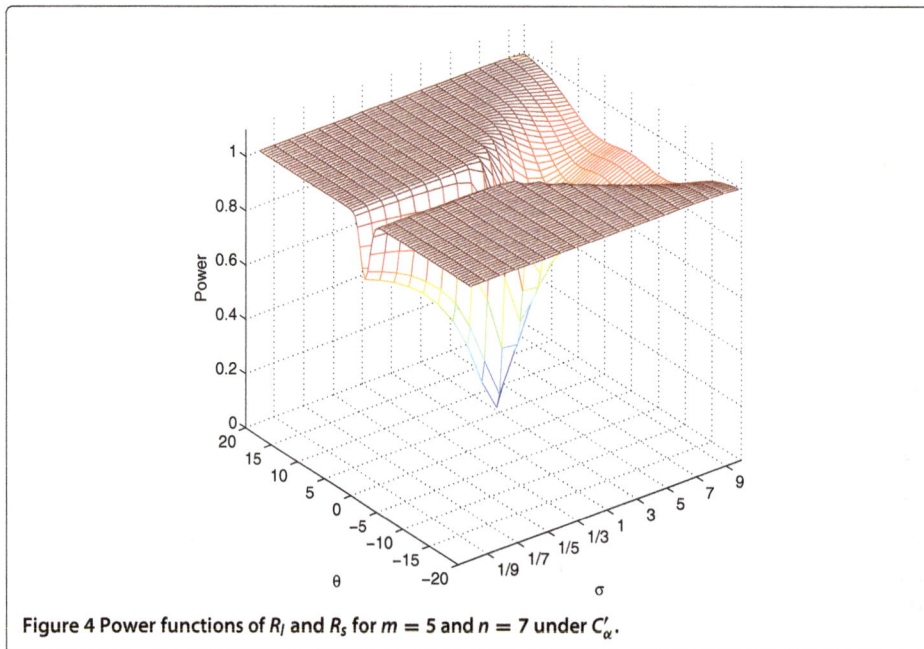

Figure 4 Power functions of R_l and R_s for $m = 5$ and $n = 7$ under C'_α.

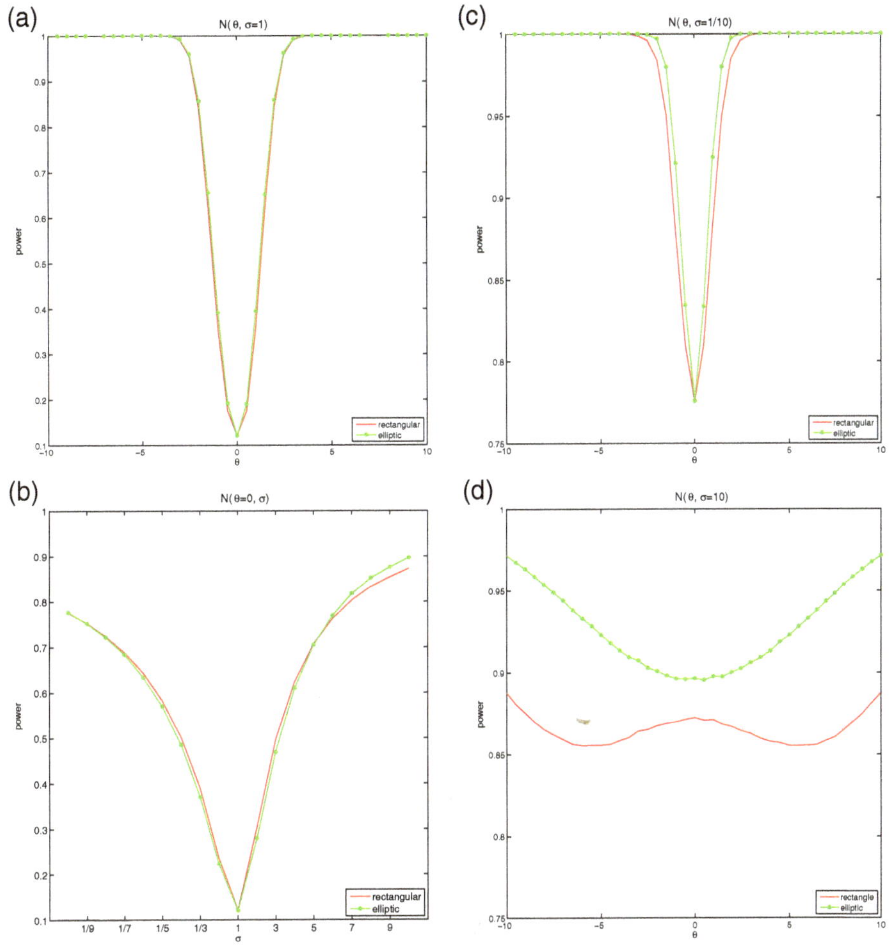

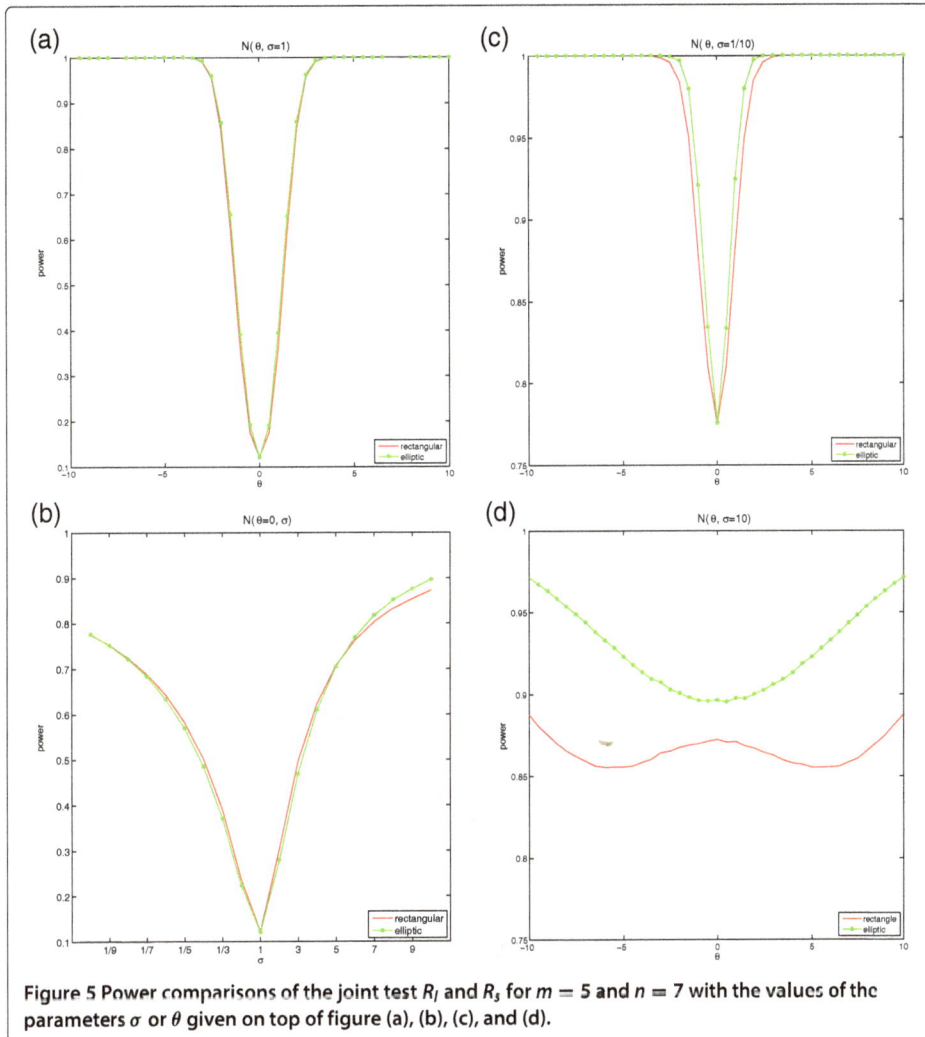

Figure 5 Power comparisons of the joint test R_l and R_s for $m = 5$ and $n = 7$ with the values of the parameters σ or θ given on top of figure (a), (b), (c), and (d).

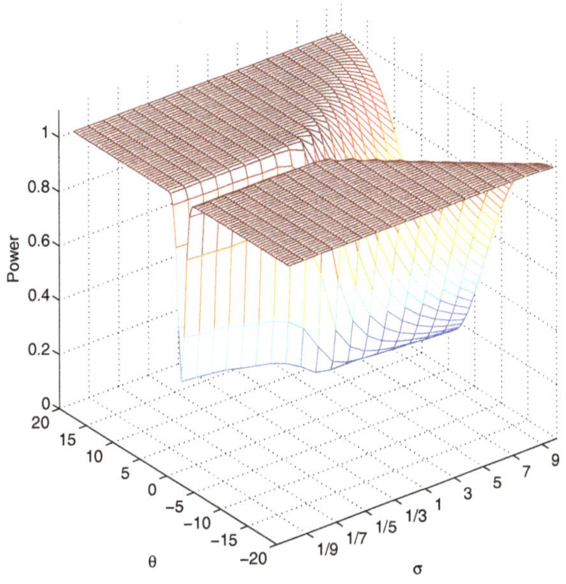

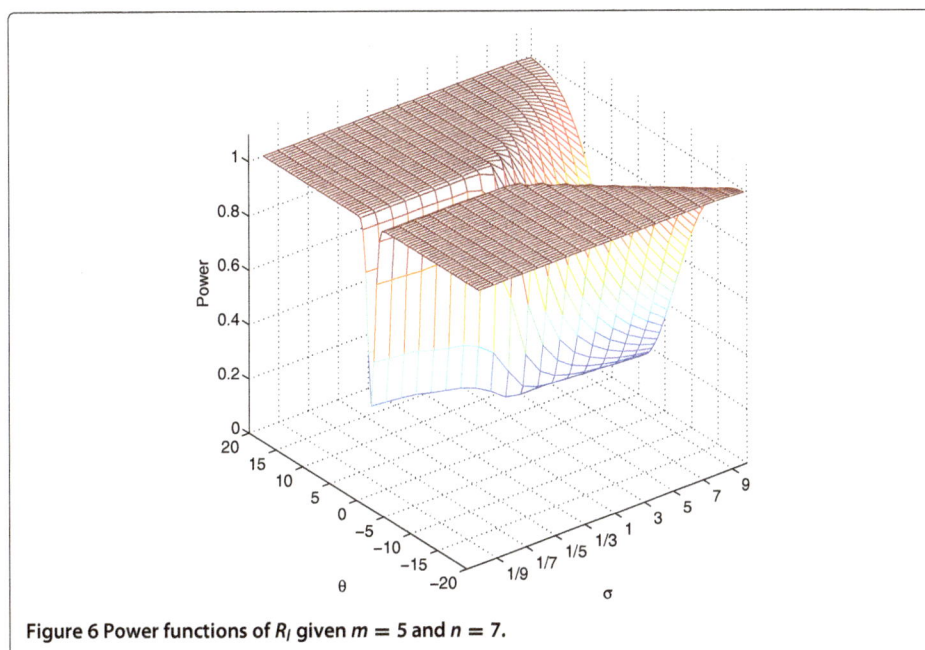

Figure 6 Power functions of R_l given $m = 5$ and $n = 7$.

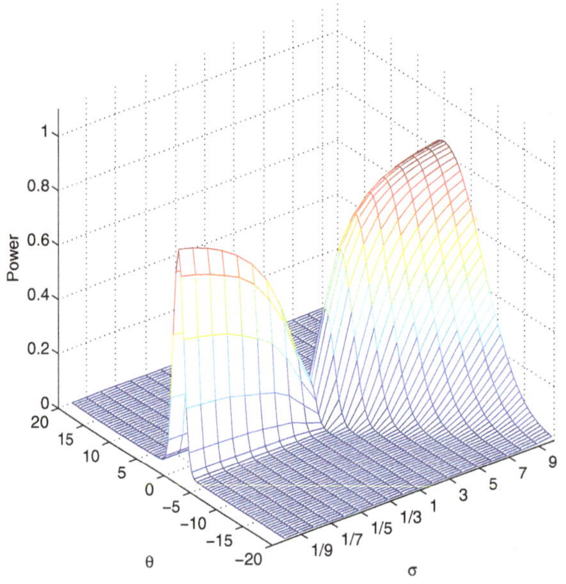

Figure 7 Power functions of R_s given $m = 5$ and $n = 7$.

Therefore, the larger the R_l is, the stronger the evidence against the null hypothesis will be. For the variation of the distribution per se, the codomain of the density function is compressed to larger numbers; therefore, in most cases, we have $Var(X_k) < Var(X)$. We then propose to reject the null hypothesis when R_s is large. For example, given $F \sim U(0, 1)$ and $G = F^k$, it is easy to see

$$\frac{E_{k+1}(X)}{E_k(X)} = \frac{(k+1)^2}{k(k+2)} > 1 \tag{18}$$

and

$$\frac{Var_{k+1}(X)}{Var_k(X)} = \frac{(k+1)^3}{k(k+2)(k+3)} < 1 \tag{19}$$

for all k. We first find the marginal and joint distributions of the ranks R_l and R_s in order to define critical regions for R_l and R_s individually and simultaneously. Due to the properties

Table 1 Power comparisons for a one-sided rank test $H_0 : F(x; \theta_o, \sigma_o) = G(x; \theta_a, \sigma_a)$ v.s. $H_a : F^k(x; \theta_o, \sigma_o) = G(x; \theta_a, \sigma_a)$

		$m = 6$ $n = 10$				$m = 10$ $n = 10$				$m = 10$ $n = 20$			
F	Test	$\beta(F)$	$\beta(F^2)$	$\beta(F^3)$	$\beta(F^6)$	$\beta(F)$	$\beta(F^2)$	$\beta(F^3)$	$\beta(F^6)$	$\beta(F)$	$\beta(F^2)$	$\beta(F^3)$	$\beta(F^6)$
$U(0,1)$	R_l	.090	.411	.647	.900	.096	.496	.761	.967	.099	.591	.845	.984
	R_s	.080	.152	.193	.218	.076	.137	.149	.123	.100	.236	.370	.638
	R_l&R_s	.100	.452	.699	.934	.100	.531	.799	.981	.100	.622	.878	.992
$t(3)$	R_l	0.090	.412	.639	.897	0.096	.493	.756	.965	0.099	.574	.841	.987
	R_s	0.080	.150	.197	.217	0.076	.137	.152	.121	0.100	.234	.367	.634
	R_l&R_s	0.100	.453	.696	.932	0.100	.528	.798	.980	0.100	.606	.874	.993
$Exp(1)$	R_l	0.090	.411	.650	.899	0.096	.490	.764	.967	0.099	.579	.841	.987
	R_s	0.080	.149	.195	.217	0.076	.140	.152	.122	0.100	.232	.376	.641
	R_l&R_s	0.100	.451	.702	.933	0.100	.525	.805	.982	0.100	.607	.875	.993

Note: A sectorial critical region is chosen for a simultaneous testing.

of the mean and variance of the alternative distribution, as shown in Equations (17), (18) and (19), we are cautious to define the critical regions. Table 1 provides powers for the tests as we choose uniform, standard Normal, student-t with 3 degrees of freedom, exponential distributions for the hypothesized distribution, a couple of different settings for sample sizes m and n, and 2, 3, 6 for k. Clearly, a joint test considering both R_l and R_s for the equality of distributions is best suited in comparison with tests considering only one of the rank statistics.

4 Conclusion

Our proposed algorithm provides a solution for finding the power of distribution equivalence tests considering the shift and scale parameters, respectively and simultaneously. Numerical studies show that a joint test should be adopted for the test homogeneity of distributions as well as under Lehmann alternatives. Also an elliptic critical region is a better choice rather than a rectangular one for a joint test. In practice, it is reasonable to have neither the normality assumption nor equal mean/variance of the interested distributions. However, our algorithm highly depends on the technology equipments as the possible states in Ω_n grow rapidly when the sample sizes increase. Therefore, we can, so far, only target small sample sizes in our work.

Competing interests

The author declares that she has no competing interests.

Acknowledgments

The author would like to thank James C. Fu and anonymous referee whose comments led to significant improvements of this manuscript.

References

Ansari, AR, Bradley, RA: Rank-Sum Tests for Dispersions. Ann. Math. Stat. **31**, 1174–1189 (1960)
Collings, BJ, Hamilton, MA: Estimating the power of the two sample Wilcoxon Test for location shift. Biometrics **44**, 847–860 (1988)
Klotz, J: Nonparametric test for scale. Ann. Math. Stat. **33**, 498–512 (1962)
Lehmann, EL: The power for rank tests. Ann. Math. Stat. **24**, 23–43 (1953)
Lehmann, EL: Nonparametrics: Statistical Methods Based on Ranks. Revised ed. Prentice-Hall, New Jersey (1998)
Mann, HB, Whitney, DR: On a test of whether one of two random variables is stochastically larger than the other. Ann. Math. Stat. **18**, 50–60 (1947)
Mood, AM: On the asymptotic efficiency of certain nonparametric two-sample tests. Ann. Math. Stat. **25**, 514–522 (1954)
Rosner, B, Glynn, RJ: Power and sample size estimation for the Wilcoxon rank sum test with application to comparisons of C statistics from alternative prediction models. Biometrics **65**, 188–197 (2009)
Shieh, G, Jan, SL, Randles, RH: On power and sample size determinations for the Wilcoxon-Mann-Whitney test. Nonparametric Stat. **18**, 33–43 (2006)
Siegel, S, Tukey, JW: A nonparametric sum of ranks procedure for relative spread in unpaired samples. J. Am. Stat. Assoc. **55**, 429–445 (1960)
Wilcoxon, F: Individual comparisons by ranking methods. Biometrics **1**, 80–83 (1945)

6

Geometric disintegration and star-shaped distributions

Wolf-Dieter Richter

Correspondence:
wolf-dieter.richter@uni-rostock.de
Institute of Mathematics, University of Rostock, Ulmenstraße 69, Haus 3, 18057 Rostock, Germany

Abstract

Geometric and stochastic representations are derived for the big class of p-generalized elliptically contoured distributions, and (generalizing Cavalieri? s and Torricelli? s method of indivisibles in a non-Euclidean sense) a geometric disintegration method is established for deriving even more general star-shaped distributions. Applications to constructing non-concentric elliptically contoured and generalized von Mises distributions are presented.

Keywords: p-generalized elliptically contoured distributions; Non-concentric elliptically contoured distributions; Star-generalized von Mises distributions; p-generalized ellipsoids and ellipsoidal coordinates; Stochastic random vector representation; Geometric measure representation; Star-generalized surface measure; Star-generalized uniform distribution; Intersection-proportion function; Ball number function

AMS subject classification: Primary 60E05; 60D05; secondary 28A50; 28A75; 51F99

1 Introduction

The needs of statistical practice and challenging probabilistic questions in the interplay of measure theory and several other mathematical disciplines stimulate the development of statistical distribution theory. In part, well established mathematical strategies are followed to enlarge known families of distributions, and partly new types of distributions are derived by new methods.

Numerous studies on multivariate probability distributions with a view towards their statistical applications are closely connected with notions like decomposition or disintegration of probability laws, invariant measures on groups and related manifolds, and cross sections. Among the basic references in this field, we refer to (Barndorff-Nielsen et al. 1989; Eaton 1983, 1989; Farrell 1976, 1985; Koehn 1970; Muirhead 1982; Nachbin 1976; Wijsman 1967, 1986, 1990). In the spirit of those works, generalizations of elliptically contoured distributions where the contours are described by a positive function that is positive homogeneous or are arbitrary cross sections are discussed in (Balkema and Nolde 2010; Fernandez et al. 1995) and in (Kamiya et al. 2008; Takemura and Kuriki 1996) respectively.

Zonoid trimming for multivariate distributions is considered in (Koshevoy and Mosler 1997; Mosler 2002). In (Balkema et al. 2010; Balkema and Nolde 2010; Joenssen and Vogel

2012; Kinoshita and Resnick 1991; Mosler 2013; Nolde 2014) the authors describe the way in which star-shaped sets and, correspondingly, star-shaped distributions occur in limit set theory, in meta density analysis, power studies for goodness-of-fit tests, depth analysis and analysis of residual dependence between extreme values in the presence of asymptotic independence, respectively. Norm-contoured sets and related estimation problems of identifying structures in high-dimensional data sets are dealt with in (Scholz 2002).

The authors in (Fang et al. 1990; Kallenberg 2005; Schindler 2003) present studies of symmetric laws from different points of view. The class of $l_{n,p}$-symmetric distributions, $n \in \mathbb{N} = \{1, 2, \ldots\}$, $p \geq 1$, was introduced in (Osiewalski and Steel 1993) and studied in (Gupta and Song 1997; Song and Gupta 1997; Schechtman and Zinn 1990; Rachev and Rüschendorf 1991; Szablowski 1998). Applications of these distributions are discussed in (Nardon and Pianca 2009; Pogány and Nadarajah 2010).

Geometric measure representations for $l_{n,p}$-symmetric distributions, $n \in \mathbb{N}, p > 0$, and for heteroscedastic Gaussian distributions were derived in (Richter 2009, 2013), respectively.

An extension of the class of $l_{n,p}$-symmetric distributions to the class of skewed $l_{n,p}$-symmetric distributions has been derived in (Arellano-Valle and Richter 2012). A general approach to geometric representations of skewed $l_{n,2}$-symmetric distributions can be found for $n = 2$ in (Günzel et al. 2012) and for arbitrary n in (Richter and Venz 2014). Another definition of power exponential distributions than the one used here was given in (Gómez et al. 1998), where a special case of elliptically contoured distributions is dealt with. Densities of p-generalized elliptically contoured distributions and of more general star-shaped distributions have been considered in (Balkema and Nolde 2010; Fernandez et al. 1995).

In the present paper, geometric and stochastic representations are derived for the big class of p-generalized elliptically contoured distributions. Generalizing Cavalieri?s and Torricelli?s method of indivisibles in a non-Euclidean sense, a geometric disintegration method is established for deriving even more general star-shaped distributions. Basic properties of these distributions are studied, applications of the new representations to constructing non-concentric elliptically contoured distributions and to generalizing the von Mises distribution are discussed, and the necessary background from non-Euclidean metric geometry is developed.

Many authors use iterated integration in distribution theory by first integrating with respect to (w.r.t.) a radius variable and then w.r.t. certain directional coordinates. In the present paper, we shall use basically the inverse order of integration. This way, we shall make use of the star-sphere intersection-proportion function (ipf) of a given set. The ipf is essentially based upon a suitably defined non-Euclidean surface content on a star sphere. The latter notion needs therefore the most effort in the present work. Areas from probability theory and mathematical statistics where the ipf successfully applies are surveyed in (Richter 2012). Applying this function allows to study the contours of mass concentration of a probability distribution independently from the tail behavior of the distribution, and often leads to a numerical stabilization of the evaluation of probability integrals. Further, the ipf allows a non-Euclidean surface measure interpretation of certain sector measures considered in the literature.

The paper is organized as follows. After quoting some preliminary facts in Section 2, we deal with the notion of a star-generalized surface content in Section 3. This notion will be studied both based upon a local definition and in terms of an integral in Section 3.1. The latter definition makes use of a preliminary system of coordinates which moreover enables a generalization of the method of indivisibles. Then Section 3.2. deals exclusively with the new surface measure on p-generalized ellipsoids. After much technical work, Theorem 5 finally proves that the local approach to the star generalized surface content results in the same quantity as a suitably defined non-Euclidean surface content in terms of an integral defined using a modified standard approach of differential geometry. To this end, some more coordinate systems are introduced and exploited. This includes consideration of star-generalized trigonometric functions and several Jacobians. In Section 4, star-shaped distributions are introduced in several steps and some of their basic properties are studied. The most explicit results are presented for the class of p-generalized elliptically contoured distributions in Section 4.7. For this specific class, all of the more general results of the preceding parts of Section 4, including the main results in Theorems 7 and 8, allow an additional interpretation which in each case is based upon a suitable non-Euclidean geometry. Moreover, the ball number function will be extended in Section 4.5 and characteristic functions are discussed in Section 4.6. In the two-dimensional case, some consequences from the preceding sections concerning the new class of non-concentric elliptically contoured distributions and a star generalization of the von Mises distribution are drawn in Sections 5.1 and 5.2, respectively. The paper ends with some concluding remarks in Section 6, basically indicating some possible future work.

2 Preliminaries

The main considerations of this paper are most easily understood by making use of a relatively easy coordinate transformation. For showing the deeper meaning of several results derived this way, we shall make use, however, of different rather technical systems of coordinates which will be introduced in later sections. Here, we begin with some preliminary notions, including the preliminary coordinate system mentioned in the Introduction.

Throughout this paper, $K \subset \mathbb{R}^n$ denotes a star body, i.e. a nonempty star-shaped set that is compact and is equal to the closure of its interior, having the origin 0_n in its interior. Its topological boundary will be denoted by S. The functional $h_K : \mathbb{R}^n \to [0, \infty)$ defined by $h_K(x) = \inf\{\lambda > 0 : x \in \lambda K\}$, $x \in \mathbb{R}^n$ where $\lambda K = \left\{(\lambda x_1, \ldots, \lambda x_n)^T : (x_1, \ldots, x_n)^T \in K\right\}$ is known as the Minkowski functional of the star body K. We assume that h_K is positive-homogeneous of degree one, i.e. $h_K(\lambda x) = \lambda h_K(x), \lambda > 0$, which is the case if, e.g., h_K is a norm or an antinorm. For the latter notion, we refer to (Moszyńska and Richter 2012), and for the role which homogeneous functionals generally play in stochastics, we refer to (Hoffmann-Jorgensen 1994).

Let us consider $K(r) = rK = \{x \in \mathbb{R}^n : h_K(x) \leq r\}$ and its boundary $S(r) = rS$ as the star ball and star sphere of Minkowski radius or star radius $r > 0$, respectively. A countable collection $\mathfrak{F} = \{C_1, C_2, ...\}$ of pairwise disjoint cones C_j with vertex being the origin 0_n and $\mathbb{R}^n = \bigcup_j C_j$ will be called a fan. By $\mathfrak{B}_n$ we denote the Borel-σ-field in $\mathbb{R}^n$. We put $S_j = S \cap C_j$, $S_j \cap \mathfrak{B}_n = \mathfrak{B}_{S,j}$ and $\mathfrak{B}_S = \sigma\{\mathfrak{B}_{S,1}, \mathfrak{B}_{S,2}, ...\}$. We shall consider only star bodies K and sets $A \in \mathfrak{B}_S$ satisfying the following condition.

Assumption 1. *The star body K and the set $A \in \mathfrak{B}_S$ are chosen such that for every j the set*

$$G(A \cap S_j) = \left\{ \vartheta \in \mathbb{R}^{n-1} : \exists \eta \text{ with } \theta = \left(\vartheta^T, \eta\right)^T \in A \cap S_j \right\}$$

is well defined and such that for every $\vartheta = (\vartheta_1, \ldots, \vartheta_{n-1})^T \in G(A \cap S_j)$ there is a uniquely determined $\eta > 0$ satisfying $h_K\left((\vartheta_1, \ldots, \vartheta_{n-1}, \eta)^T\right) = 1$.

The latter quantity will be denoted by $\eta_j(\vartheta)$, $j = 1, 2, \ldots$.

For every $x \in \mathbb{R}^n, x \neq 0$ there are uniquely determined $r > 0$ and $\theta \in S$ such that $x = r\theta$. For $x \in rS_j$, we have $x = r\left(\vartheta^T, \eta_j(\vartheta)\right)^T$, and we will write $r\eta_j(\vartheta) = y_j(\vartheta)$. Consequently, $h_K(x) = h_K(r\theta) = rh_K\left(\left(\vartheta^T, \eta_j(\vartheta)\right)^T\right) = r$.

For $j = 1, 2, \ldots$ the star spherical coordinate transformation $StSph_j :[0, \infty) \times G(S_j) \to C_j$ is defined by $x_i = r\vartheta_i, i = 1, \ldots, n-1, x_n = y_j(\vartheta)$. The equations $r = h_K(x), \vartheta_i = x_i/r$, $i = 1, \ldots, n-1$ define a.e. uniquely the inverse map of $StSph_j$.

Note that if K is convex or an axes aligned p-generalized ellipsoid, $p > 0$, see Section 3.2.1, one may assume the sets $S \cap C_{1(2)}$ to be the upper and lower hemi-spheres, $S^{+(-)} = \left\{\theta = (\theta_1, \ldots, \theta_n)^T \text{with } \theta_n > (<) 0\right\}$, respectively.

Lemma 1. *The absolute value of the Jacobian of the star-spherical coordinate transformation is $J(r, \vartheta) = r^{n-1} J_j^*(\vartheta)$, with $J_j^*(\vartheta) = |\eta_j(\vartheta) - \sum_{i=1}^{n-1} \vartheta_i \frac{\partial}{\partial \vartheta_i} \eta_j(\vartheta)|$ for every $r > 0, \vartheta \in G(S_j), j = 1, 2, \ldots$*

Proof. The formula for $J(r, \vartheta) = |\frac{d(x_1,\ldots,x_n)}{d(r,\vartheta)}|$ given in the lemma can be checked immediately by determining all partial derivatives and evaluating the resulting determinant. □

The coordinate system introduced here will be the basis of our considerations in Section 3.1 dealing with a general local notion of surface content. A specific integral notion of surface content dealt with in Section 3.2.3 will make use of another system of coordinates which will be introduced in Section 3.2.2. For the comparison study of the two seemingly rather different two approaches to measuring surfaces in Section 3.2.4, we will consider again suitable coordinates.

An essential part of the message of Lemma 1 is that the Jacobian allows a factorization into a term not depending on the radius coordinate and one that is independent of the directional coordinates.

Later in this paper, the restriction of the star spherical coordinate transformation to the case $r = 1$ will be denoted by $StSph^*$.

3 The star-generalized surface measure

3.1 Basics

The results in (Richter 2009, 2013) reflect the basic role which a suitable notion of non-Euclidean surface content plays for the study of non-spherical distributions. Here, we give first formally a local definition of a generalized surface measure which allows us to derive geometric and stochastic representations of star-shaped distributions and correspondingly distributed random vectors, respectively. For a more advanced understanding of the notions and results, we refer to Remark 6 below.

For $A \in \mathfrak{B}_S$, we introduce the central projection cone $CPC(A) = \{x \in \mathbb{R}^n : x/h_K(x) \in A\}$ and the star sector of star radius ϱ, $sector(A,\varrho) = CPC(A) \cap K(\varrho)$. We are now ready to introduce the first basic notion of this paper. To this end, let μ be the Lebesgue measure in $\mathbb{R}^n$.

Definition 1. *The star-generalized surface measure is defined on* $\varrho \cdot \mathfrak{B}_S$ *by* $\mathfrak{O}_S(A) = f'(\varrho)$ *where* $f(\varrho) = \mu(sector(A,\varrho))$.

If K is the Euclidean unit ball, and thus S is the Euclidean unit sphere, then $\mathfrak{O}_S$ equals the usual Euclidean surface content measure. The equation in Definition 1 should be well known for this case, but, astonishing enough, numerous authors do not make this very clear to their readers.

In contrast to the usual differential geometric definition of the notion of surface content in terms of an integral, the approach in Definition 1 is based upon a derivative. The equation

$$\mu(sector(A,R)) = \int_0^R \mathfrak{O}_S(rA)dr, A \in \mathfrak{B}_S \tag{1}$$

is an immediate consequence of the fundamental theorem of calculus and might seem therefore to be of no special interest, here. If, however, non-trivial explanations for $\mathfrak{O}_S$ are available as in (Richter 2009, 2013) where K is an $l_{n,p}-$ball or an ellipsoidal ball, respectively, then things change noticably. In both cases, a particular non-Euclidean geometry was identified such that the correspondingly modified integral notion of surface content based upon this non-Euclidean geometry coincides with the locally defined surface measure $\mathfrak{O}_S$. This allows a non-Euclidean interpretable extension of Cavalieri?s and Torricelli?s method of indivisibles, see (Richter 1985, 2009). Later in this paper, we shall observe this for a bigger class of star bodies. Moreover, we remark that $\mathfrak{O}_S(A) = n\mu(sector(A,1)), \forall A \in \mathfrak{B}_S$, meaning much more than just $\mathfrak{O}_S(S) = n\mu(K)$.

Theorem 1. *For sets* $A \in \mathfrak{B}_S$ *satisfying Assumption 1 in Section 2, the star-generalized surface measure allows the representation*

$$\mathfrak{O}_S(A) = \sum_j \int_{G(A \cap S_j)} J_j^*(\vartheta)d\vartheta.$$

Proof. Using star-spherical coordinates, and that $G(A \cap S_j) = StSph^{*-1}(A \cap S_j)$, we get according to Lemma 1

$$\mu(sector(A,\varrho)) = \int_{sector(A,\varrho)} dx = \int_0^\varrho \sum_j \int_{StSph^{*-1}(A \cap S_j)} r^{n-1} J_j^*(\vartheta)d(\vartheta, r).$$

Definition 1 applies. □

Remark 1. *(a) With the notations* $\mathfrak{O}_S(A) = \int\limits_A \mathfrak{O}_S(d\theta)$, *and*

$$\sum_j \int\limits_{G(A\cap S_j)} J_j^*(\vartheta)d\vartheta = \int\limits_{G(A)} J^*(\vartheta)d\vartheta,$$

an alternative expression of Theorem 1 is

$$\int\limits_A \mathfrak{O}_S(d\theta) = \int\limits_{G(A)} J^*(\vartheta)d\vartheta.$$

If f is integrable then we write $\int\limits_A f(\theta)\mathfrak{O}_S(d\theta) = \int\limits_{G(A)} f\left(\left(\vartheta^T, \eta(\vartheta)\right)^T\right) J^*(\vartheta)d\vartheta$.

(b) The sector measure on $\mathfrak{B}_S$*, i.e. the measure* $sm_K(A) = \frac{\mu(sector(A,1))}{\mu(K)}$*, satisfies the representation* $sm_K(A) = \frac{\mathfrak{O}_S(A)}{\mathfrak{O}_S(S)}, A \in \mathfrak{B}_S$.

(c) A class of examples where Theorem 1 applies is given by all star bodies K corresponding to norms or antinorms for which there exist countably many pairwise disjoint sets A_j *satisfying Assumption 1 and* $S = \bigcup\limits_j A_j$.

The following consequence of Theorem 1 follows using Fubini?s theorem and can be read in the special case $f = 1$ as a disintegration formula for the Lebesgue measure. For a certain survey of such formulas, see (Richter 2012). These representations may also be considered as closely connected with a generalized method of indivisibles with the latter being defined as the intersections of a Borel set B with the star spheres $S(r), r > 0$. Constructions of such type are called cross sections by several authors, see (Eaton 1983; Farrell 1976, 1985; Koehn 1970; Wijsman 1967, 1986, 1990) and (Takemura and Kuriki 1996).

Corollary 1. *Let the star body K satisfy Assumption 1. Then*

(a) For $B \in \mathfrak{B}_n$ *and integrable f,* $\int\limits_B f(x)dx = \int\limits_0^\infty \left[r^{n-1} \int\limits_{[\frac{1}{r}B]\cap S} f(r\theta)\mathfrak{O}_S(d\theta) \right] dr.$

(b) For bounded measurable B, $\int\limits_B dx = \int\limits_0^\infty \mathfrak{O}_S(B\cap S(r))dr.$

Proof. Changing from Cartesian to star spherical coordinates yields

$$\int\limits_B f(x)dx = \int\limits_0^\infty \left[r^{n-1} \sum_j \int\limits_{G([\frac{1}{r}B]\cap S_j)} f(StSph_j(r,\vartheta))J_j^*(\vartheta)d\vartheta \right] dr$$

$$= \int\limits_0^\infty \left[r^{n-1} \sum_j \int\limits_{[\frac{1}{r}B]\cap S_j} f(r\theta)\mathfrak{O}_S(d\theta) \right] dr.$$

The rest follows with $f = 1$ and the notation in Remark 1 □

Corollary 1 may be rewritten using the following second basic notion of this paper.

Definition 2. *The star sphere intersection-proportion function (ipf) of the set* $B \in \mathfrak{B}_n$ *is defined as* $\mathfrak{F}_S(B, r) = \mathfrak{O}_S\left(\left[\frac{1}{r}B\right]\cap S\right)/\mathfrak{O}_S(S)$, $r > 0$.

The ipf was first introduced in (Richter 1985, 1987, 1991) for Gaussian and spherical distributions, respectively, i.e. for cases where S is the Euclidean unit sphere, and generalized later in (Richter 2007) to the case that S is an $l_{n,p}-$sphere. Moreover, the ipf corresponding to an asymmetric sphere S was considered for the case that K is the shifted positive part of an $l_{n,1}-$ ball, i.e. a simplex, and for the case that K is a, possibly asymmetric, polygon or Platonic body, respectively.

Corollary 2. *If the conditions of Corollary 1(b) are satisfied,*

$$\int\limits_B dx = \mathfrak{O}_S(S) \int\limits_0^\infty r^{n-1} \mathfrak{F}_S(B,r) dr.$$

Proof. It follows from Corollary 1 that

$$\int\limits_B dx = \int\limits_0^\infty r^{n-1} \mathfrak{O}_S\left(\left[\frac{1}{r}B\right] \cap S\right) dr.$$

The rest follows by Definition 2. □

Remark 2. *According to Remark 1(b) and Definition 2, the ipf allows the sector measure interpretation* $\mathfrak{F}_S(B,r) = sm_K\left(\left[\frac{1}{r}B\right] \cap S\right)$, $r > 0$.

Whether one prefers the interpretation of the ipf according to the definition of the sector measure sm_K in terms of volumes or according to Definition 2 in terms of star-generalized surface contents may depend on several aspects. The authors in (Barthe et al. 2003; Naor 2007; Schechtman and Zinn 1990) use the notion of cone measure in similar situations.

As already mentioned in the first part of the present section, one is naturally interested in a fully differential geometric explanation of the star-generalized surface measure $\mathfrak{O}_S$ in terms of an integral. Such an explanation will be given in Section 3.2.3 when K is an element of a class of generalized ellipsoids which are star-shaped but not necessarily convex.

3.2 The star-generalized surface content of *p*-generalized ellipsoids

3.2.1 Volumes of p-generalized ellipsoids

Because the notion of the star-generalized surface content is derived from that of volumes, we first study volumes of p-generalized ellipsoids in this section. To this end, let $\mathfrak{b} = \{\mathfrak{b}_1, \ldots, \mathfrak{b}_n\}$ be any orthonormal basis (onb) in $\mathbb{R}^n$ and put $x = \sum\limits_{i=1}^{n} \xi_i \mathfrak{b}_i$ for $x \in \mathbb{R}^n$. Moreover, let $a = (a_1, \ldots, a_n)^T$ be an arbitrary vector having positive components, p a positive real number, $|.|_{a,p} : \mathbb{R}^n \rightarrow [0, \infty)$ the function defined by $|x|_{a,p} = \left(\sum\limits_1^n |\frac{\xi_i}{a_i}|^p\right)^{1/p}$, $x \in \mathbb{R}^n$ and $B_{a,p} = \left\{x \in \mathbb{R}^n : |x|_{a,p} \leq 1\right\}$ the corresponding unit ball w.r.t. $\mathfrak{b}$. Its topological boundary $E_{a,p}$ is a generalized ellipsoid having form parameter p and main axes being aligned with the coordinate axes and having lengths $2a_i, i = 1, \ldots, n$. One may consider $E_{a,p}$ also as a sphere w.r.t. the function $|.|_{a,p}$ which is a norm if $p \geq 1$ and an antinorm if $0 < p < 1$.

The $\mathfrak{b}_i$-axis may be interpreted in the sense of main axis from principal component analysis. For a discussion of these notions in connection with that of correlation, we refer to (Dietrich et al. 2013). The set $B_{a,p}(R) = RB_{a,p}$ will be called a p-generalized ellipsoidal ball, or simply p-generalized ellipsoid, of $|.|_{a,p}$-radius R, $R > 0$, and w.r.t. the basis $\mathfrak{b}$.

The evaluation of the volume of $B_{a,p}(R)$ may be immediately reduced to that of an $l_{n,p}$-ball having a suitable p-radius. To this end, we denote the $l_{n,p}$-ball of p-radius R by $K_{n,p}(R) = RK_{n,p}$ where $K_{n,p} = B_{\mathbb{1},p}, \mathbb{1} = (1, \ldots, 1)^T \in \mathbb{R}^n$, and its topological boundary, the $l_{n,p}$−sphere of p-radius R, by $S_{n,p}(R) = RS_{n,p}$. Moreover, we put $a_i^* = \prod_{j=1, j\neq i}^{n} a_j$, $i = 1, \ldots, n$ and let $diag\left(a_1^*, \ldots, a_n^*\right)$ denote a diagonal $n \times n$-matrix whose diagonal entries are $a_2 \cdot \ldots \cdot a_n, \ \ldots, a_1 \cdot \ldots \cdot a_{n-1}$, respectively. If $\mathfrak{b}$ is the standard onb in $\mathbb{R}^n$, then $diag\left(a_1^*, \ldots, a_n^*\right) B_{a,p}(R) = K_{n,p}(a_1 \ldots a_n R)$. Changing variables $u = diag\left(a_1^*, \ldots, a_n^*\right) x$ in the integral $\mu(B_{a,p}(R)) = \int\limits_{\{x \in R^n : |x|_{a,p} \leq R\}} dx$ gives

$$\mu(B_{a,p}(R)) = \int\limits_{K_{n,p}(Ra_1 \ldots a_n)} \frac{du}{(a_1 \ldots a_n)^{n-1}}.$$

Hence, $\mu(B_{a,p}(R)) = \frac{\mu(K_{n,p}(Ra_1 \ldots a_n))}{(a_1 \ldots a_n)^{n-1}} = a_1 \ldots a_n \frac{\omega_{n,p}}{n} R^n$ where, in accordance with (Richter 2009), $\omega_{n,p} = \frac{2^n \left(\Gamma(\frac{1}{p})\right)^n}{p^{n-1}\Gamma(\frac{n}{p})} = O_{p,q}(S_{n,p})$ with $\frac{1}{p} + \frac{1}{q} = 1$ is the $l_{n,q}$-surface content of the $l_{n,p}$−unit sphere $S_{n,p}$. The following theorem has thus been proved.

Theorem 2. *The p-generalized ellipsoid of* $|.|_{a,p}$*-radius R has the volume*

$$\mu(B_{a,p}(R)) = a_1 \ldots a_n \frac{\omega_{n,p}}{n} R^n.$$

Corollary 3. *The star-generalized surface content of a p-generalized ellipsoid with axes of lengths* $2a_i, i = 1, \ldots, n$ *is* $\mathfrak{O}_S(E_{a,p}) = a_1 \ldots a_n \omega_{n,p}$.

This corollary is an immediate consequence of Definition 1.

Notice that this formula for the star-generalized surface content of $E_{a,p}$ proves that the parameters p and a have separate influence on the result. Moreover, it makes no use of elliptic integrals, whereas the Euclidean surface content of $E_{a,p}$ does.

Similarly, as Equation (1), the equation

$$\mu(B_{a,p}(R)) = \int\limits_0^R \mathfrak{O}_S(E_{a,p}(r))dr \tag{2}$$

where $E_{a,p}(r) = rE_{a,p}$, might seem to be of no special interest, at this stage of our study. We shall show, however, later in this paper that $\mathfrak{O}_S$ allows a non-trivial interpretation as the surface measure w.r.t. a well defined, non-Euclidean, metric geometry. This allows us to re-define $\mathfrak{O}_S$ in a well established differential geometric approach. This will be done in the next but one section. We shall make use of a specific coordinate system which will be defined in the next section.

Following the notation in (Richter 2013), we will call the star-generalized surface measure alternatively the $E_{a,p}$-generalized surface measure if K is a p-generalized ellipsoid with axes of lengths $2a_i, i = 1, \ldots, n$.

3.2.2 The p-generalized ellipsoidal coordinates

We recall that $l_{n,p}$-generalized and ellipsoidal generalized trigonometric functions and coordinates have been shown in (Richter 2007, 2009) and (Richter 2011b, 2013) to be powerful tools for studying $l_{n,p}$-symmetric and elliptically contoured distributions, respectively. The coordinates which we define in this section are in some sense combinations and generalizations of the aforementioned ones. They will be used in Section 3.2.4 for showing the equivalence of two approaches to the star-generalized surface measure $\mathfrak{O}_S$: the local one presented already in Definition 1, and an integral one which will be given later.

Let us assume for a moment that $n = 2$ and $(x, y)^T = x\mathfrak{b}_1 + y\mathfrak{b}_2$ where $\{\mathfrak{b}_1, \mathfrak{b}_2\}$ is an onb in $\mathbb{R}^2$. In the following definition, ϕ can be interpreted as the angle between $pos(x, 0)$ and $pos(x, y)$ where $pos(\xi, \eta) = \{(\gamma\xi, \gamma\eta)^T : \gamma > 0\}$.

Definition 3. *The $E_{a,b;p}$-generalized trigonometric functions are defined as*

$$\cos_{(a,b;p)}(\phi) = \frac{(\cos\phi)/a}{N_{a,b;p}(\phi)} \text{ and } \sin_{(a,b;p)}(\phi) = \frac{(\sin\phi)/b}{N_{a,b;p}(\phi)},\ \phi \in [0, 2\pi)$$

for positive a, b, p and where $N_{a,b;p}(\phi) = (|(\cos\phi)/a|^p + |(\sin\phi)/b|^p)^{1/p}$.

Remark 3. *These generalized trigonometric functions may be extended to functions on the whole real line with period 2π. Basic analytical and geometric interpretations of these functions follow from the representations*

$$\cos_{(a,b;p)}(\phi) = \frac{x/a}{(|x/a|^p + |y/b|^p)^{1/p}},\ |\cos_{(a,b;p)}(\phi)| = \frac{|(\cos_{(a,b)}(\phi), 0)|_p}{|(\cos_{(a,b)}(\phi), \sin_{(a,b)}(\phi))|_p},$$

and

$$\sin_{(a,b;p)}(\phi) = \frac{y/b}{(|x/a|^p + |y/b|^p)^{1/p}},\ |\sin_{(a,b;p)}(\phi)| = \frac{|(0, \sin_{(a,b)}(\phi))|_p}{|(\cos_{(a,b)}(\phi), \sin_{(a,b)}(\phi))|_p},$$

where $|.|_p = |.|_{\mathbb{1},p}$ and $\sin_{(a,b)}$ and $\cos_{(a,b)}$ are defined in (Richter 2011b).

Remark 4. *Euler's formula is generalized by* $|\cos_{(a,b;p)}(\phi)|^p + |\sin_{(a,b;p)}(\phi)|^p = 1$.

Remark 5. *For every ϕ,*

$$\cos'_{(a,b;p)}(\phi) = -\frac{\sin_{a,b;p}(\phi)|\sin_{a,b;p}(\phi)|^{p-2}}{abN^2_{(a,b;p)}(\phi)},\ \sin'_{(a,b;p)}(\phi) = \frac{\cos_{a,b;p}(\phi)|\cos_{a,b;p}(\phi)|^{p-2}}{abN^2_{(a,b;p)}(\phi)}.$$

We assume again that $x = x_1\mathfrak{b}_1 + ... + x_n\mathfrak{b}_n$, $x \in \mathbb{R}^n$.

Definition 4. *The p-generalized ellipsoidal coordinate transformation $T^E_{a,p} = T^E_{a,p}(n)$, $T^E_{a,p} : M_n \to \mathbb{R}^n$, with $M_n = [0, \infty) \times M^*_n, M^*_n = [0, \pi)^{\times(n-2)} \times [0, 2\pi)$ is defined by*

$$x_1 = a_1 r \cos_{(a_1,a_2;p)}(\phi_1), x_2 = a_2 r \sin_{(a_1,a_2;p)}(\phi_1)\cos_{(a_2,a_3;p)}(\phi_2), \ldots,$$

$$x_{n-1} = a_{n-1} r \sin_{(a_1,a_2;p)}(\phi_1) ... \sin_{(a_{n-2},a_{n-1};p)}(\phi_{n-2})\cos_{(a_{n-1},a_n;p)}(\phi_{n-1}),$$

$$x_n = a_n r \sin_{(a_1,a_2;p)}(\phi_1) ... \sin_{(a_{n-2},a_{n-1};p)}(\phi_{n-2})\sin_{(a_{n-1},a_n;p)}(\phi_{n-1}).$$

Theorem 3. *The map $T^E_{a,p}$ is almost one-to-one, its inverse is a.e. given by*

$$r = \left(\sum_1^n \left(\left|\frac{x_i}{a_i}\right|\right)^p\right)^{1/p}, \phi_j = \arccos_{(a_j,a_{j+1};p)}\left(\frac{x_j/a_j}{\left(\sum_{i=j}^n |x_i/a_i|^p\right)^{1/p}}\right), \textit{and, if } x_{n-1} \neq 0,$$

$\arctan\left|\frac{x_n}{x_{n-1}}\right| = \phi_{n-1}$ *if* $(x_{n-1}, x_n) \in Q_1$, $= \pi - \phi_{n-1}$ *in* Q_2, $= -\pi + \phi_{n-1}$ *in* Q_3, *and* $= 2\pi - \phi_{n-1}$ *in* Q_4.

Here, $\arccos_{(a_j,a_{j+1};p)}$ *denotes the function inverse to* $\cos_{(a_j,a_{j+1};p)}$ *and* Q_1 *up to* Q_4 *denote anti-clockwise enumerated quadrants from* $\mathbb{R}^2$.

Proof. The proof of this theorem is quite similar to that of Theorem 1 in (Richter 2007) and is therefore omitted. □

Theorem 4. *The Jacobian of the coordinate transformation $T^E_{a,p}$ is*

$$J\left(T^E_{a,p}\right)(r, \phi_1, \ldots, \phi_{n-1}) = r^{n-1} J^*\left(T^E_{a,p}\right)(\phi_1, \ldots, \phi_{n-1}),$$

$$J^*\left(T^E_{a,p}\right)(\phi_1, \ldots, \phi_{n-1}) = \frac{1}{a_2 \cdot \ldots \cdot a_{n-1}} \prod_{i=1}^{n-1} \frac{\left(\sin_{(a_i,a_{i+1};p)}(\phi_i)\right)^{n-1-i}}{N^2_{(a_i,a_{i+1};p)}(\phi_i)}.$$

Proof. The proof will be given in four steps. First, we change variables $\frac{x_i}{a_i} = y_i$, $i = 1, \ldots, n$. The Jacobian of this transformation is $\left|\frac{D(x_1,\ldots,x_n)}{D(y_1,\ldots,y_n)}\right| = a_1 \cdot \ldots \cdot a_n$.

Next, we change variables $y_1 = \tilde{r}\mu_1, y_2 = \tilde{r}\left(1 - |\mu_1|^p\right)^{1/p} \mu_2, \ldots,$

$$y_{n-1} = \tilde{r}\left(1 - |\mu_1|^p\right)^{1/p} \cdot \ldots \cdot \left(1 - |\mu_{n-2}|^p\right)^{1/p} \mu_{n-1},$$

$$y_n = +(-)\tilde{r}\left(1 - |\mu_1|^p\right)^{1/p} \cdot \ldots \cdot \left(1 - |\mu_{n-2}|^p\right)^{1/p} \left(1 - |\mu_{n-1}|^p\right)^{1/p}.$$

As it was shown in the proof of Theorem 2 in the afore mentioned paper, the Jacobian of this transformation is $\left|\frac{D(y_1,\ldots,y_n)}{D(\tilde{r},\mu_1,\ldots,\mu_{n-1})}\right| = \tilde{r}^{n-1} \prod_{i=1}^{n-1} \left(1 - |\mu_i|^p\right)^{(n-p-i)/p}$.

Third, we change variables $\tilde{r} = r, \mu_i = \cos_{(a_i,a_{i+1};p)}(\phi_i), i = 1, \ldots, n-1$. The Jacobian of this transformation is

$$\left|\frac{D(\tilde{r}, \mu_1, \ldots, \mu_{n-1})}{D(r, \phi_1, \ldots, \phi_{n-1})}\right| = \left|\det diag\left(1, \frac{d}{d\phi_1}\cos_{(a_1,a_2;p)}(\phi_1), \ldots, \frac{d}{d\phi_{n-1}}\cos_{(a_{n-1},a_n;p)}(\phi_{n-1})\right)\right|.$$

It follows from the properties of the $E_{a,b;p}$-generalized trigonometric functions that

$$\left|\frac{D(\tilde{r}, \mu_1, \ldots, \mu_{n-1})}{D(r, \phi_1, \ldots, \phi_{n-1})}\right| = \prod_{i=1}^{n-1} \frac{\sin_{(a_i,a_{i+1};p)}(\phi_i) |\sin_{(a_i,a_{i+1};p)}(\phi_i)|^{p-2}}{a_i a_{i+1} N^2_{(a_i,a_{i+1};p)}(\phi_i)}.$$

On combining all three transformations, we get finally

$$J\left(T^E_{a,p}\right)(r, \phi_1, \ldots, \phi_{n-1}) = a_1 \ldots a_n \cdot r^{n-1} \prod_{i=1}^{n-1} \frac{\left(\sin_{(a_i,a_{i+1};p)}(\phi_i)\right)^{d-i-1}}{a_i a_{i+1} N^2_{(a_i,a_{i+1};p)}(\phi_i)}.$$

□

Corollary 4. *If* $n = 2$ *then* $J\left(T^E_{a,p}\right)(r, \phi) = \frac{r}{N^2_{(a_1,a_2;p)}(\phi)}$, *and if* $n = 3$ *then* $J\left(T^E_{a,p}\right)(r, \phi_1, \phi_2) = \frac{r^2 \sin \phi_1}{a_2^2 N^2_{(a_2,a_3;p)}(\phi_2) N^3_{(a_1,a_2;p)}(\phi_1)}$.

For the corresponding results in the case $p = 2$, we refer to (Richter 2011b, 2013).

3.2.3 Integral approach to the star-generalized surface measure on p-generalized ellipsoids

Let us recall that measuring the Euclidean surface content of $E_{a,p}(R)$ necessarily involves certain elliptic integrals. In this paper, however, we make use of a non-Euclidean definition of surface content which avoids such integrals. To this end, we shall consider the ellipsoid $E_{a,p}(R)$ as a subset of the generalized Minkowski space $\left(\mathbb{R}^n, |.|_{\frac{1}{a},q}\right)$ where $\frac{1}{a} = \left(\frac{1}{a_1}, \ldots, \frac{1}{a_n}\right)^T$ and p and q are connected with each other by the equation $\frac{1}{p} + \frac{1}{q} = 1$. We will introduce now the notion of the $|.|_{\frac{1}{a},q}$-surface content of $E_{a,p}(R)$ in a similar way as the notion of the $l_{n,q}$-surface content was introduced in (Richter 2009) for $l_{n,p}$-spheres. Notice that effects coming from scaling axes with the help of the parameter vector a and effects being due to the form parameter p are dealt with here in a separate way when introducing the function $|.|_{\frac{1}{a},q}$.

Let $\mathfrak{b}$ be the standard onb, and let y be defined as the positive solution of $\sum\limits_{i=1}^{n-1} |x_i/a_i|^p + |y/a_n|^p = R^p$ where $\sum\limits_{i=1}^{n-1} |x_i/a_i|^p < R^p$. At $(x_1, \ldots, x_{n-1}, y)^T$, $y > 0$, the vector normal to the upper half $E^+_{a,p}(R)$ of the ellipsoid $E_{a,p}(R)$ is $N(x_1, \ldots, x_{n-1}) = (-1)^n \left(\sum\limits_{i=1}^{n-1} \frac{\partial y}{\partial x_i} e_i - e_n\right)$. Since it always will be clear how to deal with the case $y < 0$, we will not further mention this case.

Definition 5. *Let $A \subset E^+_{a,p}(R) \cap \mathfrak{B}_n$. The integral (or $|.|_{\frac{1}{a},q}$-) surface content of the set A is defined by* $O_{a,p,q}(A) = \int\limits_{G(A)} |N(x_1, \ldots, x_{n-1})|_{\frac{1}{a},q}\, dx_1 \ldots dx_{n-1}$ *where* $G(A) = \left\{ (x_1, \ldots, x_{n-1})^T : (x_1, \ldots, x_n)^T \in A \right\}$.

Later in this paper, this definition will be called the integral approach to the notion of star-generalized surface content. This will be justified in the next section. Let us mention that if $a = \mathbb{1}$ then the surface measure $O_{a,p,q}$ which is based upon the geometry of the ellipsoid $E_{\frac{1}{a},q}$ equals the surface measure $O_{p,q}$ in (Richter 2009) being based upon the geometry of the $l_{n,q}$-ball $K_{n,q}$ which is dual to $K_{n,p}$. Two-dimensional special cases of Definition 5 were dealt with, e.g., in (Richter 2011a, 2011b) for arbitrary star discs and ellipses, respectively.

Lemma 2. *The $|.|_{\frac{1}{a},q}$-surface content of the whole generalized ellipsoid $E_{a,p}(R)$ of $|.|_{a,p}$-radius R is* $O_{a,p,q}\left(E_{a,p}(R)\right) = a_1 \ldots a_n \omega_{n,p} R^{n-1}$.

Proof. It follows from $\frac{\partial y}{\partial x_j} = -\frac{a_n |x_j|^{p-1}}{a_j^p \left(R^p - \sum\limits_{i=1}^{n-1} (|x_i/a_i|)^p\right)^{(p-1)/p}}$, $j = 1, \ldots, n-1$

and with $q = p/(p-1)$ that

$$|N(x_1, \ldots, x_{n-1})|^q_{\frac{1}{a},q} = \sum_{j=1}^{n-1} \frac{a_n^q |x_j/a_j|^p}{R^p - \sum\limits_{i=1}^{n-1} |x_i/a_i|^p)} + a_n^q = \frac{a_n^q R^p}{R^p - \sum\limits_{i=1}^{n-1} |x_i/a_i|^p}.$$

Hence, because of symmetry,

$$O_{a,p,q}(E_{a,p}(R)) = 2a_n R^{p-1} \int\limits_{G(E^+_{a,p}(R))} \frac{d(x_1...x_{n-1})}{\left(R^p - \sum\limits_{i=1}^{n-1} |x_i/a_i|^p\right)^{(p-1)/p}}.$$

For suitably transforming this integral, we shall introduce now another system of coordinates. Let the p-generalized $(n-1)$-dimensional standard elliptical coordinate transformation

$$T_{a,p} : M_{n-1} \to \mathbb{R}^{n-1}, M_{n-1} = [0,\infty) \times M^*_{n-1}, M^*_{n-1} = [0,\pi)^{\times(n-3)} \times [0, 2\pi)$$

be defined by

$$x_1 = a_1 r \cos_p(\phi_1), x_2 = a_2 r \sin_p(\phi_1)\cos_p(\phi_2), \ldots,$$
$$x_{n-2} = a_{n-2} r \sin_p(\phi_1) \ldots \sin_p(\phi_{n-3})\cos_p(\phi_{n-2}),$$
$$x_{n-1} = a_{n-1} r \sin_p(\phi_1) \ldots \sin_p(\phi_{n-3})\sin_p(\phi_{n-2})$$

where the p-generalized trigonometric functions $\sin_p$ and $\cos_p$ are defined in (Richter 2007). If $a = \mathbb{1} \in \mathbb{R}^{n-1}$ then this transformation coincides with the $l_{n-1,p}$-spherical coordinate transformation $SPH^{(n-1)}_p$, the Jacobian of which is given in (Richter 2007). If we write $J(T)$ for the Jacobian of a coordinate transformation T then $J(T_{a,p})(r,\phi) = \left|\frac{d(x_1,\ldots,x_{n-1})}{d(r,\phi_1,\ldots,\phi_{n-2})}\right| = a_1 \cdot \ldots \cdot a_{n-1} J\left(SPH^{(n-1)}_p\right)(r,\phi)$.

Moreover, let $J^*\left(SPH^{(n-1)}_p\right)(\phi) = J\left(SPH^{(n-1)}_p\right)(1,\phi)$ be the restriction of the Jacobian of $SPH^{(n-1)}_p$ to the sphere defined by $r = 1$.

Changing from Cartesian to p-generalized standard elliptical coordinates gives

$$O_{a,p,q}(E_{a,p}(R)) = 2a_1 \ldots a_n R^{p-1} \int\limits_0^R \frac{r^{n-2}}{(R^p - r^p)^{(p-1)/p}} dr$$

$$\times \int\limits_0^\pi \ldots \int\limits_0^\pi \int\limits_0^{2\pi} J^*\left(SPH^{(n-1)}_p\right)(\phi_1,\ldots,\phi_{n-2}) d\phi_{n-2} \ldots d\phi_1.$$

Because of

$$\int\limits_0^R \frac{r^{n-2}dr}{(R^p - r^p)^{(p-1)/p}} = R^{n-p} \int\limits_0^1 \frac{t^{n-2}dt}{(1-t^p)^{(p-1)/p}} = \frac{1}{p} R^{n-p} B\left(\frac{1}{p}, \frac{n-1}{p}\right)$$

and $\frac{1}{p}B\left(\frac{1}{p}, \frac{n-1}{p}\right)\omega_{n-1,p} = \frac{1}{2}\omega_{n,p}$, it follows that $O_{a,p,q}(E_{a,p}(R)) = a_1 \ldots a_n \omega_{n,p} R^{n-1}$ □

Hence, for the specific sets $E_{a,b}(R)$, the local and the integral approaches to the star-generalized surface content lead to the same result. In the next section, we will generalize this result. When doing this, we will again make use of a modified coordinate system.

3.2.4 Comparing the local and integral approaches to generalized surface measures on p-generalized ellipsoids

In Section 3.2.3, the surface measure $O_{a,p,q}$ was used for measuring the whole p-generalized ellipsoid $E_{a,p}$ following a differential geometric, or integral or global approach. In the present section, however, we compare it for arbitrary $A \in \mathfrak{B}^E_{a,p} = \mathfrak{B}^n \cap E_{a,p}$ with the alternative local approach which makes use of derivatives and which was introduced in Definition 1. In this sense, we continue to follow the general method of analyzing the non-Euclidean geometry underlying a multivariate probability distribution which was developed in (Richter 2009, 2013). The following theorem says that the star-generalized surface measure coincides with the integral surface measure. For a comparison of these surface measures, it is sufficient to consider them for sets $A \in \mathfrak{B}^E_{a,p}$.

Theorem 5. *With* $S = E_{a,p}$ *and* $1/p + 1/q = 1$, $\mathfrak{O}_S(A) = O_{a,p,q}(A), \forall A \in \mathfrak{B}^E_{a,p}$.

Proof. W.l.o.g., we restrict our consideration to sets $A \in E^+_{a,p} \cap \mathfrak{B}^n$ and start from a slight generalization of the first result in the proof of Lemma 2,

$$O_{a,p,q}(A) = a_n \int\limits_{G(A)} \frac{d(x_1,\ldots,x_{n-1})}{\left(1 - \sum\limits_{i=1}^{n-1} |x_i/a_i|^p\right)^{(p-1)/p}}.$$

We change from Cartesian to p-generalized ellipsoidal coordinates in $(n-1)$ dimensions, $T^E_{a,p}(n-1)\colon (x_1,\ldots,x_{n-1}) \longrightarrow (r,\phi_1,\ldots,\phi_{n-2})$. Because of $\sum\limits_{i=1}^{n-1} |x_i/a_i|^p = r^p$,

$$O_{a,p,q}(A) = a_n \int\limits_{\left(T^E_{a,p}(n-1)\right)^{-1}(G(A))} \frac{r^{n-2}}{(1-r^p)^{(p-1)/p}} J^* \, d(r,\phi_1,\ldots,\phi_{n-2}),$$

$$J^* = J^*\left(T^E_{a,p}(n-1)\right)(\phi_1,\ldots,\phi_{n-2}) = a_{n-1} \prod_{i=1}^{n-2} \frac{\left(\sin_{(a_i,a_{i+1};p)}(\phi_i)\right)^{n-1-i}}{a_{i+1} N^2_{(a_i,a_{i+1};p)}(\phi_i)}.$$

If $A = A(r_1,r_2,M^*) = \left\{ \left(y_1,\ldots,y_{n-1}, \left(1 - \sum\limits_{i=1}^{n-1} |x_i/a_i|^p\right)^{1/p}\right)^T : \right.$

$$\left. (y_1,\ldots,y_{n-1})^T = T^E_{a,p}(n-1)([r_1,r_2) \times M^*) \right\},$$

with $M^* = \{(\phi_1,\ldots,\phi_{n-2}) : \phi_{il} \le \phi_i \le \phi_{iu}, i = 1,\ldots,n-2\}$

$$\subset [0,\pi)^{\times(n-3)} \times [0,2\pi) = M^*_{n-1}$$

then $$O_{a,p,q}(A) = a_n a_{n-1} \int\limits_{r1}^{r_2} \frac{r^{n-2}}{(1-r^p)^{(p-1)/p}} dr$$

$$\times \int\limits_{M^*} J^*\left(T^E_{a,p}(n-1)\right)(\phi_1,\ldots,\phi_{n-2}) d(\phi_1,\ldots,\phi_{n-2}).$$

In what follows, we use the coordinate transformation $\tilde{T}_{a,p} : (R,r,\phi) \to z[R,r,\phi], \phi = (\phi_1,\ldots,\phi_{n-2})$ defined by

$z_1 = a_1 R r \cos_{(a_1,a_2;p)}(\phi_1), z_2 = a_2 R r \sin_{(a_1,a_2;p)}(\phi_1) \cos_{(a_2,a_3;p)}(\phi_2), \ldots,$

$z_{n-2} = a_{n-2} R r \sin_{(a_1,a_2;p)}(\phi_1) \cdot \ldots \cdot \sin_{(a_{n-3},a_{n-2};p)}(\phi_{n-3}) \cos_{(a_{n-2},a_{n-1};p)}(\phi_{n-2})$

$z_{n-1} = a_{n-1} R r \sin_{(a_1,a_2;p)}(\phi_1) \cdot \ldots \cdot \sin_{(a_{n-3},a_{n-2};p)}(\phi_{n-3}) \sin_{(a_{n-2},a_{n-1};p)}(\phi_{n-2}),$

$z_n = a_n R (1 - r^p)^{1/p}.$

This transformation allows the representations

$A(r_1,r_2,M^*) = \tilde{T}_{a,p}(1,[r_1,r_2),M^*) = \{z[R,r,\phi] : R = 1, r \in [r_1,r_2), \phi \in M^*\}$ and

$sector(A(r_1,r_2,M^*),\rho) = \tilde{T}_{a,p}([0,\rho) \times [r_1,r_2) \times M^*)$

$= \{z[R,r,\phi] : 0 \le R < \rho, r \in [r_1,r_2), \phi \in M^*\}.$

The volume

$$\mu(sector(A(r_1, r_2, M^*), \rho)) = \int\limits_{sector(A(r_1,r_2,M^*),\rho)} dz$$

may therefore be written as

$$\mu(sector(A(r_1, r_2, M^*), \rho)) = \int\limits_{R=0}^{\rho} \int\limits_{r=r_1}^{r_2} \int\limits_{\phi \in M^*} J(\tilde{T}_{a,p})(R, r, \phi) dR dr d\phi.$$

Here, $J(\tilde{T}_{a,p})(R, r, \phi) = \frac{D(z_1,\dots,z_n)}{D(R,r,\phi_1,\dots,\phi_{(n-2)})}$ can be evaluated as in (Richter 2013) where the case $p = 2$ was dealt with:

$$\begin{aligned} J(\tilde{T}_{a,p})(R, r, \phi) &= |z_{nr} a_{n-1} (rR)^{n-2} J^* r - z_{nR} a_{n-1} (rR)^{n-2} J^* R| \\ &= \left| \left(-\frac{a_n R r^{p-1}}{(1 - r^p)^{(p-1)/p}} (rR)^{n-2} r - a_n (1 - r^p)^{1/p} (rR)^{n-2} R \right) J^* a_{n-1} \right| \\ &= a_{n-1} a_n J^* (T^E_{a,p}(n-1))(\phi) R^{n-1} \frac{r^{n-2}}{(1 - r^p)^{1-1/p}}. \end{aligned}$$

It follows that

$$\lambda\big(sector\big(A\left(r_1, r_2, M^*\right), \rho\big)\big) = a_{n-1} a_n \int\limits_0^{\rho} R^{n-1} dR \int\limits_{r_1}^{r_2} \frac{r^{n-2}}{\sqrt{1 - r^2}} dr \int\limits_{M^*} J^* \left(T^E_{a,p}(n-1)\right)(\phi) d\phi.$$

We consider now the local approach to the non-Euclidean surface content,

$$\frac{d}{d\rho} \mu(sector(A(r_1, r_2, M^*), \rho))|_{\rho=1} = a_{n-1} a_n \int\limits_{r_1}^{r_2} \frac{r^{n-2}}{\sqrt{1 - r^2}} dr \int\limits_{M^*} J^* \left(T^E_{a,p}(n-1)\right)(\phi) d\phi,$$

and observe that $\mathfrak{O}_S(A(r_1, r_2, M^*)) = O_{a,p,q}(A(r_1, r_2, M^*))$.

The measures $\mathfrak{O}_S$ and $O_{a,p,q}$ coincide on the semi-algebra which is generated by the sets of the type $A(r_1, r_2, M^*)$. It follows from the measure extension theorem that these measures coincide on the whole Borel-σ-field $\mathfrak{B}^E_{a,p}$ on $E_{a,p}$, too. □

Remark 6. ***Reformulating the results of Section 3.1***

In the special case that $K = B_{a,p}, S = E_{a,p}$, just considered here, all the statements of Equations (1) and (2), Theorem 1, Corollaries 1-3 and Remarks 1 and 2 remain valid if the integral surface measure $O_{a,p,q}$ is used instead of the star-generalized surface measure $\mathfrak{O}_S$. The same is true for all those statements quoted below which are using the local notions of Section 2.

4 Star-shaped distributions and geometric disintegration

4.1 Star-shaped uniform distributions

In this section, we extend the method of indivisibles which was used so far for the Lebesgue measure to a class of probability laws which contains the families of elliptically contoured and $l_{n,p}$-symmetric distributions as special cases. This method was originally developed in (Richter 1985, 1987) for proving large deviation limit theorems for the multivariate standard Gaussian law.

We continue to use the notations from Section 2. Note that the following general consideration always covers the very well interpretable specific case that $S = E_{a,p}$ and thus $\mathfrak{O}_S = O_{a,p,q}$.

Definition 6. *The star-generalized uniform probability distribution on the Borel σ-field $\mathfrak{B}_S$ is defined as $\omega_S(A) = \mathfrak{O}_S(A)/\mathfrak{O}_S(S)$.*

Remark 7. *(a) Let $A \in \mathfrak{B}_S$. Then $\omega_S(A) = sm_K(A)$.*
(b) Let $B \in \mathfrak{B}_n$. Then $\omega_S\left(\left[\frac{1}{r}B\right] \cap S\right) = \mathfrak{F}_S(B, r)$, $r > 0$.

Let $(\Omega, \mathfrak{A}, P)$ be a probability space and $Y : \Omega \to \mathbb{R}^n$ a random vector being uniformly distributed on K, i.e.

$$P(Y \in B) = \mu(B)/\mu(K), \ \forall B \in K \cap \mathfrak{B}_n.$$

The a.s. defined normalized random vector $U_S = Y/h_K(Y)$ takes its values in $S, P(h_K(U_S) = 1) = 1$. Let us further put $R = h_K(Y)$.

Theorem 6. *(a) U_S follows the star-generalized uniform distribution, $U_S \sim \omega_S$.*
(b) The pdf of R is $f_R(r) = I_{(0,1)}(r)n \cdot r^{n-1}$.
(c) The random elements U_S and R are stochastically independent.

Proof. (a) Let $A \in \mathfrak{B}_S$. Then $P(U_S \in A) = P(Y \in sector(A, 1))$, and
$P(U_S \in A) = \frac{\mu(sector(A,1))}{\mu(K)} = \frac{1}{\mu(K)} \int\limits_0^1 r^{n-1} dr \int\limits_{StSph^{*-1}(A)} J^*(\vartheta) d\vartheta = \frac{\mathfrak{O}_S(A)}{n\mu(K)}$.
Because of $\mathfrak{O}_S(S) = n\mu(K)$, we have $P(U_S \in A) = \omega_S(A)$.
(b) For $0 < r < 1$, we consider the cumulative distribution function (cdf) of R,

$$P(R < r) = P(Y \in K(r)) = \mu(K(r))/\mu(K) = r^n I_{(0,1)}(r).$$

(c) The independence of U_S and R follows from $P(R < \varrho, U_S \in A)$
$= P(Y \in sector(A, \varrho)) = \int\limits_{sector(A,\varrho)} P^Y(dx) = \frac{1}{\mu(K)} \int\limits_{sector(A,\varrho)} d\mu$

$$= \frac{1}{\mu(K)} \int\limits_0^{\varrho} \int\limits_{G(A)} r^{n-1} J^*(\vartheta) d\vartheta dr = \frac{1}{\mu(K)} \frac{\varrho^n}{n} \mathfrak{O}_S(A) = P(R < \varrho) P(U_S \in A)$$ □

Remark 8. *(a) The pdf of R^2 is $\frac{d}{dr}P(R^2 < r) = I_{(0,1)}(r)\frac{n}{2}r^{n/2-1}$, $r \in R$.*
(b) The probability distribution of the random vector Y allows the representation

$$P(Y \in B) = \int\limits_0^{\infty} P\left(U_S \in \frac{1}{r}B | R = r\right) dP(R < r)$$

which may be considered as a reformulation of Corollary 2 with

$$\mathfrak{F}_S(B, r) = \omega_S\left(\left[\frac{1}{r}B\right] \cap S\right) = P\left(U_S \in \frac{1}{r}B | R = r\right) \quad a.s. \tag{3}$$

That is why the family of probability measures $\mathfrak{P} = \{P_r, r > 0\}$ where P_r is defined on the Borel σ-field $\mathfrak{B}_n$ by $P_r(B) = \omega_S\left(\left[\frac{1}{r}B\right] \cap S\right) = P\left(U_S \in \frac{1}{r}B | R = r\right)$, may be called a geometric disintegration of P^Y w.r.t. P^R. The family $\mathfrak{P}$ may also be considered as a regular conditional probability.

4.2 Continuous star-shaped distributions

There are different ways to introduce more general classes of star-shaped distributions than the uniform ones considered so far. One of the possibilities is to continue with star-shaped distributions having a density, to derive their most basic properties and finally to introduce the class of all star-shaped distributions having just the latter as their defining properties. This way may be considered as formally generalizing the notion of norm-contoured distributions in (Richter, W.-D.: Norm contoured distributions in R^2, submitted), as well as being statistically well motivated by comparing empirical density level sets with level sets of Minkowski functionals of suitably chosen star bodies. This way will be followed in the present and in the following two sections. An alternative possibility would be just to introduce here the general class of star-shaped distributions and to restrict consideration to special classes of distributions like continuous ones only later.

Definition 7. *Let* $g : \mathbb{R}^+ \to \mathbb{R}^+$ *satisfy the assumptions* $0 < I(g) < \infty$ *where* $I(g) = \int_0^\infty r^{n-1} g(r)dr$. *We call g a density generating function (dgf),* $\varphi_{g,K}(x) = C(g,K)g(h_K(x)), x \in \mathbb{R}^n$ *a star-shaped density and K its contour defining star body.*

A probability measure having the density $\varphi_{g,K}$ will be denoted by $\Phi_{g,K}$. Let us emphasize that according to this definition 0_n may be any point from the set of all points w.r.t. which K is star-shaped, hence K needs not to be symmetric w.r.t. 0_n. Densities of such type have been studied already in (Balkema and Nolde 2010; Fernandez et al. 1995). Our more general considerations in Sections 4.3-4.7, however, seem to be new. The following theorem deals with a geometric measure representation of continuous star-shaped distributions.

Theorem 7. *For every* $B \in \mathfrak{B}^n$, $\Phi_{g,K}(B) = C(g,K)\mathfrak{O}_S(S) \int_0^\infty r^{n-1} g(r)\mathfrak{F}_S(B,r)dr$.

Proof. Because of $\Phi_{g,K}(B) = C(g,K) \int_B g(h_K(x))dx$ it follows from Corollary 1(a) that $\Phi_{g,K}(B) = C(g,K) \int_0^\infty r^{n-1} \int_{[\frac{1}{r}B]\cap S} g(h_K(r\theta))\mathfrak{O}_S(d\theta)dr$. Hence,

$$\Phi_{g,K}(B) = C(g,K) \int_0^\infty r^{n-1} g(r)\mathfrak{O}_S\left(\left[\tfrac{1}{r}B\right] \cap S\right) dr. \qquad \square$$

Classes of dgfs are surveyed, e.g., in (Fang et al. 1990) and (Richter 2013). Numerous types of applications of special cases of the geometric measure representation in Theorem 7 are surveyed in (Richter 2009, 2012). Later applications are to be found in (Arellano-Valle and Richter 2012; Batún-Cutz et al. 2013) and (Günzel et al. 2012).

4.3 Stochastic representations

In this section, we consider that property of continuous star-shaped distributions which will serve in the next section to define a general class of star-shaped distributions.

Theorem 8. *If* $Y \sim \Phi_{g,K}$ *then Y allows the stochastic representation* $Y \stackrel{d}{=} R_g U_S$ *where* R_g *and* U_S *are stochastically independent,* $U_S \sim \omega_S$ *and* R_g *follows the density* $f(r) = \frac{1}{I(g)} r^{n-1} g(r), r > 0$.

Proof. $P(R < \varrho) = P(Y \in K(\varrho)) = C(g,K) \int_{K(\varrho)} g(h_K(x))dx$

$$= 2C(g,K) \int_0^{\rho} \int_{G(S^+)} g\left(h_K\left(r\left(\vartheta^T, \eta^+(\vartheta)\right)^T\right)\right) r^{n-1} J^*(\vartheta) d\vartheta dr$$

$$= 2C(g,K) \int_0^{\rho} r^{n-1} g(r) dr \int_{G(S^+)} J^*(\vartheta) d\vartheta$$

□

Remark 9. *The normalizing constant $C(g,K)$ in Definition 7 allows according to Theorem 7 the representation $C(g,K) = \frac{1}{\mathfrak{O}_S(S)I(g)}$ and the statement of Theorem 7 may according to Theorem 8 be written as*

$$\Phi_{g,K}(B) = \int_0^{\infty} \mathfrak{F}_S(B,r) P(R \in dr),\ B \in \mathfrak{B}_n \tag{4}$$

where $\mathfrak{F}_S(B,r)$ may be interpreted as in (3). Hence, (4) may be read as a generalization of Remark 8(b). Moreover,

$$\varphi_{g,K}(x) = \frac{g(h_K(x))}{\mathfrak{O}_S(S)I(g)},\ x \in \mathbb{R}^n. \tag{5}$$

4.4 General star-shaped distributions

The results of the previous section may serve as a starting point for defining general star-shaped distributions. We follow the way in (Fang et al. 1990) and (Richter 2009, 2013) when we use the stochastic representation from Section 4.3 for defining now a large family of star-shaped distributions.

Definition 8. *A random vector $Y : \Omega \to \mathbb{R}^n$ is said to follow a star-shaped distribution if there are a star body K having the origin as an interior point, $0_n \in int K$, a vector $\nu \in \mathbb{R}^n$, and a random variable $R : \Omega \to [0,\infty)$ such that $Y - \nu$ satisfies the stochastic representation $Y - \nu \stackrel{d}{=} R \cdot U_S$ where U_S follows the star-generalized uniform distribution on the boundary S of K, $U_S \sim \omega_S$, and R and U_S are stochastically independent. If Y has a density with dgf g then by $\Phi_{g,K,\nu}$ the distribution law of Y is denoted, $Y \sim \Phi_{g,K,\nu}$, and K is called a density contour defining star body of the star-shaped distribution $\Phi_{g,K,\nu}$.*

The set of all star-shaped distributions on $\mathfrak{B}_n$ will be denoted $StSh^{(n)}$ and its subset of continuous distributions by

$$CStSh^{(n)} = \{\Phi_{g,K,\nu} : \nu \in \mathbb{R}^n,\ K \text{ is a star body with } 0 \in \text{ int } K, g \text{ is a dgf}\}.$$

We recall that star-shaped sets are associated with multivariate stable distributions in (Molchanov 2009) to describe characteristic functions, thus playing there another role than in Definition 8. To finish this section, we remark that both the set of all star bodies having the origin as an interior point and the set $StSh^{(n)}$ are invariant w.r.t. any orthogonal transformation.

4.5 Extension of the ball number function

In (Richter 2011), the ball number function was defined for $l_{n,p}$-balls and the problem of extending it to balls being as general as possible was stated. It follows from the results in Section 3 that both the ratios $\frac{\mu(B_{a,p}(r))}{r^n}$ and $\frac{O_{a,p,q}(E_{a,p}(r))}{nr^{n-1}}$ do not depend on the star radius r, and that their constant values are one and the same number. This common value will be called the ball number $\pi(B_{a,p})$ of $B_{a,p}(r), r > 0$. Here, $\pi(B_{a,p}) = \mu(B_{a,p}) = a_1 \cdot \ldots \cdot a_n \cdot \frac{\omega_{n,p}}{n}$, hence the region where the ball number function is defined is extended here to all $B_{a,p}$ balls.

4.6 Characteristic functions

Let $Y : \Omega \to \mathbb{R}^n$ be a star-shaped distributed random vector which satisfies the stochastic representation $Y \stackrel{d}{=} R \cdot U_S$ where the non-negative random variable R is independent of the star-generalized uniformly distributed random vector U_S, and let moreover ϕ_Y and ϕ_{U_S} denote the characteristic functions (ch.f.) of the vectors Y and U_S, respectively. Further, let F_R denote the cdf of R.

Theorem 9. *The ch.f. of the star-shaped distributed random vector Y allows the integral representation* $\phi_Y(t) = \int_0^\infty \phi_{U_S}(rt) dF_R(r)$, $t \in \mathbb{R}^n$.

Proof. Because of the independence of R and U_S, Theorem 1.1.6 in (Sasvári 2013) applies. □

This theorem was proved first for spherically distributed vectors in (Schoenberg 1938) and later for $l_{n,1}$-symmetrically distributed random vectors in (Ng and Tian 2001) and for continuous $l_{n,p}$-symmetrically distributed vectors in (Kalke 2013).

Remark 10. *(a) The ch.f.* $\phi_{U_S}(t) = Ee^{it^T U_S}$, $t \in \mathbb{R}^n$, *of* U_S *allows the integral representation* $\phi_{U_S}(t) = \int_S \cos\left(t^T\theta\right) \omega_S(d\theta) + i \int_S \sin\left(t^T\theta\right) \omega_S(d\theta)$.

(b) The ch.f. ϕ_Y *of a star-shaped distributed random vector Y having a density with dgf g and contour defining star body K allows the representation*

$$\mathfrak{O}_S(S) I(g) \phi_Y(t) = \int_0^\infty \left[\sum_j \int_{G(S_j)} \cos(t, r \begin{pmatrix} \vartheta \\ y(\vartheta) \end{pmatrix}) J_j^*(\vartheta) d\vartheta \right] r^{n-1} g(r) dr$$
$$+ i \int_0^\infty \left[\sum_j \int_{G(S_j)} \sin(t, r \begin{pmatrix} \vartheta \\ y(\vartheta) \end{pmatrix}) J_j^*(\vartheta) d\vartheta \right] r^{n-1} g(r) dr$$

where $(.,.)$ *denotes the Euclidean scalar product in* $\mathbb{R}^n$.

(c) On combining the representations in (a) and (b), and taking into account Remark 1, we get an alternative direct proof of Theorem 9 if Y has density $\varphi_{g,K}$.

(d) If U_S *is symmetrically distributed w.r.t. the origin,* $U_S \stackrel{d}{=} -U_S$, *then the imaginary parts of the integral representations in (a) and (b) vanish, and both* ϕ_{U_K} *and* ϕ_Y *are symmetric w.r.t. the origin.*

4.7 The class of *p*-generalized elliptically contoured distributions

The general principle for deriving geometric and stochastic representations of star-shaped distributions developed so far will be proved in this section to successfully apply to a class of distributions considered in Section 3.5 of (Arellano-Valle and Richter 2012) and including both the $l_{n,p}$-symmetric ones, accordingly represented in (Richter 2009), and the elliptically contoured ones, analogously dealt with in (Richter 2013).

According to Definition 6, Theorem 5 and Corollary 3, the p-generalized elliptically contoured uniform distribution on $\mathfrak{B}_{E_{a,p}}$ is defined for arbitrary $p > 0$ by

$$\omega_{a,p,q}(A) = D(n,a,p)O_{a,p,q}(A) \text{ with } D(n,a,p) = \frac{p^{n-1}\Gamma\left(\frac{n}{p}\right)}{a_1 \cdot \ldots \cdot a_n 2^n \left(\Gamma\left(\frac{1}{p}\right)\right)^n}$$

and with q satisfying $1/p + 1/q = 1$.

If a random vector $Y : \Omega \to \mathbb{R}^n$ follows the uniform probability distribution on $B_{a,p}$ then $U \sim \omega_{a,p,q} = \omega_{E_{a,p}}$ where U is a.s. defined as $U = Y/R$ and is independent of $R = h_{B_{a,p}}(Y)$. The latter, non-negative, random variable has the density described in Theorem 6(b).

Let $\mathfrak{O}(n)$ denote the set of all orthogonal $n \times n$ matrices. A random vector $Y : \Omega \to \mathbb{R}^n$ is said to follow a p-generalized elliptically contoured distribution $EC_{a,p,\nu,O}$ with parameters $a = (a_1, \ldots, a_n)^T, a_i > 0, i = 1, \ldots, n,\ p > 0,\ \nu \in \mathbb{R}^n$ and $O \in \mathfrak{O}(n)$ if there exists a random variable $R : \Omega \to [0, \infty)$ such that Y satisfies the stochastic representation

$$O^T(Y - \nu) \stackrel{d}{=} R \cdot U$$

where $U \sim \omega_{E_{a,p}}$ and U and R are stochastically independent. Note that Y has a density f_Y iff R has a density. If $O^T(Y - \nu)$ has the dgf g, i.e. if

$$f_Y(y) = C(g,a,p)g\left(\left|O^T(y - \nu)\right|_{a,p}\right), y \in \mathbb{R}^n$$

with $C(g,a,p) = C(g, B_{a,p})$ then R has the density

$$f_R(r) = I(g)^{-1}r^{n-1}g(r)I_{[0,\infty)}(r).$$

In this case, we write $Y \sim \Phi_{g,a,p,\nu,O}$ and $f_Y = \varphi_{g,a,p,\nu,O}$. Note that $\Phi_{g,a,p,0_n,I_n} = \Phi_{g,B_{a,p},0_n}$ where I_n denotes the $n \times n$-unit matrix. The measure $\Phi_{g,a,p,\nu,O}$ allows the geometric representation

$$\Phi_{g,a,p,\nu,O}(B) = \int_0^\infty \mathfrak{F}_{a,p}\left(O^T(B - \nu, r)\right) dF_R(r) \tag{6}$$

where $\mathfrak{F}_{a,p}(M,r) = \omega_{E_{a,p}}\left(\left[\frac{1}{r}M\right] \cap E_{a,p}\right), r > 0.$

Example 1. *In the case of dimension $n = 2$, Figure 1 shows the density $\varphi_{g,a,p,\nu,O}$ and contours of its superlevel sets where $g(r) = \exp\left\{-\frac{r^p}{p}\right\}, a = (3,1)^T, p$ takes several values, $\nu = (0,0)^T$ and*

$$O = \begin{pmatrix} \cos\alpha & \sin\alpha \\ -\sin\alpha & \cos\alpha \end{pmatrix}, \quad \alpha = 5\pi/3.$$

The matrix O causes an anticlockwise rotation through an angle of size $\pi/3$.

Remark 11 (On independent coordinates). *Let $(R, \Phi_1, \ldots, \Phi_{n-1}) = \left(T^E_{a,p}\right)^{-1}(Y)$ be the random p-generalized ellipsoidal coordinates of Y where $Y \sim \Phi_{g,a,p,0_n,O}$. According to Theorem 3, the generalized radius is $\left(\sum_{i=1}^n |Y_i/a_i|^p\right)^{1/p} = R$, and by the density transformation formula and Theorem 4,*

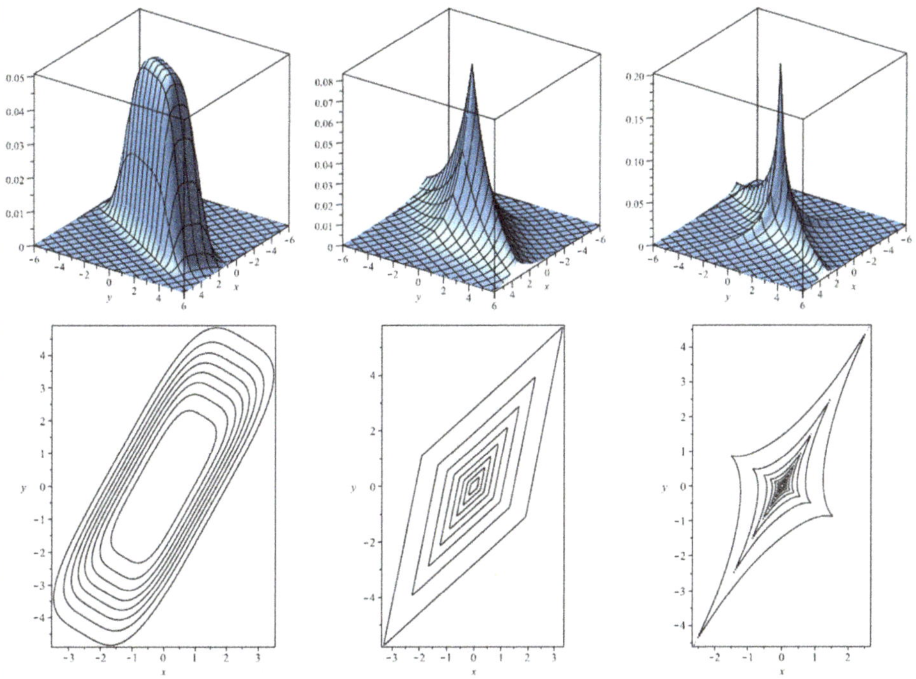

Figure 1 ***p*-generalized elliptically contoured densities for $p = 4$, 1 and 0.6, from left to right.**

$$f_{(R,\Phi_1,\dots,\Phi_{n-1})}(r,\phi_1,\dots,\phi_{n-1}) = \frac{C(g,a,p)}{a_2\cdot\ldots\cdot a_{n-1}} g(r) r^{n-1} \prod_{i=1}^{n-1} h_i(\phi_i)$$

with $h_i(\phi_i) = \frac{\left(\sin_{(a_i,a_{i+1};p)}(\phi_i)\right)^{n-i-1}}{N^2_{(a_i,a_{i+1};p)}(\phi_i)}$, $i = 1,\dots,n-1$ *being integrable functions. Thus, suitably normalized, the functions* $h_0, h_1, \dots, h_{n-1}$ *with* $h_0(r) = rg(r)I_{(0,\infty)}(r)$ *are the densities of* $R, \Phi_1, \dots, \Phi_{n-1}$*, respectively.*

Hence, the random coordinates $R, \Phi_1, \dots, \Phi_{n-1}$ *are stochastically independent.*

This result opens new perspectives for various applications which are not considered in the present paper.

5 Applications

5.1 The non-concentric elliptically contoured distribution class

The general distribution classes considered in the present paper include various interesting special cases to be studied only in detail in the future. Just to start with, let $0 < b < a$ and

$$K_{a,b} = \left\{(x,y)^T \in \mathbb{R}^2 : (x/a)^2 + (y/b)^2 \le 1\right\}.$$

A point $(e,f)^T$ is from the inner part of $K_{a,b}$ iff it satisfies the inequality $(e/a)^2 + (f/b)^2 < 1$. For such points, we put $K_{a,b,e,f} = K_{a,b} - (e,f)^T$. As because $rK_{a,b,e,f} = K_{ar,br,er,fr}$, for arbitrary dgf g, the level sets of the density

$$\varphi^*_{g,a,b,e,f}(x,y) = C\left(g, K_{a,b,e,f}\right) g\left(h_{K_{a,b,e,f}}\left((x,y)^T\right)\right), (x,y)^T \in \mathbb{R}^2$$

are the boundaries of the sets $K_{ar,br,er,fr}$, $r > 0$. Note that if $(e, 0)^T$ is a focal point of the elliptical disc $K_{a,b}$ then the origin $(0, 0)^T$ is a focal point of $E_{ar,br,er,0}$, for all $r > 0$. In this case, we call the origin a position of the density $\varphi^*_{g,a,b,e,0}$. Similarly, the origin may also be called a position of the density $\varphi^*_{g,a,b,0,f}$ if $(0,f)^T$ is a focal point of $K_{a,b}$. The distribution class

$$NCEC = \left\{\varphi^*_{g,a,b,e,f}\ ,\ g \text{ is a dgf}, 0 < b < a, (e/a)^2 + (f/b)^2 < 1\right\}$$

will be called a non-concentric elliptically contoured distribution class. It is left to the reader to derive explicit expressions for the Minkowski functional $h_{K_{a,b,e,f}}$ and the normalizing constant $C(g, K_{a,b,e,f})$.

5.2 Circular distributions

In this section, we study directional distributions on further using the results of the present work. For a recent overview on circular distributions we refer to (Pewsey et al. 2013).

In the case of dimension two, the density of $\Phi_{g,K,\nu}$ is

$$\varphi_{g,K,\nu}(x, y) = C(g, K)g\left(h_K\left(\begin{pmatrix} x \\ y \end{pmatrix} - \begin{pmatrix} \nu_1 \\ \nu_2 \end{pmatrix}\right)\right),\ (x, y)^T \in \mathbb{R}^2.$$

We recall that star-generalized trigonometric functions and random polar coordinates are defined in (Richter 2011a) by $\cos_K(\phi) = \frac{\cos\phi}{h_K(\cos\phi,\sin\phi)}$, $\sin_K(\phi) = \frac{\sin\phi}{h_K(\cos\phi,\sin\phi)}$ and $X = R\cos_K(\Phi), Y = R\sin_K(\Phi)$, respectively. The K-generalized radius coordinate is $R = h_K(X, Y)$, and the angle Φ satisfies the representation of the usual polar angle,

$$\arctan(|Y/X|) = \Phi \text{ in } Q_1,\ = \pi - \Phi \text{ in } Q_2,\ = \Phi - \pi \text{ in } Q_3, = \ 2\pi - \Phi \text{ in } Q_4.$$

The Jacobian of this transformation is $rR_S^2(\phi)$ where the function $R_S(\phi) = 1/h_K(\cos\phi, \sin\phi)$ describes the boundary S of K:

$$S = \left\{R_S(\phi)\begin{pmatrix} \cos\phi \\ \sin\phi \end{pmatrix}, 0 \le \phi < 2\pi\right\} = \left\{(x, y)^T : h_K(x, y) = 1\right\}.$$

With uniquely determined $\mu \in [0, 2\pi)$ and $\lambda > 0$, the location vector $(\nu_1, \nu_2)^T$ can be represented as

$$\begin{pmatrix} \nu_1 \\ \nu_2 \end{pmatrix} = \lambda\begin{pmatrix} \cos_K(\mu) \\ \sin_K(\mu) \end{pmatrix}.$$

Thus, the density of $(R, \Phi)^T$ is

$$f_{(R,\Phi)}(r, \phi) = C(g, K)rR_S^2(\phi)g\left(h_K\left(\begin{pmatrix} r\cos_K(\phi) - \lambda\cos_K(\mu) \\ r\sin_K(\phi) - \lambda\sin_K(\mu) \end{pmatrix}\right)\right)I_{[0,2\pi)}(\phi)I_{[0,\infty)}(r).$$

Integrating $f_{(R,\Phi)}$ w.r.t. ϕ, and dividing $f_{(R,\Phi)}$ by the latter result, gives $f_{\Phi|R}(\phi|r) = vMd_{g,K,r,\lambda,\mu}(\phi)$ where

$$vMd_{g,K,r,\lambda,\mu}(\phi) = \frac{R_S^2(\phi)g\left(h_K\left(\begin{pmatrix} r\cos_K(\phi) - \lambda\cos_K(\mu) \\ r\sin_K(\phi) - \lambda\sin_K(\mu) \end{pmatrix}\right)\right)}{\int_0^{2\pi} R_S^2(\phi)g\left(h_K\left(\begin{pmatrix} r\cos_S(\phi) - \lambda\cos_S(\mu) \\ r\sin_S(\phi) - \lambda\sin_S(\mu) \end{pmatrix}\right)\right)d\phi}. \quad (7)$$

This function will be called a star-shaped generalization of the von Mises density which itself appears as a special case for K being the Euclidean unit disc in $\mathbb{R}^2$ and $g(r) = \exp\left\{-r^2/2\right\}$, $r > 0$, c.f. (von Mises 1918).

Example 2. *For illustrating our principle of constructing generalized von Mises densities at the hand of a concrete example, let P_n denote the polygon having the n vertices $I_{n,i} = \left(\cos\left(\frac{2\pi}{n}(i-1)\right), \sin\left(\frac{2\pi}{n}(i-1)\right)\right)^T, i = 1,\ldots,n,\ n \geq 3$, and let K_n be the convex body circumscribed by P_n. The Minkowski functional of K_n has been dealt with in (Richter, W-D, Schicker, K: Circle numbers of centered regular convex polygons, submitted). Figure 2 shows, from the left to the right, in each row, the polygonally contoured density $\varphi_{g,K_n,\nu}(x,y), (x,y)^T \in \mathbb{R}^2$, the contours of this density (black shapes) together with the polygonally generalized circle consisting of all points $(x,y)^T \in \mathbb{R}^2$ satisfying the condition $h_{K_n}\left((x,y)^T\right) = r$ (blue shape), and the polygonally generalized von Mises density $vMd_{g,K_n,r,\lambda}$ where $g(\rho) = \exp\left\{-\frac{\rho^2}{2}\right\}, \rho > 0,\ r = 1, \mu = 3\pi/4$ and $(n,\lambda) = \left(3,\frac{1}{2}\right)$ in the first row, but $(n,\lambda) = \left(4,\frac{3}{2}\right)$ in the second row. Each of the arrows in the middle panel shows the way from the center of the blue drawn polygonally generalized circle to that of the black drawn ones.*

6 Concluding remarks

Exact distributions of numerous statistics like mean value statistic, Student statistic, Chi-squared statistic have been derived in a broad and well known literature under the assumption that the sample vector follows a multivariate normal law, or more generally, an elliptically contoured one. Similarly, exact statistical distributions have been derived when the sample distribution is exponential or one of its geometric generalizations. For

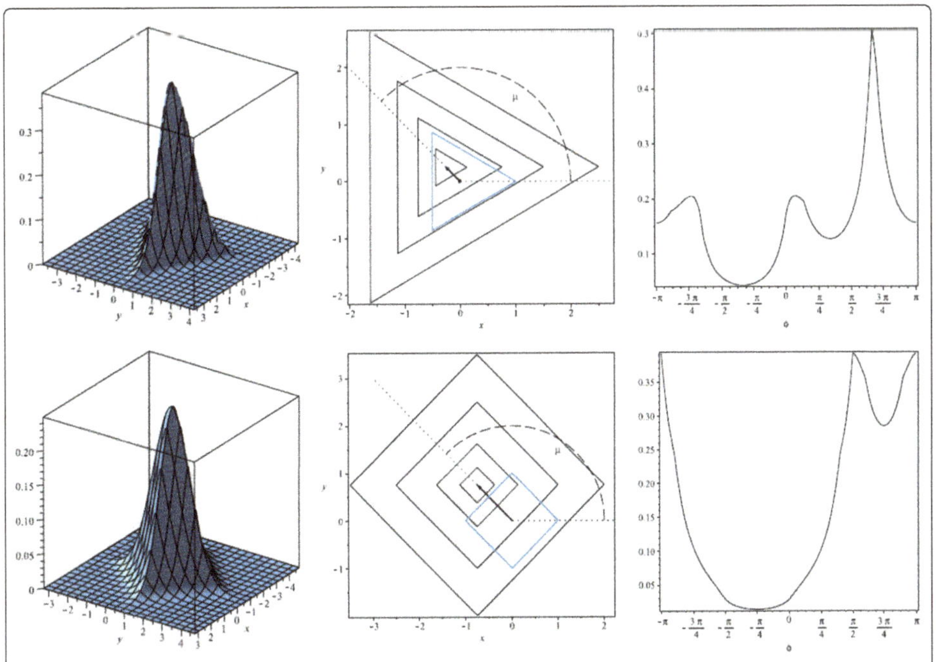

Figure 2 From left to right: polygonally contoured density, density contours with polygonally generalized circle describing the condition $h_{K_n}((x,y)) = 1$, polygonally generalized von Mises density, where $(n,\lambda) = (3,\frac{1}{2})$ in the first row and $= (4,\frac{3}{2})$ in the second row.

results of the latter type, we refer to (Henschel 2001, 2002; Henschel and Richter 2002). However, there seem, in general, not to be as many exact distributional results as in the Gaussian or elliptically contoured cases in the case of samples from any other multivariate distribution. Mathematical research without one of the above assumptions often deals with asymptotic considerations for large sample sizes. Many results of such work are again closely connected with properties of the Gaussian law which occurs often as a limit law.

The present paper provides new possibilities for deriving representations of exact statistical distributions if the sample vector follows a probability law coming from a rather big class of probability laws. It is not the place here to demonstrate in any detail all the possible applications of the present results. Instead, we refer to (Richter 2012) and several references given therein for getting a first overview. A recent example for deriving new results on distributions of functions of Gaussian vectors is given in (Richter 2014). This example might stimulate consideration of exact distributions of new classes of statistics.

Competing interests
The author declares that he has no competing interests.

Acknowledgements
The author is grateful to two Reviewers for their valuable hints which led to improvements of the manuscript, and to Thomas Dietrich for drawing the figures with *Maple 17*.

References

Arellano-Valle, RB, Richter, W-D: On skewed continuous $l_{n,p}$-symmetric distributions. Chilean J. Stat. **3**(2), 193–212 (2012)
Balkema, AA, Embrechts, P, Nolde, N: Meta densities and the shape of their sample clouds. J. Multivariate Anal. **101**, 1738–1754 (2010)
Balkema, AA, Nolde, N: Asymptotic independence for unimodal densities. Adv. Appl. Prob. **42**, 411–432 (2010)
Barndorff-Nielsen, OE, Blaesild, P, Eriksen, PS: Decomposition and Invariance of Measures, and Statistical Transformation Models, Lecture Notes in Statistics, Vol. 58. Springer, New York (1989)
Barthe, F, Csörnyei, M, Naor, A: A note on simulataneous polar and cartesian decomposition. In: Geometric Aspects of, Functional Analysis, Lecture Notes in Mathematics, vol. 1807, pp. 1–19. Springer, Berlin, (2003)
Batún-Cutz, J, Gonzáles-Farías, G, Richter, W-D: Maximum distributions for $l_{2,p}$-symmetric vectors are skewed $l_{1,p}$-symmetric distributions. Stat. Probab. Lett. **83**, 2260–2268 (2013)
Dietrich, T, Kalke, S, Richter, W-D: Stochastic representations and a geometric parametrization of the two-dimensional Gaussian law. Chilean J. Stat. **4**(2), 27–59 (2013)
Eaton, ML: Multivariate Statistics, a Vector Space Approach. Wiley, New York (1983)
Eaton, M L: Group invariance applications in statistics. In: Regional Conference, Series in Probability and Statistics, vol. 1, pp. 1–133. Institute of, Mathematical Statistics, Hayward, California, (1989)
Fang, K-T, Kotz, S, Ng, KW: Symmetric Multivariate and Related Distributions. Chapman & Hall, London (1990)
Farrell, RH: Techniques of, Multivariate Calculation. Lecture Notes in Mathematics 520. Springer, Berlin (1976)
Farrell, R H: Multivariate Calculation: Use of the Continuous Groups. Springer, Berlin (1985)
Fernandez, C, Osiewalski, J, Steel, MFJ: Modeling and inference with ν-spherical distributions. J. Amer. Stat. Assoc. **90**, 1331–1340 (1995)
Gómez, E, Gomez-Viilegas, MA, Marín, JM: A multivariate generalization of the power exponential family of distributions. Commun. Stat. Theory Methods. **27**(3), 589–600 (1998)
Günzel, T, Richter, W-D, Scheutzow, S, Schicker, K, Venz, J: Geometric approach to the skewed normal distribution. J. Stat. Plann. Inference. **142**, 3209–3224 (2012)
Gupta, A, Song, D: l_p-norm spherical distributions. J. Stat. Plann. Inference. **60**, 241–260 (1997)
Henschel, V: Exact distributions in the model of a regression line for the threshold parameter with exponential distribution of errors. Kybernetika. **37**(6), 703–723 (2001)
Henschel, V: Statistical inference in simplicially contoured sample distributions. Metrika. **56**, 215–228 (2002)
Henschel, V, Richter, W-D: Geometric generalization of the exponential law. J. Multivariate Anal. **81**, 189–204 (2002)
Hoffmann-Jorgensen, J: Probability with a View Towards Statistics II. Chapman & Hall, New York (1994)
Joenssen, DW, Vogel, J: A power study of goodness-of-fit tests for multivariate normality implemented in R. J. Stat. Comput. Simul. **75**, 93–107 (2012)
Kalke, S: Geometrische Strukturen $l_{n,p}$-symmetrischer charakteristischer Funktionen. Dissertation, Universität Rostock (2013)
Kallenberg, O: Probabilistic Symmetries and Invariance Principles. Springer, New York (2005)
Kamiya, H, Takemura, A, Kuriki, S: Star-shaped distributions and their generalizations. J. Stat. Plann. Inference. **138**, 3429–3447 (2008)
Kinoshita, K, Resnick, SI: Convergence of scaled random samples in R. Ann. Probab. **19**, 1640–1663 (1991)
Koehn, U: Global cross sections and the densities of maximal invariants. Ann. Math. Stat. **41**(6), 2045–2056 (1970)

Koshevoy, G, Mosler, K: Zonoid trimming for multivariate distributions. Ann. Stat. **9**, 1998–2017 (1997)
Molchanov, I: Convex and star-shaped sets associated with multivariate stable distributions I: Moments and densities. J. Multiv. Anal. **100**(10), 2195–2213 (2009)
Mosler, K: Multivariate Dispersion, Central, Regions and Depth: The Lift Zonoid Approach. Springer, New York (2002)
Mosler, K: Depth statistics (2013). http://arxiv.org/pdf/1207.4988.pdf
Moszyńska, M, Richter, W-D: Reverse triangle inequality. Antinorms and semi-antinorms. Studia Scientiarum Mathematicarum Hungarica. **49**(1), 120–138 (2012)
Muirhead, RJ: Aspects of Multivariate Statistical Theory. Wiley, New York (1982)
Nachbin, L: The Haar Integral. New York (1976)
Naor, A: The surface measure and cone measure on the sphere of l_p^n. Trans. Am. Math. Soc. **359**(3), 1045–1080 (2007)
Nardon, M, Pianca, P: Simulation techniques for generalized Gaussian densities. J. Stat. Comp. Simul. **11**, 1317–1329 (2009)
Ng, K-W, Tian, G: Characteristic functions of l_1-spherical and l_1-norm symmetric distributions and their applications. J. Multiv. Anal. **76**(2), 192–213 (2001)
Nolde, N: Geometric interpretation of the residual dependence coefficient. J. Multiv. Anal. **123**, 85–95 (2014)
Osiewalski, J, Steel, MFJ: Robust Basian inference in l_q-spherical models. Biometrika. **80**(2), 456–460 (1993)
Pewsey, A, Neuhäuser, M, Graeme, DR: Circular Statistics in R. Oxford (2013)
Pogány, TK, Nadarajah, S: On the characteristic function of the generalized normal distribution. C.R. Math., Acad. Sci. Paris. **203**(3), 203–206 (2010)
Rachev, ST, Rüschendorf, L: Approximate independence of distributions on spheres and their stability properties. Ann. Probab. **19**(3), 1311–1337 (1991)
Richter, W-D: Laplace Gauss integrals, Gaussian measure asymptotic behaviour and probabilities of moderate deviations. Z. Anal. Anwendungen. **4**(3), 257–267 (1985)
Richter, W-D: The Laplace integral and probabilities of moderate deviations in R^k. (In Russian). Translation: J. Math. Sci. **38**(6), 2391–2397 (1987)
Richter, W-D: Eine geometrische Methode in der Stochastik. Rostocker Mathematisches Kolloquium. **44**, 63–72 (1991)
Richter, W-D: Generalized spherical and simplicial coordinates. J. Math. Anal. Appl. **336**, 1187–1202 (2007)
Richter, W-D: Continuous $l_{n,p}$-symmetric distributions. Lithuanian Math. J. **49**, 93–108 (2009)
Richter, W-D: On the ball number function. Lithuanian Math. J. **51**(3), 440–449 (2011)
Richter, W-D: Circle numbers for star discs. ISRN Geometry. (16) (2011a). Article ID 479262
Richter, W-D: Ellipses numbers and geometric measure representations. J. Appl. Anal. **17**, 165–179 (2011b)
Richter, W-D: Exact distributions under non-standard model assumptions. AIP Conf. Proc. **1479**(442) (2012). doi:10.1063/1.4756160
Richter, W-D: Geometric and stochastic representations for elliptically contoured distributions. Commun. Stat.: Theory Methods. **42**(4), 579–602 (2013)
Richter, W-D: Classes of standard Gaussian random variables and their generalizations. In: Knif, J, Pape, B (eds.) Contributions to Mathematics, Statistics, Econometrics, and Finance Essays in Honour of Professor Seppo Pynnönen. Acta Wasaensia 296, pp. 53–69. University of Vaasa, Vaasa, (2014)
Richter, W-D, Venz, J: Geometric representations of multivariate skewed elliptically contoured distributions. Chilean J. Stat. **5**(2), 20 (2014)
Sasvári, Z: Multivariate Characteristic and Correlation Functions. De Gruyter, Berlin (2013)
Schechtman, G, Zinn, J: On the volume of the intersection of two l_p^n balls. Proc. Am. Math. Soc. **110**(1), 217–224 (1990)
Schindler, W: Measures with Symmetry Properties. Springer, Berlin-Heidelberg (2003)
Schoenberg, IJ: Metric spaces and completely monotone functions. Ann. Math. **39**, 811–841 (1938)
Scholz, SP: Robustheitskonzepte und -untersuchungen für Schätzer konvexer Körper. Dissertation, Universität Dortmund (2002)
Song, D, Gupta, AK: l_p-norm uniform distributions. Proc. Am.Math. Soc. **125**(2), 595–601 (1997)
Szablowski, PJ: Uniform distributions on spheres in finite dimensional l_α and their generalizations. J. Multivariate Anal. **64**, 103–107 (1998)
Takemura, A, Kuriki, S: Theory of cross sectionally contoured distributions and its applications. Discussion, Paper Series 96-F-15. Faculty of Economics, University of Tokyo (1996). http://www.stat.t.u-tokyo.ac.jp/~takemura/papers/dp96f15.pdf
von Mises, R: Über die Ganzzahligkeit der Atomgewichte und verwandte Fragen. Physikalische Zeitschrift. **19**, 490–500 (1918)
Wijsman, RA: Cross-sections of orbits and their application to densities of maximal invariants. In: Proc. Fifth Berkeley, Symp. on Math.Statist. and Prob., vol. 1, pp. 389–400. University of California Press, Berkeley and Los Angeles, (1967)
Wijsman, RA: Global cross sections as a tool for factorization of measures and distribution of maximal invariant. Sankhya. **48A**, 1–42 (1986)
Wijsman, R A: Invariant Measures on Groups and Their Use in Statistics, Hayward (1990)

7

Bounds on the mean residual lifetime of progressive type II right censored order statistics

Mohammad Z Raqab

Correspondence:
mraqab@stat.kuniv.edu
Department of Statistics and Operations Research, Kuwait University, P.O. Box 5969 Safat, 13060 Kuwait City, Kuwait

Abstract

In this paper, we present some sharp upper bounds on the deviations of the mean residual lifetime of progressive type II right censored order statistics from the mean residual lifetime, $m(t) = E(X_t) = E(X - t|X > t)$, for arbitrary $t > 0$. We also describe the distributional forms for which the bounds are attained. The obtained bounds are numerically evaluated and compared with other classical rough ones.

Keywords: Cauchy-Schwarz inequality; Sharp bounds; Characterization; Greatest convex minorant approximation

AMS subject classifications: 62G30; 62E15

1 Introduction

The scheme of progressive type II censored sampling is an important scheme in life-testing experiments. The experimenter can remove units from a life test at various stages during the experiments, possibly resulting in a saving of cost and of time (see Sen 1986).

Suppose N units are randomly placed on a life test; at the first failure time of one of the units, a number of R_1 surviving units is randomly withdrawn from the test; at the second failure time, R_2 surviving units are selected at random and taken out of the experiment, and so on; finally, at the time of the nth failure, the remaining $R_n = N - R_1 - \ldots - R_{n-1} - n$ objects are removed. This type of schemes is applied in clinical trials. In the context of life-testing, suppose that $X^{\tilde{R}}_{1:n} \leq \ldots \leq X^{\tilde{R}}_{n:n}$ are the lifetimes of the completely observed units to fail, and that $\tilde{R} = (R_1, \ldots, R_n)$ represents the number of units withdrawn at these failure times. The model of ordinary order statistics is contained in the above set-up by choosing $\tilde{R} = (0, \ldots, 0)$(that is $n = N$) as censoring scheme, where no withdrawals are made. The type II censored order statistics are obtained by setting $R_1 = \ldots = R_{n-1} = 0$ and $R_n = N - n$. For more details on progressive censoring, one can refer to Balakrishnan and Aggarwala (2000) and references therein.

Let X be a lifetime variable with cumulative distribution function (cdf) F, probability density function (pdf) f and finite first and second moments. Further, let $t = F^{-1}(p) > 0$

$(0 \le p < 1)$ be the 100pth percentile of F. The mean and variance of the residual lifetime (MRL) r.v. X_t are defined as

$$m(t) = \mathbb{E}(X_t) = \mathbb{E}(X - t|X > t) = \frac{1}{1-p}\int_p^1 \left[F^{-1}(u) - t\right] du,$$

$$\sigma^2(t) = \mathbb{E}_F(X_t - m(t))^2 = \frac{1}{1-p}\int_p^1 \left[F^{-1}(u) - t - m(t)\right]^2 du.$$

The mean residual lifetime $m(t)$ is of interest in many fields such as reliability, survival analysis, actuarial studies, etc. It plays an important role in studying the conditional tail measure of the lifetime data. In fact, $m(t)$ can be considered as the conditional tail measure given that an object did not fail in $(0, t)$. In actuarial studies, the insurer might ignore the losses below the deductible $t > 0$ since they do not result in insurance payment (insurer might not be aware of the losses below t). That is, the mean excess loss function or MRL function $m(t)$ does arise naturally. It is well-known that the MRL function $m(t)$ characterizes the distribution function F uniquely (see, for example, Kotz and Shanbhag (1980)). Indeed, when X is non-negative, then for $t > 0$ we have

$$\overline{F}(t) = \frac{m(0)}{m(t)} \exp\left(-\int_0^t \frac{1}{m(x)} dx\right).$$

In the spirit of order statistics, Li and Chen (2004) have discussed the aging properties of the residual life length of m-out-of-n system with independent (not necessarily identical) components given that $(n-m)$th failure has occurred at time $t > 0$. The aging properties of the parallel system have been studied by Abouammoh and El-Neweihi (1986). Poursaeed and Nematollahi (2008) have studied the mean past and mean residual life functions of the components of parallel system under double monitoring. Recently, Hashemi et al. (2010) have studied some properties of the residual lifetime of progressively type II right censored order statistics (PCOSs). For comprehensive review and applications of the mean residual function, we refer, for example, to Guess and Proschan (1988), Bairamov et al. (2002), Asadi and Bayramoglu (2005) and Asadi and Bayramoglu (2006).

In fact, there are two concepts for the MRL of PCOSs appeared in the literature. The first one represents the average of remaining waiting time after a certain time $t > 0$, to the failure of the sth item from progressively type II censored sample. Precisely,

$$M_{s,n}^{\tilde{R}}(t) = \mathbb{E}\left(X_{s:n:N}^{\tilde{R}} - t|X_{s:n:N}^{\tilde{R}} > t\right). \tag{1}$$

Another related concept of the MRL of PCOSs (Hashemi et al. (2010)) can be defined as follows. Under the condition that exactly r items, have failed at or before time t, the MRL of the sth failure time, for $0 \le r < s \le n \le N$, is

$$M_{s,n}^{*\tilde{R}}(t) = \mathbb{E}\left(X_{s:n:N}^{\tilde{R}} - t|X_{r:n:N}^{\tilde{R}} \le t < X_{r+1:n:N}^{\tilde{R}}\right). \tag{2}$$

It should be noticed here that these definitions of MRL are different from the ordinary definition of $m(t)$. The bounds for the moments of order statistics appeared frequently in the literature in the last two decades. Arnold (1985) and Rychlik (1993) presented more general sharp bounds for the maximum and arbitrary combination of order statistics, respectively, of possibly dependent samples. Raqab (2003) developed p-norm bounds for the moments of PCOSs, measured in scale units generated by absolute moments of the parent distribution of a single observation. Recently, Raqab and Rychlik (2011) developed sharp bounds for the MRL of a m-out-of-n system using the greatest convex minorant approximation combined with the Cauchy-Schwarz inequality. Balakrishnan et al. (2001) applied several methods to derive different bounds for means and variances of PCOSs.

In this paper, we establish sharp bounds for the deviations of the MRL $M^{\tilde{R}}_{s,n}(t)$ of PCOSs from the MRL $m(t)$ of X for fixed $t > 0$. The so obtained bounds here are derived based on L^2- projection approach combined with the Cauchy-Schwarz inequality. These bounds can be considered as optimal estimates of the residual lifetime of PCOSs. Our work is quite general in nature as the bounds expressed in terms of $m(t)$ and $\sigma(t)$ units and the bounds are valid for all parent distribution functions F such that $F^{-1}(t) = p, 0 < p < 1$. Evaluations of the resulting bounds for various choices schemes are also presented and compared with other classical ones.

2 Bounds for MRL of PCOSs

Let $X_1, \cdots, X_n$ be the iid random variables (r.v.'s) with cdf F, pdf f, and finite first and second moments. Assume that $t = F^{-1}(p)$ $(0 \leq p < 1)$ is the 100pth percentile of F. Now, consider the deviation of the MRL of PCOSs from the MRL function of the parent r.v. X_1 as

$$\begin{aligned} M^{\tilde{R}}_{s,n}(t) - m(t) &= \mathbb{E}\left(X^{\tilde{R}}_{s:n:N} - t | X^{\tilde{R}}_{s:n:N} > t\right) - \mathbb{E}_F(X_1 - t | X_1 > t) \\ &= \int_p^1 \left[F^{-1}(u) - t\right] \frac{f^{\tilde{R}}_{s:n:N}(u)}{1 - F^{\tilde{R}}_{s:n:N}(p)} du - \int_p^1 \left[F^{-1}(u) - t\right] \frac{du}{1-p} \\ &= \int_p^1 \left[F^{-1}(u) - t - m(t)\right] \left[\frac{f^{\tilde{R}}_{s:n:N}(u)}{1 - F^{\tilde{R}}_{s:n:N}(p)} - \frac{1}{1-p}\right] du, \end{aligned} \tag{3}$$

where $f^{\tilde{R}}_{r:n:N}$ and $F^{\tilde{R}}_{r:n:N}$ are the pdf and cdf of the *rth* PCOS from standard uniform distribution over unit interval $[0, 1]$ of the forms:

$$\begin{aligned} f^{\tilde{R}}_{r:n:N}(u) &= C_{r-1} \sum_{i=1}^{r} a_{i,r}(1-u)^{\gamma_i - 1}, \ 1 \leq r \leq n, \\ F^{\tilde{R}}_{r:n:N}(u) &= 1 - C_{r-1} \sum_{i=1}^{r} \frac{a_{i,r}}{\gamma_i}(1-u)^{\gamma_i}, \ 1 \leq r \leq n, \end{aligned}$$

with

$$C_{r-1} = \prod_{i=1}^{r} \gamma_i; \ \gamma_i = N - \sum_{j=1}^{i-1} R_j - i + 1; \gamma_1 = N, \ \text{and} \ a_{i,r} = \prod_{\substack{j=1 \\ j \neq i}}^{r} (\gamma_j - \gamma_i)^{-1},$$

(cf., e.g., Kamps and Cramer 2001). The following lemma describe the variability of functions $f^{\tilde{R}}_{s:n:N}(u)$ in $[0, 1]$.

Lemma 1 (Balakrishnan et al. (2001)). *For $n \geq 2$, $f^{\tilde{R}}_{1:n:N}(u)$ is decreasing with $f^{\tilde{R}}_{1:n:N}(0) = \gamma_1$ and $\lim_{u \nearrow 1} f^{\tilde{R}}_{1:n:N}(u) = 0$. For $\gamma_n = 1$, $f^{\tilde{R}}_{n:n:N}(u)$ is increasing on $(0, 1)$ with $f^{\tilde{R}}_{n:n:N}(0) = 0$ and $\lim_{u \nearrow 1} f^{\tilde{R}}_{n:n:N}(u) = \prod_{j=1}^{n-1} \frac{\gamma_j}{\gamma_j - 1}$. If either $(1 < s < n)$ or $(s = n \geq 2, \gamma_n > 1)$, $f^{\tilde{R}}_{s:n:N}(u)$ is first increasing starting from 0 at 0, then ultimately decreasing to 0 at 1. Moreover, each $f^{\tilde{R}}_{s:n:N}(u)$ has a unique maximum in $(0,1)$ at $\varepsilon_{s,n,N} > 0$ satisfying*

$$\sum_{i=1}^{s} a_{i,s}(\gamma_i - 1)(1 - \varepsilon_{s,n,N})^{\gamma_i - 2} = 0. \tag{4}$$

The following lemma describes the Moriguti's inequality obtained by projecting the integrable function h onto the convex cone of non-decreasing functions in $L^2([0,1], du)$.

Lemma 2 (Projection of $f^{\tilde{R}}_{s:n:N}(u)$). *For either* $(1 < s < n)$ *or* $(s = n \geq 2, \gamma_n > 1)$, *there exists a unique* $u_{s,n,N} \in (p,1)$ *defined as the solution to equation*

$$\sum_{i=1}^{s} a_{i,s}\left(\frac{\gamma_i - 1}{\gamma_i}\right)(1-u)^{\gamma_i} = 0, \quad p < u < 1, \tag{5}$$

such that for

$$\widehat{f}^{\tilde{R}}_{s:n:N}(u) = \begin{cases} f^{\tilde{R}}_{s:n:N}\left(\min\left\{u, u_{s,n,N}\right\}\right), & \text{if } 0 < p < u_{s,n,N}, \\ \frac{1-F^{\tilde{R}}_{s:n:N}(p)}{1-p} & \text{if } u_{s,n,N} \leq p < 1, \end{cases}$$

and every non-decreasing function $w \in L^1\left(\left[p,1\right], du\right)$, *we have*

$$\int_p^1 w(u) f^{\tilde{R}}_{s:n:N}(u)\, du \leq \int_p^1 w(u) \widehat{f}^{\tilde{R}}_{s:n:N}(u)\, du, \tag{6}$$

with the equality iff

$$w(u) = const, \quad u_{s,n,N} < u < 1. \tag{7}$$

∇

Proof. The result of the above lemma is specifically based on a combination of the statement of Lemma 1 with Moriguti (1953), (Lemma 1). It is enough to show that $\widehat{f}^{\tilde{R}}_{s:n:N}(u)$ is the derivative of the greatest convex minorant (GCM) $\widehat{F}^{\tilde{R}}_{s:n:N}(u)$ of the antiderivative $F^{\tilde{R}}_{s:n:N}(u)$ of $f^{\tilde{R}}_{s:n:N}(u)$. By Lemma 1, the function $F^{\tilde{R}}_{s:n:N}$ is convex increasing on $[0, \varepsilon_{r,n,N})$, and concave increasing on $(\varepsilon_{r,n,N}, 1]$. Therefore, the GCM of the antiderivative $F^{\tilde{R}}_{s:n:N}(u) - F^{\tilde{R}}_{s:n:N}(p)$ of $f^{\tilde{R}}_{s:n:N}(u), p < u < 1$, is defined as the supremum of all convex functions dominated by $F^{\tilde{R}}_{s:n:N}(u) - F^{\tilde{R}}_{s:n:N}(p), p < u < 1$ and its form is defined as follows:

$$\widehat{F}^{\tilde{R}}_{s:n:N}(u) = \begin{cases} F^{\tilde{R}}_{s:n:N}(u) - F^{\tilde{R}}_{s:n:N}(p), & \text{if } p < u \leq u_{s,n,N}, \\ F^{\tilde{R}}_{s:n:N}(u_{s,n,N}) - F^{\tilde{R}}_{s:n:N}(p) & \text{if } u_{s,n,N} < u < 1, \\ +f^{\tilde{R}}_{s:n:N}(u_{s,n,N})(u - u_{s,n,N}), & \end{cases}$$

for a unique $u = u_{s,n,N} \in (p, \varepsilon_{s,n,N}) > 0$ satisfying

$$1 - F^{\tilde{R}}_{s:n:N}(u) = f^{\tilde{R}}_{s:n:N}(u)(1-u).$$

or equivalently Eq. (5). If $p \geq u_{s,n,N}$, then

$$\widehat{F}^{\tilde{R}}_{s:n:N}(u) = \frac{1 - F^{\tilde{R}}_{s:n:N}(p)}{1-p}(u-p), \quad p < u < 1.$$

Differentiating the function $\widehat{F}^{\tilde{R}}_{s:n:N}$, we complete the proof. Function $\widehat{f}^{\tilde{R}}_{s:n:N}(u)$ is called the projection of $f^{\tilde{R}}_{s:n:N}(u)$ onto the convex cone of nondecreasing functions in $L^2([0,1], du)$.

In fact, the deviation $M^{\tilde{R}}_{s,n,N}(t) - m(t)$ in (3) can be evaluated using the inequalities of the integral of the product of two proportional functions. The Moriguti projection method (see Moriguti (1953)) can be used effectively in developing improved bounds of the statistical functional over general families of distributions. Applying the Cauchy-Schwarz inequality to (3), we have for either $(1 < s < n)$ or $(s = n \geq 2, \gamma_n > 1)$,

$$M^{\tilde{R}}_{s,n,N}(t) - m(t) \leq \left\{ \int_p^1 \left[F^{-1}(u) - t - m(t)\right]^2 \frac{du}{1-p} \right\}^{1/2} \times \left\{ (1-p) \int_p^1 \left[\frac{f^{\tilde{R}}_{s:n:N}(u)}{1 - F^{\tilde{R}}_{s:n:N}(p)} - \frac{1}{1-p} \right]^2 du \right\}^{1/2} = \sigma(t)\, B^{\tilde{R}}_0(s,n,N;p). \tag{8}$$

Generally, the right hand side of (8) is not monotonic and then the bound $B^{\tilde{R}}_0(s,n,N;p)$ is not attainable. By (3), (6), and the Cauchy-Schwarz inequality, we get

$$M^{\tilde{R}}_{s,n,N}(t) - m(t) \leq \int_p^1 \left[F^{-1}(u) - t - m(t)\right] \left[\frac{\widehat{f}^{\tilde{R}}_{s:n:N}(u)}{1 - F^{\tilde{R}}_{s:n:N}(p)} - \frac{1}{1-p} \right] du, \leq B^{\tilde{R}}_1(s,n,N;p)\, \sigma(t), \tag{9}$$

with

$$B^{\tilde{R}}_1(s,n,N;p) = \left\{ (1-p) \int_p^{u_{s,n,N}} \frac{\left[f^{\tilde{R}}_{s:n:N}(u)\right]^2}{\left[1 - F^{\tilde{R}}_{s:n:N}(p)\right]^2} + \frac{(1-p)(1-u_{s,n,N})\left[f^{\tilde{R}}_{s:n:N}(u)\right]^2}{\left[1 - F^{\tilde{R}}_{s:n:N}(p)\right]^2} - 1 \right\}^{1/2}. \tag{10}$$

The latter inequality in Eq. (9) becomes equality iff

$$F^{-1}(u) - t - m(t) = \alpha \left(\frac{\widehat{f}^{\tilde{R}}_{s:n:N}(u)}{1 - F^{\tilde{R}}_{s:n:N}(p)} - \frac{1}{1-p} \right), \quad p \leq u < 1, \quad \text{for some } \alpha > 0. \tag{11}$$

Note that for $p < u_{s,n,N}$, the right-hand side is nonnegative and nondecreasing as desired. From (7), the former inequality of (9) becomes equality iff the right-hand side of Eq. (11) is a constant on $(u_{s,n,N}, 1)$. The moment condition

$$\int_p^1 \left[F^{-1}(u) - t - m(t)\right]^2 \frac{du}{1-p} = \sigma^2(t),$$

forces $\alpha = (1-p)\sigma(t)/B^{\tilde{R}}_1(s,n,N;p)$. Plugging it in Eq. (11) and using the fact that $F^{-1}(u) \leq t,\ 0 < u < p$, we readily form the cdf that attains the bound in (10). The non-negativity jump at t leads to the additional condition given below in (13). Thus we have proven the following theorem. □

Theorem 1. *For a parent distribution F with finite mean μ, variance σ^2 and either $(1 < s < n)$ or $(s = n \geq 2, \gamma_n > 1)$ with $p < u_{s,n,N}$, we have*

$$\frac{M^{\tilde{R}}_{s,n,N}(t) - m(t)}{\sigma(t)} \leq B^{\tilde{R}}_1(s,n,N;p), \tag{12}$$

where $B^{\tilde{R}}_1(s,n,N;p)$ is defined in Eq. (10) with unique $u_{s,n,N} \in (p, \varepsilon_{s,n,N})$ solving Eq. (5). Under the additional condition

$$m(t) \geq \left(1 - \frac{(1-p) f^{\tilde{R}}_{s:n:N}(p)}{1 - F^{\tilde{R}}_{s:n:N}(p)}\right) \frac{\sigma(t)}{B^{\tilde{R}}_1(s,n,N;p)}, \tag{13}$$

the equality in (12) *is attained in limit by the distribution function*

$$F(x) = \begin{cases} F_0(x), & \text{if } x < t, \\ p, & \text{if } t \le x < b_1, \\ f_{s:n:N}^{\tilde{R}-1}\left(\frac{1-F_{s:n:N}^{\tilde{R}}(p)}{1-p}\left[B_1^{\tilde{R}}(s,n,N;p)\left(\frac{x-t-m(t)}{\sigma(t)}\right)+1\right]\right), & \text{if } b_1 \le x < b_2, \\ 1, & \text{if } x \ge b_2, \end{cases} \quad (14)$$

where F_0 *is an arbitrary continuous function with* $F_0(t) = p$, *and*

$$b_1 = t + m(t) - \left(1 - \frac{(1-p) f_{s:n:N}^{\tilde{R}}(p)}{1 - F_{s:n:N}^{\tilde{R}}(p)}\right) \frac{\sigma(t)}{B_1^{\tilde{R}}(s,n,N;p)},$$
$$b_2 = t + m(t) + \left(\frac{(1-p) f_{s:n:N}^{\tilde{R}}(u_{s,N,p})}{1 - F_{s:n:N}^{\tilde{R}}(p)} - 1\right) \frac{\sigma(t)}{B_1^{\tilde{R}}(s,n,N;p)}.$$

∇

Remark 1. Distribution function (14) is a location-scale family of distributions consisting any arbitrary continuous function supported on the left of t with probability p, constant supported on (t, b_1), has a smooth component on finite support $[b_1, b_2]$, and one atom of measure $(1 - u_{s,n,N})$ at the right end of the support interval. By the monotonicity of $f_{s:n:N}^{\tilde{R}}(u)$ on $(p, u_{s,n,N})$, then the right-hand side of (13) is positive. The arbitrary distribution function F_0 with $F_0(t) = p$ defined on the left tail can be assumed in all other distributions attaining the bounds for the MRL function of PCOSs. In a consequence of that, the distribution function F is not absolutely continuous function. However, it can be approximated by sequences of absolutely continuous functions attaining the bound asymptotically.

Let us consider the extreme progressive type II censored order statistics (i) ($s = n$, $\gamma_n = 1$) (ii) $s = 1$ (iii) $1 < s < n$ or ($s = n \ge 2, \gamma_n > 1$) with $p \ge u_{s,n,N}$. In the former case, it is easily checked that $\left[F_{n:n:N}^{\tilde{R}}(u)\right]'' = (1-u)^{-1} f_{n-1:n:N}^{\tilde{R}}(u) \ge 0$, and then the antiderivative $F_{s:n:N}^{\tilde{R}}(u)$ of $f_{s:n:N}^{\tilde{R}}(u)$ is convex on $[0, 1]$. Therefore, the projection of $f_{s:n:N}^{\tilde{R}}(u)$ is the function itself. Arguments similar to those in the above lines provide the corresponding bound and the distribution function for which the bound is attained. For the second case, the pdf is decreasing and its anti-derivative is concave for $(p, 1)$. Therefore, the GCM for the latter two cases amounts to the linear function

$$\widehat{F}_{1:n:N}^{\tilde{R}}(u) = (1-p)^{N-1}(u-p), \; p < u < 1,$$

and then the projection of $f_{s:n:N}^{\tilde{R}}(u)$ is just constant. This turns out that the right-hand side of the 2nd inequality amounts to 0. In this case, Moriguti's method does not lead to an improved bound and the equality becomes equality iff $F^{-1}(u) - t - m(t)$ is constant on $(p, 1)$. Combining this with the moment condition and the fact that $F(t) = p$, we immediately obtain the distribution function attaining the bound. The so obtained bounds for the extreme cases and their respective marginal distributions are summarized in two subsequent theorems.

Theorem 2. *For $\gamma_n = 1$, the bound on $\left(M^{\tilde{R}}_{n,n,N}(t) - m(t)\right)/\sigma(t)$ is*

$$B^{\tilde{R}}_1(n,n,N;p) = \left\{(1-p)\int_p^1 \frac{\left[f^{\tilde{R}}_{n:n:N}(u)\right]^2}{\left[1-F^{\tilde{R}}_{n:n:N}(p)\right]^2} - 1\right\}^{1/2}.$$

Under Condition (13) with $(s = n, \gamma_n = 1)$, the bound is attained in the limit by continuous distribution functions converging to the following cdf

$$F(x) = \begin{cases} F_0(x), & \text{if } x < t, \\ p, & \text{if } t \le x < c_1, \\ f^{\tilde{R}-1}_{n:n:N}\left(\frac{1-F^{\tilde{R}}_{n:n:N}(p)}{1-p}\left[B^{\tilde{R}}_1(n,n,N;p)\left(\frac{x-t-m(t)}{\sigma(t)}\right)+1\right]\right), & \text{if } c_1 \le x < c_2, \\ 1, & \text{if } x \ge c_2, \end{cases} \tag{15}$$

where F_0 is an arbitrary continuous function with $F_0(t) = p$, and

$$c_1 = t + m(t) - \left(1 - \frac{(1-p)f^{\tilde{R}}_{n:n:N}(p)}{1-F^{\tilde{R}}_{n:n:N}(p)}\right)\frac{\sigma(t)}{B^{\tilde{R}}_1(n,n,N;p)},$$

$$c_2 = t + m(t) + \left(\frac{(1-p)\prod_{j=1}^{n-1}\frac{\gamma_j}{\gamma_j - 1}}{1-F^{\tilde{R}}_{n:n:N}(p)} - 1\right)\frac{\sigma(t)}{B^{\tilde{R}}_1(n,n,N;p)}.$$

∇

Theorem 3. *If either $s = 1$ or $(1 < s < n)$ with $p \ge u_{s,n,N}$) or $(s = n \ge 2, \gamma_n > 1)$ with $p \ge u_{s,n,N}$, then the bound on $\left(M^{\tilde{R}}_{s,n,N}(t) - m(t)\right)/\sigma(t)$ is 0, which is attained in the limit by continuous distribution function converging to the cdf of the form*

$$F(x) = \begin{cases} F_0(x), & \text{if } x < t, \\ p, & \text{if } t \le x < t + m(t), \\ 1, & \text{if } x \ge t + m(t). \end{cases} \tag{16}$$

∇

Now we use the integrand maximization-based approach to derive another bound (Papadatos 1997). This bound depends only $m(t)$ and then the previous assumption of finiteness of the second moment can be removed here.

The MRL of PCOSs can be rewritten as

$$\begin{aligned} M^{\tilde{R}}_{s,n,N}(t) &= \mathbb{E}\left(X^{\tilde{R}}_{s:n:N} - t | X^{\tilde{R}}_{s:n:N} > t\right) \\ &= \int_t^\infty \frac{1-F^{\tilde{R}}_{s:n:N}(F(x))}{1-F^{\tilde{R}}_{s:n:N}(p)} dx \\ &= \frac{1-p}{1-F^{\tilde{R}}_{s:n:N}(p)} \int_{p<F(x)<1} \frac{1-F^{\tilde{R}}_{s:n:N}(F(x))}{1-F(x)} \frac{1-F(x)}{1-p} dx \\ &\le \frac{1-p}{1-F^{\tilde{R}}_{s:n:N}(p)} \sup_{p<u<1}\left(\frac{1-F^{\tilde{R}}_{s:n:N}(u)}{1-u}\right) m(t). \end{aligned} \tag{17}$$

The function to be maximized in (17) represents the slopes of the straight lines joining the points of the graph of $F^{\tilde{R}}_{s:n:N}$ at u and 1. It increases on $(0, u_{s,n,N})$ and then decreases on $(u_{s,n,N}, 1)$ with a maximal value at $u = u_{s,n,N}$. In the spirit of these lines, we have

$$\sup_{p<u<1}\left(\frac{1-F^{\tilde{R}}_{s:n:N}(u)}{1-u}\right) = \begin{cases} (1-p)^{N-1}, & \text{if } s=1,\\ \prod_{i=1}^{n-1}\frac{\gamma_i}{\gamma_i-1}, & \text{if } (s=n, \gamma_n=1)\\ f^{\tilde{R}}_{s:n:N}(u_{s,n,N}), & \text{if } (1<s<n) \text{ or } (s=n\geq 2, \gamma_n>1)\\ & \quad\text{with } p<u_{s,n,N}\\ \frac{1-F^{\tilde{R}}_{s:n:N}(p)}{1-p}, & \text{if } (1<s<n) \text{ or } (s=n\geq 2, \gamma_n>1)\\ & \quad\text{with } p\geq u_{s,n,N}. \end{cases}$$

Plugging the extreme values into (17), we readily establish the new bound. For either $(1 < s < n)$ or $(s = n \geq 2, \gamma_n > 1)$ with $p < u_{s,n,N}$, the inequality becomes equality if $F(x) = u_{s,n,N}$ or $F(x) = 1$ for $x > t$. That is, we have two jumps $(u_{s,n,N} - p)$ and $(1 - u_{s,n,N})$ at t and some point $\eta > t$, respectively. Using the moment condition, the value of η is found to be

$$\eta = t + \frac{1-p}{1-u_{s,n,N}}\, m(t).$$

If either $(1 < s < n)$ with $p \geq u_{s,n,N}$ or $(s = n \geq 2, \gamma_n > 1)$ with $p \geq u_{s,n,N}$ or $s = 1$, the bounds here are the same as the ones in Theorem 3. So the attainability conditions in Theorem 3 are applied here. For $(s = n, \gamma_n = 1)$, we should take the distribution that have values $F(x)$ close to 1 if $x > t$ and $F(x) < 1$. We may assume $F(x) = 1 - \vartheta$ for $\{x : p < F(x) < 1\}$. The moment condition implies that the distribution function has the jump from $1 - \vartheta$ to 1 at $\eta = t + \frac{1-p}{\vartheta} m(t)$. The bounds obtained as well as the marginal distributions attaining these bounds are described in Theorem 4 below.

Theorem 4. *For any continuous parent distribution F having a finite mean μ, we have for every point $t > 0$ from the support of F,*

$$\frac{M^{\tilde{R}}_{s,n,N}(t)}{m(t)} \leq B^{\tilde{R}}_2(s,n,N;p)$$
$$= \begin{cases} \frac{(1-p)}{1-F^{\tilde{R}}_{s:n:N}(p)}\prod_{i=1}^{n-1}\frac{\gamma_i}{\gamma_i-1}, & \text{if } (s=n, \gamma_n=1),\\ \frac{(1-p)f^{\tilde{R}}_{s:n:N}(u_{s,n,N})}{1-F^{\tilde{R}}_{s:n:N}(p)}, & \text{if } (1<s<n) \text{ or } (s=n\geq 2, \gamma_n>1)\\ & \quad\text{with } p<u_{s,n,N},\\ 1, & \text{if } s=1 \text{ or } (1<s<n) \text{ with } p\geq u_{s,n,N}\\ & \quad\text{or } (s=n\geq 2, \gamma_n>1) \text{ with } p\geq u_{s,n,N}. \end{cases} \tag{18}$$

The bounds are attained in the limit by continuous distribution functions converging to the following distribution functions. If $1 < s < n$ and $p < u_{s,n,N}$, the limiting distribution is of the form

$$F(x) = \begin{cases} F_0(x), & \text{if } x<t,\\ u_{s,n,N}, & \text{if } t\leq x< t+\frac{1-p}{1-u_{s,n,N}}m(t),\\ 1, & \text{if } x\geq t+\frac{1-p}{1-u_{s,n,N}}m(t). \end{cases} \tag{19}$$

If $(s = n, \gamma_n = 1)$, *then the bound is attained in limit by continuous approximations of the family of distribution functions*

$$F(x) = \begin{cases} F_0(x), & \text{if } x < t, \\ 1-\vartheta, & \text{if } t \leq x < t + \frac{1-p}{\vartheta}m(t), \\ 1, & \text{if } x \geq t + \frac{1-p}{\vartheta}m(t), \end{cases} \tag{20}$$

as ϑ *tends to 0. Finally, if either* $s = 1$ *or* $(1 < s < n)$ *with* $p \geq u_{s,n,N}$ *or* $(s = n \geq 2, \gamma_n > 1)$ *with* $p \geq u_{s,n,N}$, *then the limiting distribution is* (16).

∇

3 Numerical experiments and discussion

In this section, we have conducted a numerical study to evaluate the resulting bounds on the MRL of the PCOSs $\left(M^{\tilde{R}}_{s,n,N}(t) - m(t)\right)/\sigma(t)$. The bounds are evaluated for $s = 3, 5$, $n = 5$ and $N = 10, 20$ with different censoring schemes and $p = 0, 0.1, 0.2, 0.3$. Clearly, large values of p need not to be considered here since if p is getting large $(p \geq u_{s,n,N})$, the bounds are reduced to the trivial minimal values.

The numbers $\varepsilon_{s,n,N}$'s and $u_{s,n,N}$'s can be obtained by solving (4) and (5). The solutions of (4) and (5) are determined numerically by means of the Newton-Raphson method. The solutions are plugged in (10) and (18) and obtain the final results. For the extreme cases, the bounds do not depend on $u_{s,n,N}$'s and their evaluations can be performed directly. The results are presented in Tables 1 and 2. In Table 1, we compare the bounds $B^{\tilde{R}}_1(s, n, N; p)$ presented in Theorems 1, 2 and 3 with the rough ones $B^{\tilde{R}}_0(s, n, N; p)$. The bounds are identical for $(s = n, \gamma_n = 1)$ [See, for example, the bounds for $(N = 10, s = 5, \tilde{R} = (5, 0, 0, 0, 0))$ and $(N = 20, s = 5, \tilde{R} = (15, 0, 0, 0, 0))$]. The bounds combined with the Moriguti's method $B^{\tilde{R}}_1(n, n, N; p)$ compare very well with the classical bounds derived by direct application of the Cauchy-Schwarz inequality, $B^{\tilde{R}}_0(n, n, N; p)$. For fixed censoring

Table 1 Values of $B^{\tilde{R}}_0(s, 5, N; p)$ and $B^{\tilde{R}}_1(s, 5, N; p)$ for different censoring schemes and p

			$p = 0$		$p = 0.1$		$p = 0.2$		$p = 0.3$	
N	s	$\tilde{R}$	$B^{\tilde{R}}_0(.)$	$B^{\tilde{R}}_1(.)$	$B^{\tilde{R}}_0(.)$	$B^{\tilde{R}}_1(.)$	$B^{\tilde{R}}_0(.)$	$B^{\tilde{R}}_1(.)$	$B^{\tilde{R}}_0(.)$	$B^{\tilde{R}}_1(.)$
10	3	(2,0,0,0,3)	0.9689	0.2148	0.9389	0.0000	1.0455	0.0000	1.1663	0.0000
		(5,0,0,0,0)	0.6591	0.3130	0.5722	0.1235	0.5711	0.0009	0.6218	0.0000
		(1,1,1,1,1)	0.9976	0.2069	0.9750	0.0000	1.0924	0.0000	1.2170	0.0000
		(0,0,3,0,2)	1.1087	0.1848	1.1130	0.0000	1.2702	0.0000	1.4162	0.0000
10	5	(2,0,0,0,3)	0.8555	0.5342	0.7486	0.3976	0.6468	0.2287	0.6017	0.0628
		(5,0,0,0,0)	1.2341	1.2341	1.1273	1.1273	1.0103	1.0103	0.8830	0.8830
		(1,1,1,1,1)	0.7593	0.6325	0.6477	0.5105	0.5256	0.3651	0.4203	0.2102
		(0,0,3,0,2)	0.7957	0.5426	0.6865	0.4077	0.5800	0.2429	0.5253	0.0808
20	3	(5,4,3,2,1)	1.3773	0.1325	1.4928	0.0000	1.7098	0.0000	1.8312	0.0000
		(15,0,0,0,0)	0.6463	0.2671	0.5758	0.0629	0.6014	0.0000	0.6571	0.0000
		(3,3,3,3,3)	1.5385	0.1182	1.7214	0.0000	1.9888	0.0000	2.1359	0.0000
		(5,0,5,0,5)	1.5285	0.1199	1.7054	0.0000	1.9769	0.0000	2.1345	0.0000
20	5	(5,4,3,2,1)	0.6666	0.4871	0.5504	0.3407	0.4493	0.1700	0.4206	0.0312
		(15,0,0,0,0)	1.1842	1.1842	1.0780	1.0780	0.9618	0.9618	0.8363	0.8363
		(3,3,3,3,3)	0.9126	0.3631	0.8155	0.1675	0.8067	0.0000	0.8982	0.0000
		(5,0,5,0,5)	1.0145	0.3474	0.9206	0.1413	0.9340	0.0000	1.05954	0.0000

Table 2 Values of $B_2^{\tilde{R}}(s, 5, N; p)$ for different censoring schemes and p

		$B_2^{\tilde{R}}(s, 5, N; p)$				
N	s	$\tilde{R}$	$p = 0$	$p = 0.1$	$p = 0.2$	$p = 0.3$
10	3	(2,0,0,0,3)	1.0611	1.0000	1.0000	1.0000
		(5,0,0,0,0)	1.1355	1.0377	1.0000	1.0000
		(1,1,1,1,1)	1.0566	1.0000	1.0000	1.0000
		(0,0,3,0,2)	1.0448	1.0000	1.0000	1.0000
10	5	(2,0,0,0,3)	1.3572	1.2222	1.0992	1.0169
		(5,0,0,0,0)	4.4444	4.0001	3.5575	3.1231
		(1,1,1,1,1)	1.5260	1.3737	1.2282	1.1064
		(0,0,3,0,2)	1.3733	1.2366	1.1117	1.0249
20	3	(5,4,3,2,1)	1.0228	1.0000	1.0000	1.0000
		(15,0,0,0,0)	1.0994	1.0143	1.0000	1.0000
		(3,3,3,3,3)	1.0180	1.0000	1.0000	1.0000
		(5,0,5,0,5)	1.0185	1.0000	1.0000	1.0000
20	5	(5,4,3,2,1)	1.3062	1.1769	1.0676	1.0071
		(15,0,0,0,0)	4.2105	3.7896	3.3712	2.9631
		(3,3,3,3,3)	1.1607	1.0497	1.0000	1.0000
		(5,0,5,0,5)	1.1458	1.0374	1.0000	1.0000

scheme, as p increases, the bounds decrease in p, and become zero if $p \geq u_{s,n,n}$. It is not surprising to observe that while we can't see a regular behavior for $B_0^{\tilde{R}}(s, n, N; p)$, the Moriguti's bound increases when s increases.

Table 2 presents numerical values of bounds $B_2^{\tilde{R}}(n, n, N; p)$ of Theorem 4. The bounds of PCOSs here are represented in the scale units m(t). The values of the bounds take the maximized values when $p = 0$ and then decrease to the minimum value when p gets large. It is observed that the large bound occurs when $(s = n, \gamma_n = 1)$.

It is of interest to point out that the two estimates of the MRL of PCOSs $M_{s,n,N}^{\tilde{R}}(t)$ presented in Tables 1 and 2 are quite different. For comparison purposes, we have to rewrite both bounds as follows:

$$M_{s,n,N}^{\tilde{R}}(t) \leq B_{11}^{\tilde{R}}(s, n, N; p) = m(t) + B_1^{\tilde{R}}(s, n, N; p)\ \sigma(t),$$

$$M_{s,n,N}^{\tilde{R}}(t) \leq B_{22}^{\tilde{R}}(s, n, N; p) = B_2^{\tilde{R}}(s, n, N; p)\ m(t).$$

The former bound competes the latter one iff

$$m(t) \geq \frac{B_1^{\tilde{R}}(s, n, N; p)}{B_2^{\tilde{R}}(s, n, N; p) - 1}\ \sigma(t).$$

This means that the mean-variance bound $B_{11}^{\tilde{R}}(s, n, N; p)$ compares well with respect to $B_{22}^{\tilde{R}}(s, n, N; p)$ when the residual mean $m(t)$ is large with respect to the residual standard deviation. Let illustrate that via a specific example. We calculate bounds $B_{11}^{\tilde{R}}(s, n, N; p)$ and $B_{22}^{\tilde{R}}(s, n, N; p)$ in the cases $(m(t) = 0.5, \sigma(t) = 1)$ and $(m(t) = 3, \sigma(t) = 0.25)$. In the first case, we obtain $B_{11}^{\tilde{R}}(3, 5, 10; 0.1) = 0.6235 > B_{22}^{\tilde{R}}(3, 5, 10; 0.1) = 0.5189$, whereas for the latter case, we have $B_{11}^{\tilde{R}}(3, 5, 10; 0.1) = 3.0309 < B_{22}^{\tilde{R}}(3, 5, 10; 0.1) = 3.1133$. It assures the superiority of Moriguti projection method over maximization-based method

[Papadatos (1997)] when $m(t)$ is large with respect to $\sigma(t)$. The result is reversed when $m(t)$ tends to be small when compared to $\sigma(t)$.

4 Concluding remarks

In this paper, we have developed the maximized evaluations of the MRL function of PCOSs of order statistics, measured in location and scale units of the residual life random variable $X_t = (X - t|X > t)$ and expressed in terms of percentile points $t = F^{-1}(p), 0 < p < 1$. The projection based approach as well as the maximum functional based approach method are used to obtain two different bounds. It is shown that the mean-variance bound derived based on the projection method competes well when compared with the ordinary bounds obtained via Cauchy-Schwarz inequality. For comparing the two so obtained bounds, we recommend choosing the mean-variance bound if we can assume that the mean residual life is large with respect to the residual variance.

Similar results can be done for the second concept of the MRL of PCOSs, $M^{*\tilde{R}}_{s,n}(t) = \mathbb{E}\left(X^{\tilde{R}}_{s:n:N} - t|X^{\tilde{R}}_{r:n:N} \leq t < X^{\tilde{R}}_{r+1:n:N}\right)$. Specifically, the bounds for $M^{*\tilde{R}}_{s,n}(t)$ are identical to the optimal bounds for the unconditional expectation $E\left(X^{\tilde{S}}_{s-r:n-r:\gamma_{r+1}}\right)$, where $\tilde{S} = (R_{r+1}, \ldots, R_n)$ (see, Lemma 2.1, Hashemi et al. (2010)). Accordingly, the optimal bounds for $M^{*\tilde{R}}_{s,n}(t)$ will be free of the parameter p $(0 < p < 1)$. Therefore, the results for the alternative concept of MRL of PCOSs $M^{*\tilde{R}}_{s,n}(t)$ are similar to those of progressive type II censored order statistics established in Raqab (2003). While the bounds do not depend on p, the probability distributions characterized via these bounds are described in terms of p, $m(t)$, and $\sigma(t)$.

Acknowledgement
The author would like to thank the referees for their suggestions and comments on the first draft of this manuscript.

References

Abouammoh, A, El-Neweihi, E: Closure of the NBUE and DMRL classes under formation of parallel systems. Statist. Probab. Lett. **4**, 223–225 (1986)

Arnold, BC: *p*-Norm bounds on the expectation of the maximum of a possibly dependent sample. J. Multivariate Anal. **17**, 316–332 (1985)

Asadi, M, Bayramoglu, I: A note on the mean residual life function of a parallel system. Commun. Stat. Theory Methods **34**(2), 475–485 (2005)

Asadi, M, Bayramoglu, I: The mean residual life function of a *k*-out-of-*n* structure at the system level. IEEE Trans. Reliability **55**(2), 314–318 (2006)

Bairamov, I, Ahsanullah, M, Akhundov, I: A Residual life function of a system having parallel or series structures. J. Stat. Theory Appl. **1**(2), 119–132 (2002)

Balakrishnan, N, Aggarwala, R: Progressive Censoring. Birkhäuser, Boston (2000)

Balakrishnan, N, Cramer, E, Kamps, U: Bounds for means and variances of progressive type II censored order statistics. Statist. Probab. Lett. **54**, 301–315 (2001)

Guess, F, Proschan, F: Mean residual life: theory and applications. In: Krishnaiha, PR, Rao, CR (eds.) Handbook of Statistics, vol. 7, pp. 215–224. Elsevier, Netherlands, (1988)

Hashemi, M, Tavangar, M, Asadi, M: Some properties of the residual lifetime of progressively Type-II right censored order statistics. Statist. Probab. Lett. **80**, 848–859 (2010)

Kamps, U, Cramer, E: On distributions of generalized order statistics. Statistics **35**, 269–280 (2001)

Kotz, S, Shanbhag, DN: Some new approaches in probability distributions. Adv. Appl. Probability **12**, 903–921 (1980)

Li, X, Chen, J: Aging properties of the residual life of *k*-out-of-*n* systems with independent but non-identical components. Appl. Stochastic Models Business Ind. **20**, 143–153 (2004)

Moriguti, S: A Modification of Schwarz's inequality with applications to distributions. Ann. Math. Statist. **24**, 107–113 (1953)

Papadatos, N: Exact bounds for the expectations of order statistics from non-negative populations. Ann. Inst. Stat. Math. **49**(4), 727–736 (1997)

Poursaeed, MH, Nematollahi, AR: On the mean past and the mean residual life under double monitoring. Commun. Stat. Theory Meth. **37**, 1119–1133 (2008)

Raqab, MZ: P-Norm bounds for moments of progressive type II censored order statistics. Statist. Probab. Lett. **64**, 393–402 (2003)

Raqab, MZ, Rychlik, T: Bounds for the mean residual life function of a *k*-out-of-*n* system. Metrika **74**(3), 361–380 (2011)

Rychlik, T: Sharp bounds on *L*-estimates and their expectations for dependent samples. Commun. Statist. Theory Meth. **22**, 1053–1068 (1993)

Sen, PK: Progressive Censoring Schemes. In: Kotz, S, Johnson, NL (eds.) Encyclopedia of Statistical Sciences, vol. 7, pp. 296–299. Wiley, New York, (1986)

T-normal family of distributions: a new approach to generalize the normal distribution

Ayman Alzaatreh[1*†], Carl Lee[2†] and Felix Famoye[2†]

*Correspondence:
alzaatreha@apsu.edu
†Equal contributors
[1]Department of Mathematics and Statistics, Austin Peay State University, Clarksville, TN 37044, USA
Full list of author information is available at the end of the article

Abstract

The idea of generating skewed distributions from normal has been of great interest among researchers for decades. This paper proposes four families of generalized normal distributions using the T-X framework. These four families of distributions are named as T-normal families arising from the quantile functions of (i) standard exponential, (ii) standard log-logistic, (iii) standard logistic and (iv) standard extreme value distributions. Some general properties including moments, mean deviations and Shannon entropy of the T-normal family are studied. Four new generalized normal distributions are developed using the T-normal method. Some properties of these four generalized normal distributions are studied in detail. The shapes of the proposed T-normal distributions can be symmetric, skewed to the right, skewed to the left, or bimodal. Two data sets, one skewed unimodal and the other bimodal, are fitted by using the generalized T-normal distributions.

AMS 2010 Subject Classification: 60E05; 62E15; 62P10

Keywords: T-X distributions; Hazard function; Quantile function; Shannon entropy

1 Introduction

The normal distribution is perhaps the most commonly used probability distribution in both statistical theory and applications. The normal distribution was first used by de Moivre (1733) in the literature as an approximation to the binomial distribution. However, the development of the normal distribution by Gauss (1809, 1816) became the standard used in the modern statistics. Hence, the normal distribution is also commonly known as the Gaussian distribution. Properties of the normal distribution have been well developed (e.g., see Johnson et al. 1994; Patel and Read 1996). The distribution also plays an important role in generating new distributions.

Methods for developing generalized normal distributions seemed very limited until Azzalini (1985). A random variable X_λ is said to follow the skewed normal distribution $SN(\lambda)$ if the probability density function (PDF) of X_λ is $g(x|\lambda) = 2\phi(x)\Phi(\lambda x)$, where $\phi(x)$ and $\Phi(x)$ are $N(0,1)$ PDF and cumulative distribution function (CDF) respectively. Various extensions of $SN(\lambda)$ have been proposed and studied (e.g., Arellano-Valle et al. 2004; Arnold and Beaver 2002; Arnold et al. 2007; Choudhury and Abdul 2011; Balakrishnan 2002; Gupta and Gupta 2004; Sharafi and Behboodian 2008; Yadegari et al. 2008). For reviews on skewed normal and its generalization, one may refer to Kotz and Vicari (2005) and Lee et al. (2013). Pourahmadi (2007) showed that the skewed normal distribution

$SN(\lambda)$ approaches half-normal as $\lambda \to \infty$. This explains why skewed normal distribution is limited in fitting real data. In order to allow for fitting diverse magnitudes of skewness, various works have been done by introducing different methods to capture the magnitude of the skewness.

Fernández and Steel (1998) introduced a two-piece PDF as $g(x) = \begin{cases} cf(\alpha x), x \geq 0, \\ cf(x/\alpha), x < 0. \end{cases}$, $c > 0$ and $\alpha > 0$, where f is a symmetric PDF defined on $\Re$, which is unimodal and symmetric around 0. When f is normal, it is a generalized skewed normal. Kotz and Vicari (2005) suggested that α and $1/\alpha$ be replaced by α_1 and α_2 respectively, in order to have more flexibility of controlling skewness. Another general framework that introduces a skew mechanism to symmetric distributions was given by Ferreira and Steel (2006). The corresponding skew family is $g(x|f, q) = f(x)q[F(x)]$, $x \in \Re$. The PDF $g(x|f, q)$ is a weighted version of $f(x)$, with the weight function given by $q[F(x)]$. If q follows the uniform distribution, then, $g = f$. When f is normal, this is a general framework for developing skewed normal distributions.

Eugene et al. (2002) introduced the beta-generated family of distributions with CDF

$$G(x) = \int_0^{F(x)} b(t)dt, \tag{1.1}$$

where $b(t)$ is the PDF of the beta random variable and $F(x)$ is the CDF of any random variable. The corresponding PDF to (1.1) is given by

$$g(x) = \frac{1}{B(\alpha, \beta)} f(x)F^{\alpha-1}(x)(1 - F(x))^{\beta-1}, \quad \alpha, \beta > 0. \tag{1.2}$$

If F is Φ, the CDF of the normal distribution, equation (1.2) defines the beta-normal distribution. If α and β are integers, (1.2) is the αth order statistic of the random sample of size $(\alpha + \beta - 1)$.

The beta-normal distribution can be unimodal or bimodal and it has been applied to fit a variety of real data including bimodal cases (Famoye et al. 2004). The main distinction between the method of skewed normal and the beta-generated normal is that the skewed normal method introduces a skewing mechanism into the normal distribution to generate skewed normal family. The skewness of the distribution is estimated by the skewing parameter. On the other hand, the beta-normal distribution is generated by adding more parameters using beta distribution as the generator. Thus, the skewness is not directly defined by a specific parameter; rather it is the combination of all shape parameters that play the role of measuring skewness. For detailed review about the methods for generating continuous distributions, including the normal distribution, one may refer to Lee et al. (2013).

Alzaatreh et al. (2013) extended the beta generated family and defined the $T\text{-}X(W)$ family. The CDF of the $T\text{-}X(W)$ distribution is $G(x) = \int_a^{W(F(x))} r(t)dt$, where $r(t)$ is the PDF of the random variable T with support (a, b) for $-\infty \leq a < b \leq \infty$. The function $W(F(x))$ is monotonic and absolutely continuous. Aljarrah et al. (2014) took $W(F(x))$ to be the quantile function of a random variable Y and defined the $T\text{-}X\{Y\}$ family as

$$G(x) = \int_a^{Q_Y(F(x))} r(t)dt = R(Q_Y(F(x))), \tag{1.3}$$

where $Q_Y(p)$ is the quantile function of the random variable Y. In (1.1), X is used as a random variable having CDF $F(x)$ and then as a random variable having CDF $G(x)$ which may be confusing. This article first gives a unified notation to re-define the T-$X\{Y\}$ as T-$R\{Y\}$ and proposes several different generalizations of the normal distribution using the T-$R\{Y\}$ framework.

Section 2 gives the unified definition of T-$R\{Y\}$ family and defines several new generalized normal families. Section 3 gives some general properties of the proposed generalized normal families. Section 4 defines some new generalized normal distributions and studies some of their properties. Section 5 provides some applications to numerical data sets and the paper ends with a short summary and conclusions.

2 *T*-normal families of distributions

Let T, R and Y be random variables with CDF $F_T(x) = P(T \leq x), F_R(x) = P(R \leq x)$ and $F_Y(x) = P(Y \leq x)$. The corresponding quantile functions are $Q_T(p)$, $Q_R(p)$ and $Q_Y(p)$, where the quantile function is defined as $Q_Z(p) = \inf\{z : F_Z(z) \geq p\}$, $0 < p < 1$. If densities exist, we denote them by $f_T(x), f_R(x)$ and $f_Y(x)$. Now assume the random variable $T \in (a, b)$ and $Y \in (c, d)$, for $-\infty \leq a < b \leq \infty$ and $-\infty \leq c < d \leq \infty$. Following the technique proposed by Aljarrah et al. (2014), the CDF of the random variable X is defined as

$$F_X(x) = \int_a^{Q_Y(F_R(x))} f_T(t)dt = P[T \leq Q_Y(F_R(x))] = F_T(Q_Y(F_R(x))). \tag{2.1}$$

Note that (2.1) is an alternative expression to (1.3) without using X in two different situations. Hereafter, the family of distributions in (2.1) will be called the T-$R\{Y\}$ family of distributions.

Remark 1. If X follows the distribution in (2.1), it is easy to see that

(i) $X \stackrel{d}{=} Q_R(F_Y(T))$,
(ii) $Q_X(p) = Q_R(F_Y(Q_T(p)))$,
(iii) If $T \stackrel{d}{=} Y$ then $X \stackrel{d}{=} R$ and
(iv) If $Y \stackrel{d}{=} R$ then $X \stackrel{d}{=} T$.

The corresponding PDF associated with (2.1) is

$$f_X(x) = f_T(Q_Y(F_R(x))) \times Q'_Y(F_R(x)) \times f_R(x), \tag{2.2}$$

where $Q'_Y(F_R) = \frac{d}{dF_R} Q_Y(F_R)$. Using the fact that $Q_Y(F_Y(x)) = x$, it follows that $Q'_Y(F_Y(x)) \times f_Y(x) = 1$ so that $Q'_Y(p) = 1/f_Y(Q_Y(p))$. By taking $p = F_R(x)$, (2.2) reduces to

$$f_X(x) = f_R(x) \times \frac{f_T(Q_Y(F_R(x)))}{f_Y(Q_Y(F_R(x)))}. \tag{2.3}$$

From (2.1) and (2.3), the hazard function of the random variable X can be written as

$$\begin{aligned} h_X(x) &= \frac{f_X(x)}{1-F_X(x)} \\ &= \frac{f_R(x)}{1-F_R(x)} \times \frac{1-F_Y(Q_Y(F_R(x)))}{f_Y(Q_Y(F_R(x)))} \times \frac{f_T(Q_Y(F_R(x)))}{1-F_T(Q_Y(F_R(x)))} \\ &= h_R(x) \times \frac{h_T(Q_Y(F_R(x)))}{h_Y(Q_Y(F_R(x)))}. \end{aligned} \tag{2.4}$$

One can see from (2.3) and (2.4) that

$$\frac{f_X(x)}{f_R(x)} = \frac{f_T(Q_Y(F_R(x)))}{f_Y(Q_Y(F_R(x)))} \quad \text{and} \quad \frac{h_X(x)}{h_R(x)} = \frac{h_T(Q_Y(F_R(x)))}{h_Y(Q_Y(F_R(x)))}.$$

Some general properties of the T-$R\{Y\}$ family were recently studied in the literature, for more details see Aljarrah et al. (2014). Equivalent expressions to (2.2) - (2.4) are given in Aljarrah et al. (2014) by using the T-$X\{Y\}$ notation. Table 1 gives some distributions of the T-$R\{Y\}$ families based on quantile functions of some standard forms of distribution and some commonly used random variables T. The explicit expression of a T-$R\{Y\}$ family can be obtained using (2.3) for different combinations of random variables T, R, and Y.

Several extensions from Table 1 can be made. First, one can use the quantile function of non-standard distributions, such as non-standard exponential, log-logistic, logistic, extreme value, and Weibull. For example, the quantile function of log-logistic is $Q_Y(p) = \alpha(p/(1-p))^{1/\beta}$, $\alpha, \beta > 0$. By using this Q_Y function, two additional parameters corresponding to the log-logistic distribution may be added to the T-R\{log-logistic\} family. Aljarrah et al. (2014) gave a more detailed list of T-$R\{Y\}$ distributions based on quantile functions of non-standard distributions. Secondly, one can introduce exponentiated and scale parameters by replacing $F_T(x)$ by $F_T^{\delta}(\alpha x)$, $\alpha, \delta > 0$ as well as for $F_R(x)$.

If R is a normal random variable with PDF $f_R(x) = \phi(x)$ and CDF $F_R(x) = \Phi(x)$, then (2.1) gives the T-normal$\{Y\}$ family of distributions as

$$F_X(x) = \int_a^{Q_Y(\Phi(x))} f_T(t)dt = F_T(Q_Y(\Phi(x))). \tag{2.5}$$

The corresponding PDF associated with (2.5) is

$$f_X(x) = f_T(Q_Y(\Phi(x))) \times Q'_Y(\Phi(x)) \times \phi(x) = \phi(x) \times \frac{f_T(Q_Y(\Phi(x)))}{f_Y(Q_Y(\Phi(x)))}. \tag{2.6}$$

Table 1 Families of T-$R\{Y\}$ distributions based on different choices for the random variables Y* and T

Y	$Q_Y(p)$	T	$F_T(x)$ or $f_T(x)$
(a) Uniform	p	(1) Exponentiated-Exponential	$F_T(x) = (1-e^{-\lambda x})^{\alpha}$
(b) Exponential	$-\log(1-p)$	(2) Weibull	$F_T(x) = 1-e^{-(x/\gamma)^c}$
(c) Log-logistic	$p/(1-p)$	(3) Logistic	$F_T(x) = 1-(1+e^{-\lambda x})^{-1}$
(d) Logistic	$\log(p/(1-p))$	(4) Log-logistic	$F_T(x) = 1-[1+(x/\alpha)^{\beta}]^{-1}$
(e) Extreme value	$\log(-\log(1-p))$	(5) Pareto	$F_T(x) = 1-(\alpha/x)^{\beta}$
		(6) Cauchy	$f_T(x) = \{\pi\beta(1+[(x-\alpha)/\beta]^2)\}^{-1}$
		(7) Pascal	$f_T(x) = 0.5\lambda e^{-\lambda\lvert x\rvert}$
		(8) Gamma	$f_T(x) = x^{\alpha-1}e^{-x/\beta}/[\beta^{\alpha}\Gamma(\alpha)]$

*standard random variable Y.

The hazard function of the T-normal$\{Y\}$ family is given by $h_X(x) = h_\phi(x) \times \frac{h_T(Q_Y(\Phi(x)))}{h_Y(Q_Y(\Phi(x)))}$, where $h_\phi(x) = \phi(x)/(1 - \Phi(x))$.

The T-normal$\{Y\}$ family is a general framework for generating many different generalizations of the normal distribution. Various existing generalizations of normal distributions can be obtained based on this framework. The beta normal (Eugene et al. 2002), Kumaraswamy normal (Cordeiro and de Castro 2011), and generalized beta-generated normal (Alexander et al. 2012) belong to the T-normal{standard uniform} families. The gamma-normal distribution studied by Alzaatreh, et al. (2014) is a member of T-normal{standard exponential} family. For distribution "parsimony", we will focus on the quantile functions of standard distributions in order to limit the number of parameters. Generalizations from using non-standard quantile functions or adding exponentiated and/or scale parameters can be derived in a straightforward manner. In the following, we define the families of generalized normal (GN) distributions, T-normal$\{Y\}$, using the standard quantile functions (b)-(e) defined in Table 1.

2.1 Family of GN distributions from the quantile function of exponential distribution (*T-N*{exponential})

By using the quantile function (b) in Table 1: $Q_Y(\Phi(x)) = -\log(1 - \Phi(x))$, the corresponding CDF to (2.5) is

$$F_X(x) = F_T\left\{-\log(1 - \Phi(x))\right\} = F_T\left(H_\phi(x)\right), \tag{2.7}$$

and the corresponding PDF is

$$f_X(x) = \frac{\phi(x)}{1 - \Phi(x)} f_T\left(-\log\left(1 - \Phi(x)\right)\right) = h_\phi(x) f_T\left(H_\phi(x)\right), \tag{2.8}$$

where $h_\phi(x)$ and $H_\phi(x) = -\log[1 - \Phi(x)]$ are the hazard and cumulative hazard functions for the normal distribution, respectively. Thus, this family of GN distributions is denoted as T-N{exponential}, which arises from the "hazard function of the normal distribution".

2.2 Family of GN distributions from the quantile function of log-logistic distribution (*T-N*{log-logistic})

By using the quantile function (c) in Table 1: $Q_Y(\Phi(x)) = \Phi(x)/(1 - \Phi(x))$, the corresponding CDF to (2.5) is

$$F_X(x) = F_T\left\{\Phi(x)/(1 - \Phi(x))\right\}, \tag{2.9}$$

and the corresponding PDF is

$$f_X(x) = \frac{\phi(x)}{(1 - \Phi(x))^2} f_T\left(\frac{\Phi(x)}{1 - \Phi(x)}\right). \tag{2.10}$$

The family of GN distributions in (2.9) is denoted as T-N{log-logistic}, which arises from the "odds of the normal distribution".

2.3 Family of GN distributions from the quantile function of logistic distribution (*T-N*{logistic})

By using the quantile function (d) in Table 1: $Q_Y(\Phi(x)) = \log(\Phi(x)/(1 - \Phi(x)))$, the corresponding CDF to (2.5) is

$$F_X(x) = F_T\left\{\log[\Phi(x)/(1 - \Phi(x))]\right\}, \tag{2.11}$$

and the corresponding PDF is

$$f_X(x) = \frac{\phi(x)}{\Phi(x)(1-\Phi(x))} f_T\left(\log\{\Phi(x)/(1-\Phi(x))\}\right) = \frac{h_\phi(x)}{\Phi(x)} f_T\left(\log\left\{\frac{\Phi(x)}{1-\Phi(x)}\right\}\right). \tag{2.12}$$

The family of GN distributions in (2.11) is denoted as T-N\{logistic\}, which arises from the "logit function of the normal distribution".

2.4 Family of GN distributions from the quantile function of extreme value distribution (T-N\{extreme value\})

By using the quantile function (e) in Table 1: $Q_Y(\Phi(x)) = \log(-\log(1-\Phi(x))$, the corresponding CDF to (2.5) is

$$F_X(x) = F_T\left\{\log[-\log(1-\Phi(x))]\right\}, \tag{2.13}$$

and the corresponding PDF is

$$f_X(x) = \frac{\phi(x)}{-(1-\Phi(x))\log(1-\Phi(x))} f_T\left\{\log\left(-\log(1-\Phi(x)\right)\right\} = \frac{h_\phi(x)}{H_\phi(x)} f_T(\log(H_\phi(x)). \tag{2.14}$$

The family of GN distributions in (2.13) is denoted as T-N\{extreme value\}, which arises from the "extreme value function of the normal distribution".

3 Some properties of the T-normal family of distributions

In this section, some of the general properties of the T-normal family will be discussed.

Lemma 1 (Transformation). For any random variable T with PDF $f_T(x)$, then the random variable

(i) $X = \Phi^{-1}(1-e^{-T})$ follows the distribution of T-N\{exponential\} family in (2.7).
(ii) $X = \Phi^{-1}(T/(1+T))$ follows the distribution of T-N\{log-logistic\} family in (2.9).
(iii) $X = \Phi^{-1}\left(e^T/(1+e^T)\right)$ follows the distribution of T-N\{logistic\} family in (2.11).
(iv) $X = \Phi^{-1}\left(1-e^{-e^T}\right)$ follows the distribution of T-N\{extreme value\} family in (2.13).

Proof. The result follows immediately from Remark 1(i). □

Lemma 1 gives the relationships between the random variable X and the random variable T. These relationships can be used to generate random samples from X by using T. For example, one can simulate the random variable X which follows the distribution of T-N\{exponential\} family in (2.7) by first simulating random variable T from the PDF $f_T(x)$ and then computing $X = \Phi^{-1}\left(1-e^{-T}\right)$, which has the CDF $F_X(x)$.

Lemma 2. The quantile functions for the (i) T-N\{exponential\}, (ii) T-N\{log-logistic\}, (iii) T-N\{logistic\}, and (iv) T-N\{extreme value\} distributions, are respectively,

(i) $Q_X(p) = \Phi^{-1}\left\{1-e^{-Q_T(p)}\right\}$,
(ii) $Q_X(p) = \Phi^{-1}\left\{Q_T(p)/\left(1+Q_T(p)\right)\right\}$,

(iii) $Q_X(p) = \Phi^{-1}\left\{e^{Q_T(p)}/\left(1 + e^{Q_T(p)}\right)\right\}$,

(iv) $Q_X(p) = \Phi^{-1}\left\{1 - e^{-e^{Q_T(p)}}\right\}$.

Proof. The result follows directly from Remark 1(ii). □

Theorem 1. The mode(s) of the T-$N\{Y\}$ family are the solutions of the equation

$$x = \mu + \sigma^2\phi(x)\left\{\frac{Q''_Y(\Phi(x))}{Q'_Y(\Phi(x))} + \frac{f'_T(Q_Y(\Phi(x)))}{f_T(Q_Y(\Phi(x)))}Q'_Y(\Phi(x))\right\}. \tag{3.1}$$

Proof. One can show the result in (3.1) by setting the derivative of the equation (2.6) to zero and then using the fact that $\phi'(x) = -\sigma^{-2}(x-\mu)\phi(x)$. □

Corollary 1. The mode(s) of the (i) T-N\{exponential\}, (ii) T-N\{log-logistic\}, (iii) T-N\{logistic\}, and (iv) T-N\{extreme value\} distributions, respectively, are the solutions of the equations

$$\text{(i) } x = \mu + \sigma^2 h_\phi(x)\left\{1 + \frac{f'_T\left(H_\phi(x)\right)}{f_T\left(H_\phi(x)\right)}\right\}, \tag{3.2}$$

$$\text{(ii) } x = \mu + \sigma^2 h_\phi(x)\left\{2 + \frac{f'_T\left(\Phi(x)/(1-\Phi(x))\right)}{(1-\Phi(x))f_T\left(\Phi(x)/(1-\Phi(x))\right)}\right\},$$

$$\text{(iii) } x = \mu + \sigma^2\frac{h_\phi(x)}{\Phi(x)}\left\{\frac{f'_T\left(\log\{\Phi(x)/(1-\Phi(x))\}\right)}{f_T\left(\log\{\Phi(x)/(1-\Phi(x))\}\right)} + 2\Phi(x) - 1\right\},$$

$$\text{(iv) } x = \mu + \sigma^2\frac{h_\phi(x)}{H_\phi(x)}\left\{\frac{f'_T\{\log(H_\phi(x))\}}{f_T\{\log(H_\phi(x))\}} + H_\phi(x) - 1\right\}.$$

Note that the results in Theorem 1 do not imply that the mode is unique. It is possible that there is more than one mode for some of these GN distributions. For example, the logistic-N\{logistic\} distribution given in section 4 is a bimodal distribution. If T follows the gamma distribution with parameters α and β, equation (3.2) can be simplified to

$$x = \mu + \sigma^2 h_\phi(x)\left\{1 + \frac{f'_T\left(H_\phi(x)\right)}{f_T\left(H_\phi(x)\right)}\right\} = \mu + \sigma^2 h_\phi(x)\left[(\alpha-1)/H_\phi(x) - \beta^{-1} + 1\right].$$

This agrees with the result obtained by Alzaatreh et al. (2014) for the gamma-normal distribution.

The entropy of a random variable X is a measure of variation of uncertainty (Rényi 1961). Shannon's entropy for a random variable X with PDF $g(x)$ is defined as $E\left\{-\log\left(g(X)\right)\right\}$.

Theorem 2. The Shannon's entropies for the T-$N\{Y\}$ family is given by

$$\eta_X = \eta_T + E\left(\log f_Y(T)\right) + \log(\sigma\sqrt{2\pi}) + \frac{1}{2\sigma^2}E(X-\mu)^2. \tag{3.3}$$

Proof. Since $X \overset{d}{=} Q_R(F_Y(T))$, this implies that $T \overset{d}{=} Q_Y(F_R(X))$. Hence, from (2.3) we have $f_X(x) = \frac{f_T(t)}{f_Y(t)} \times f_R(x)$. This implies

$$\eta_X = \eta_T + E\left(\log f_Y(T)\right) - E\left(\log f_R(X)\right). \tag{3.4}$$

For the T-$N\{Y\}$ family we have $f_R(x) = \phi(x)$, so

$$\log(\phi(x)) = -\log(\sigma\sqrt{2\pi}) - [(x-\mu)/\sigma]^2/2. \tag{3.5}$$

The result in (3.3) follows from (3.4) and (3.5). □

Corollary 2. The Shannon's entropies for the (i) T-N\{exponential\}, (ii) T-N\{log-logistic\}, (iii) T-N\{logistic\}, and (iv) T-N\{extreme value\} distributions, respectively, are given by

$$\text{(i)}\quad \eta_X = \log(\sigma\sqrt{2\pi}) - \mu_T + \eta_T + E(X-\mu)^2/(2\sigma^2), \tag{3.6}$$

$$\text{(ii)}\quad \eta_X = \log(\sigma\sqrt{2\pi}) - 2E(\log(1+T)) + \eta_T + E(X-\mu)^2/(2\sigma^2),$$

$$\text{(iii)}\quad \eta_X = \log(\sigma\sqrt{2\pi}) - 2E(\log(1+e^T)) + \mu_T + \eta_T + E(X-\mu)^2/(2\sigma^2),$$

$$\text{(iv)}\quad \eta_X = \log(\sigma\sqrt{2\pi}) - E(e^T) + \mu_T + \eta_T + E(X-\mu)^2/(2\sigma^2).$$

Proof. The results in (i)-(iv) can be easily shown using (3.3) and the fact that $f_Y(T) = e^{-T}$, $(1+T)^{-2}$, $e^T(1+e^T)^{-2}$ and $e^T e^{-e^T}$ for exponential, log-logistic, logistic and extreme value, respectively. □

Theorem 3. The rth non-central moments of the (i) T-N\{exponential\}, (ii) T-N\{log-logistic\}, (iii) T-N\{logistic\}, and (iv) T-N\{extreme value\} distributions, respectively, can be expressed as

$$\text{(i)}\quad E(X^r) = \sum_{j=0}^{r} \sum_{k_1,\cdots,k_j=0}^{\infty} \sum_{i=0}^{2s_j+j} 2^{j/2}\sigma^j\mu^{r-j}A(\underline{k})\binom{r}{j}\binom{2s_j+j}{i}(-2)^i M_T(-i), \tag{3.7}$$

$$\text{(ii)}\quad E(X^r) = \sum_{j=0}^{r} \sum_{k_1,k_2,\cdots,k_j=0}^{\infty} 2^{j/2}\sigma^j\mu^{r-j}A(\underline{k})\binom{r}{j}E\left\{\left(\frac{T-1}{T+1}\right)^{2s_j+j}\right\}, \tag{3.8}$$

$$\text{(iii)}\quad E(X^r) = \sum_{j=0}^{r} \sum_{k_1,k_2,\cdots,k_j=0}^{\infty} 2^{j/2}\sigma^j\mu^{r-j}A(\underline{k})\binom{r}{j}E\left\{\left(\frac{e^T}{1+e^T}\right)^{2s_j+j}\right\}, \tag{3.9}$$

$$\text{(iv)}\quad E(X^r) = \sum_{j=0}^{r} \sum_{k_1,\cdots,k_j=0}^{\infty} \sum_{i=0}^{2s_j+j} 2^{j/2}\sigma^j\mu^{r-j}A(\underline{k})\binom{r}{j}\binom{2s_j+j}{i}(-2)^i M_{e^T}(-i), \tag{3.10}$$

where $A(\underline{k}) = A(k_1,k_2,\cdots,k_j) = (\sqrt{\pi}/2)^{2s_j+j}a_{k_1}a_{k_2}\cdots a_{k_j}$, $s_j = k_1+k_2+\cdots+k_j$, $M_T(-i) = E(e^{-iT})$, $a_k = \frac{c_k}{2k+1}$, $c_k = \sum_{j=0}^{k-1}\frac{c_j c_{k-1-j}}{(j+1)(2j+1)}$, and $c_0 = 1$.

Proof. We first show (3.7). By using Lemma 1, the rth moments for the T-N\{exponential\} distribution can be written as $E(X^r) = E\big(\Phi^{-1}(1 - e^{-T})\big)^r$. Since

$$\Phi^{-1}(1 - e^{-T}) = \sqrt{2}\,\sigma\,\mathrm{erf}^{-1}(1 - 2e^{-T}) + \mu,$$

the rth moments can be written as

$$E(X^r) = E\Big(\sqrt{2}\,\sigma\,\mathrm{erf}^{-1}(1 - 2e^{-T}) + \mu\Big)^r = \sum_{j=0}^{r} \binom{r}{j} 2^{j/2}\sigma^j\, E\left\{\left(\mathrm{erf}^{-1}(1 - e^{-T})\right)^j\right\} \mu^{r-j}. \tag{3.11}$$

On using the series representation for $\mathrm{erf}^{-1}(1 - 2e^{-T})$ (Wolfram.com, 2014), we get $\mathrm{erf}^{-1}(1 - 2e^{-T}) = \sum_{k=0}^{\infty} a_k(\sqrt{\pi}/2)^{2k+1}(1 - 2e^{-T})^{2k+1}$, where $a_k = \frac{c_k}{2k+1}$, $c_k = \sum_{j=0}^{k-1} \frac{c_j c_{k-1-j}}{(j+1)(2j+1)}$, and $c_0 = 1$. This implies

$$\left(\mathrm{erf}^{-1}(1 - 2e^{-T})\right)^j = \sum_{k_1, k_2, \cdots, k_j = 0}^{\infty} A(k_1, k_2, \cdots, k_j)(1 - 2e^{-T})^{2s_j + j}, \tag{3.12}$$

where $A(k_1, k_2, \cdots, k_j) = (\sqrt{\pi}/2)^{2s_j+j} a_{k_1} a_{k_2} \cdots a_{k_j}$ and $s_j = k_1 + k_2 + \cdots + k_j$. By using the binomial expansion on $(1 - 2e^{-T})^{2s_j+j}$, (3.12) can be written as

$$\left(\mathrm{erf}^{-1}(1 - 2e^{-T})\right)^j = \sum_{k_1, k_2, \cdots, k_j = 0}^{\infty} \sum_{i=0}^{2s_j+j} A(k_1, k_2, \cdots, k_j) \binom{2s_j + j}{i} (-2)^i e^{-iT}. \tag{3.13}$$

The result of (3.7) follows by using equation (3.13) in equation (3.11). The results of (3.8)-(3.10) can be obtained by applying the same technique for (3.7). □

If T follows the gamma distribution with parameters α and β for the T-N\{exponential\}, we obtain the term $M_T(-i) = (1 + \beta i)^{-\alpha}$ in (3.7). Thus, (3.7) gives the rth non-central moment of gamma-N\{exponential\} distribution as shown in Alzaatreh et al. (2014).

The deviation from the mean and the deviation from the median are used to measure the dispersion and the spread in a population from the center. The mean deviation from the mean is denoted by $D(\mu)$, and the mean deviation from the median M is denoted by $D(M)$.

Theorem 4. $D(\mu)$ and $D(M)$ for each of (i) T-N\{exponential\}, (ii) T-N\{log-logistic\}, (iii) T-N\{logistic\}, and (iv) T-N\{extreme value\} distributions, respectively, are

$$\text{(i)}\quad D(\mu) = \sqrt{2}\sigma \sum_{k=0}^{\infty} \sum_{i=0}^{2k+1} A(k) \binom{2k+1}{i} (-2)^{i+1} S_{e^{-u}}(\mu, 0, i), \tag{3.14}$$

$$D(M) = \sqrt{2}\sigma \sum_{k=0}^{\infty} \sum_{i=0}^{2k+1} A(k) \binom{2k+1}{i} (-2)^{i+1} S_{e^{-u}}(M, 0, i), \tag{3.15}$$

where $S_\xi(c, a, \alpha) = \int_a^{Q_Y(\Phi(c))} \xi^\alpha f_T(u)du$ and $Q_Y(\Phi(c)) = -\log(1 - \Phi(c))$.

$$\text{(ii)}\quad D(\mu) = -\sqrt{8}\sigma \sum_{k=0}^{\infty} A(k) S_{\frac{u-1}{u+1}}(\mu, 0, 2k+1), \tag{3.16}$$

$$D(M) = -\sqrt{8}\sigma \sum_{k=0}^{\infty} A(k) S_{\frac{u-1}{u+1}}(M, 0, 2k+1), \tag{3.17}$$

where $Q_Y(\Phi(c)) = \Phi(c)/(1-\Phi(c))$.

$$\text{(iii)}\quad D(\mu) = -\sqrt{8}\sigma\sum_{k=0}^{\infty}A(k)S_{\frac{e^u}{1+e^u}}(\mu,-\infty,2k+1), \tag{3.18}$$

$$D(M) = -\sqrt{8}\sigma\sum_{k=0}^{\infty}A(k)S_{\frac{e^u}{1+e^u}}(M,-\infty,2k+1), \tag{3.19}$$

where $Q_Y(\Phi(c)) = \log\{\Phi(c)/(1-\Phi(c))\}$.

$$\text{(iv)}\quad D(\mu) = \sqrt{2}\sigma\sum_{k=0}^{\infty}\sum_{i=0}^{2k+1}A(k)\binom{2k+1}{i}(-2)^i S_{e^{-e^u}}(\mu,-\infty,i), \tag{3.20}$$

$$D(M) = \sqrt{2}\sigma\sum_{k=0}^{\infty}\sum_{i=0}^{2k+1}A(k)\binom{2k+1}{i}(-2)^i S_{e^{-e^u}}(M,-\infty,i), \tag{3.21}$$

where $Q_Y(\Phi(c)) = \log\{-\log(1-\Phi(c))\}$.

Proof. By definitions of $D(\mu)$ and $D(M)$,

$$D(\mu) = \int_{-\infty}^{\mu}(\mu-x)f_X(x)dx + \int_{\mu}^{\infty}(x-\mu)f_X(x)dx = 2\int_{-\infty}^{\mu}(\mu-x)f_X(x)dx$$

$$= 2\mu F_X(\mu) - 2\int_{-\infty}^{\mu}xf_X(x)dx. \tag{3.22}$$

$$D(M) = \int_{-\infty}^{M}(M-x)f_X(x)dx + \int_{M}^{\infty}(x-M)f_X(x)dx$$

$$= 2\int_{-\infty}^{M}(M-x)f_X(x)dx + E(X) - M$$

$$= \mu - 2\int_{-\infty}^{M}xf_X(x)dx. \tag{3.23}$$

We first proof the results (3.14) and (3.15) for the T-N\{exponential\} family. Defining the integral

$$I_c = \int_{-\infty}^{c}xf_X(x)dx = \int_{-\infty}^{c}\frac{x\phi(x)}{1-\Phi(x)}f_T\left\{-\log(1-\Phi(x)\right\}dx, \tag{3.24}$$

and using the substitution $u = -\log(1-\Phi(x))$, (3.24) can be written as

$$I_c = \int_{0}^{-\log(1-\Phi(c))}\Phi^{-1}(1-e^{-u})f_T(u)du. \tag{3.25}$$

By using similar approach as in Theorem 3, the equation (3.25) can be written as

$$I_c = \mu F_X(c) + \sqrt{2}\sigma\sum_{k=0}^{\infty}\sum_{i=0}^{2k+1}A(k)\binom{2k+1}{i}(-2)^i S_{e^{-u}}(c,0,i), \tag{3.26}$$

where $A(k)$ is defined in the proof of Theorem 3, $S_{\xi}(c,a,\alpha) = \int_a^{Q_Y(\Phi(c))}\xi^{\alpha}f_T(u)du$ and $Q_Y(\Phi(c)) = -\log(1-\Phi(c))$. The results in (3.14) and (3.15) follow by using (3.26) in (3.22) and (3.23). Applying the same techniques of showing (3.14) and (3.15), one can show the results of (3.16) and (3.17) for (ii), (3.18) and (3.19) for (iii), and (3.20) and (3.21) for (iv). □

4 Some examples of GN families of distributions with different *T* distributions

In this section different T distributions are used to generate different GN distributions. In the following subsections, we present four new GN distributions namely, Weibull-N{exponential}, exponential-N{log-logistic}, logistic-N{logistic} and logistic-N{extreme value}. For illustrative purposes, we study some properties of the Weibull-N{logistic} distribution. To conserve space, properties of other GN distributions are not given. One can follow the same method to study the properties of other GN distributions.

4.1 The Weibull-*N*{exponential} distribution

If a random variable T follows the Weibull distribution with parameters c and γ, then $f_T(x) = c\gamma^{-1}(\frac{x}{\gamma})^{c-1}e^{-(\frac{x}{\gamma})^c}$, $c, \gamma > 0$. From (2.8), the PDF of the Weibull-N{exponential} is defined as

$$f_X(x) = \frac{c}{\gamma}\frac{\phi(x)}{1-\Phi(x)}\left\{\frac{-\log(1-\Phi(x))}{\gamma}\right\}^{c-1}\exp\left(-\left\{\frac{-\log(1-\Phi(x))}{\gamma}\right\}^{c}\right). \quad (4.1)$$

Remark 2.

(i) When $c = 1$, the Weibull-N{exponential} reduces to the exponential-normal distribution with $\theta = 1/\gamma$.

(ii) When $c = \gamma = 1$, the Weibull-N{exponential} reduces to the normal distribution.

(iii) When $c = 1$ and $\gamma^{-1} = n \in N$, the PDF in (4.1) reduces to the distribution of the minimum order statistics, $x_{(1)}$, from a normal random sample of size n.

By using (2.7), the CDF of the Weibull-N{exponential} is given by

$$F_X(x) = 1 - \exp\left(-\{-(1/\gamma)\log(1-\Phi(x))\}^c\right).$$

In Figures 1 and 2, various graphs of $f_X(x)$ when $\mu = 0, \sigma = 1$ and for various values of c and γ are provided. These Figures indicate that the Weibull-N{exponential} PDF can be left skewed, right skewed, or symmetric. Also, the Weibull-N{exponential} is left skewed whenever $\gamma > 1$ and right skewed whenever $\gamma < 1$. For fixed γ, the peak increases as c increases.

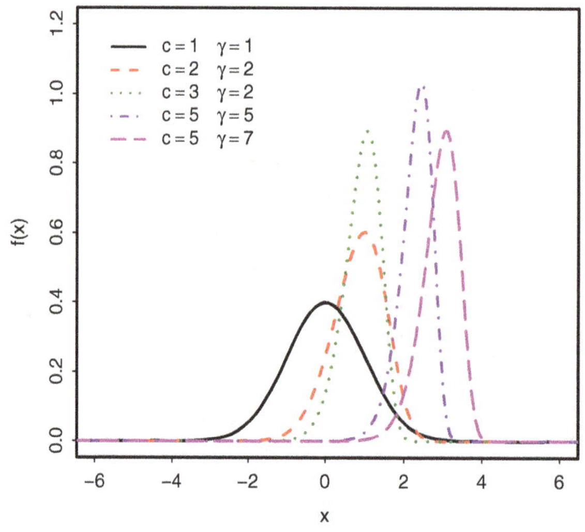

Figure 1 The PDF of Weibull-*N*{exponential} for various values of *c* and *γ*.

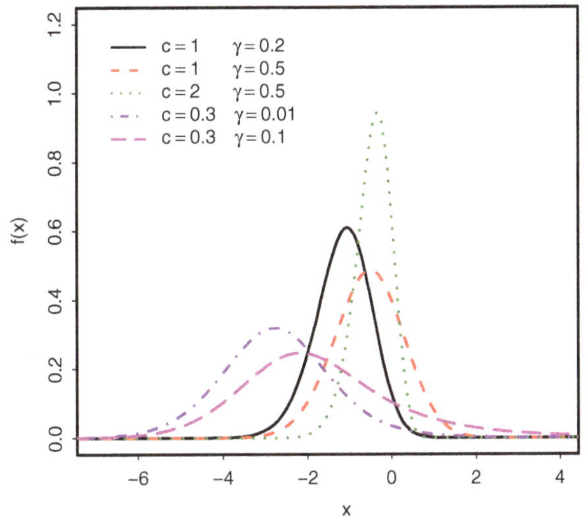

Figure 2 The PDF of Weibull-*N*{exponential} for various values of *c* and *γ*.

Some properties of the Weibull-N\{exponential\} are obtained in the following by using the general properties discussed in section 3.

(1) Quantile function: By using Lemma 2, the quantile function of the Weibull-N\{exponential\} distribution is given by $Q_X(p) = \Phi^{-1}\left\{1 - \exp(-\gamma(-\log(1-p))^{1/c})\right\}$.

(2) Mode: By using Corollary 1, the mode of Weibull-N\{exponential\} distribution is the solution of the following equation

$$x = \mu + \sigma^2 h_\phi(x)\left\{\frac{c-1}{H_\phi(x)} - c\gamma^{-c}(H_\phi(x))^{c-1} + 1\right\}.$$

(3) Shannon entropy: By using Corollary 2 and the fact that $\mu_T = \gamma\Gamma(1+1/c)$ and $\eta_T = 1 + \xi(1-1/c) + \log(\gamma/c)$ (see Song 2001), one can easily obtain the Shannon entropy of Weibull-N\{exponential\} distribution as

$$\eta_X = \log(\sigma\sqrt{2\pi}) - \gamma\Gamma(1+1/c) + \xi(1-1/c) + \log(\gamma/c) + E(X-\mu)^2/(2\sigma^2) + 1.$$

(4) Moments: By using Theorem 3, a series representation of the rth moments of the Weibull-N\{exponential\} distribution can be obtained by replacing $M_T(-i)$ with

$$\sum_{k=0}^{\infty} \frac{(-1)^k \gamma^k}{k!} \Gamma\left(1 + \frac{k}{c}\right)$$

in equation (3.7).

(5) Mean deviations: By using Theorem 4, the mean deviation from the mean and the mean deviation from the median of Weibull-N\{exponential\} distribution can be obtained by replacing $S(\mu, 0, i)$ and $S(M, 0, i)$ with $\frac{c}{i^c\gamma^c} \sum_{k=0}^{\infty} \frac{(-1)^k}{\gamma^{ck} i^{ck} k!} \Gamma[c(k+1), -i \log(1-\Phi(\mu))]$ and $\frac{c}{i^c\gamma^c} \sum_{k=0}^{\infty} \frac{(-1)^k}{\gamma^{ck} i^{ck} k!} \Gamma[c(k+1), -i\log(1-\Phi(M))]$ in equations (3.14) and (3.15) respectively, where $\Gamma(\alpha, x) = \int_0^x u^{\alpha-1} e^{-u} du$ is the incomplete gamma function.

4.2 The exponential-*N* {log-logistic} distribution

If a random variable T follows the exponential distribution with parameter λ, then $f_T(x) = \lambda e^{-\lambda x},\ \lambda > 0$. From (2.10), the PDF of the exponential-N\{log-logistic\} is defined as

$$f_X(x) = \frac{\lambda\phi(x)}{(1-\Phi(x))^2}\exp\left[\frac{-\lambda\Phi(x)}{1-\Phi(x)}\right]. \tag{4.2}$$

From (2.9), the CDF of (4.2) is given by $F_X(x) = 1 - \exp\left[\frac{\lambda\Phi(x)}{1-\Phi(x)}\right]$.

In Figure 3, various graphs of $f_X(x)$ when $\mu = 0, \sigma = 1$ and for various values of λ are provided. These graphs indicate that the exponential-N\{log-logistic\} distribution is always left skewed. Also, the skewness increases as λ decreases.

4.3 The logistic-*N*{logistic} distribution

If a random variable T follows the logistic distribution with parameter λ, then $f_T(x) = \lambda e^{-\lambda x}(1+e^{-\lambda x})^{-2},\ \lambda > 0$. From (2.12), the PDF of logistic-N\{logistic\} distribution is defined as

$$f_X(x) = \frac{\lambda\phi(x)\Phi^{\lambda-1}(x)(1-\Phi(x))^{\lambda-1}}{[\Phi^{\lambda}(x)+(1-\Phi(x))^{\lambda}]^2}. \tag{4.3}$$

From (2.11), the CDF of (4.3) is given by $F_X(x) = \frac{\Phi^{\lambda}(x)}{\Phi^{\lambda}(x)+(1-\Phi(x))^{\lambda}}$.

When $\lambda = 1$, (4.3) reduces to the normal distribution. In Figure 4, various graphs of $f_X(x)$ when $\mu = 0, \sigma = 1$ and for various values of λ are provided. These graphs indicate that the PDF of logistic-N\{logistic\} can be bimodal and the bimodality occurs for small values of λ. Also, it is easy to see from the PDFs in (4.3) that the distribution is symmetric for all values of λ.

4.4 The logistic-*N* {extreme value} distribution

If a random variable T follows the logistic distribution with parameter λ, then $f_T(x) = \lambda e^{-\lambda x}(1+e^{-\lambda x})^{-2},\ \lambda > 0$. From (2.14), the PDF of the logistic-N\{extreme value\} distribution is defined as

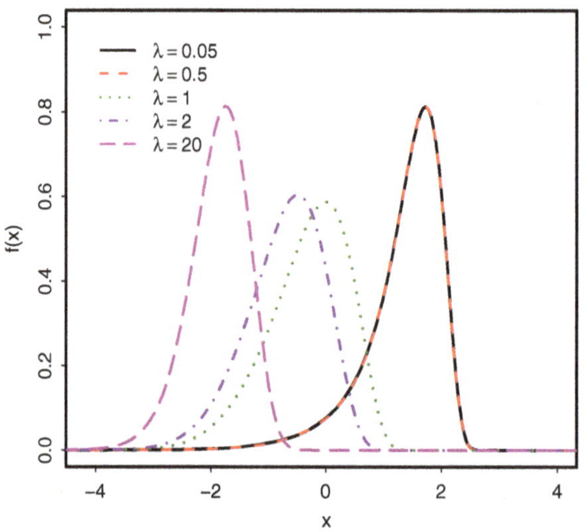

Figure 3 PDF of the exponential-*N*{log-logistic} distribution for various values of λ.

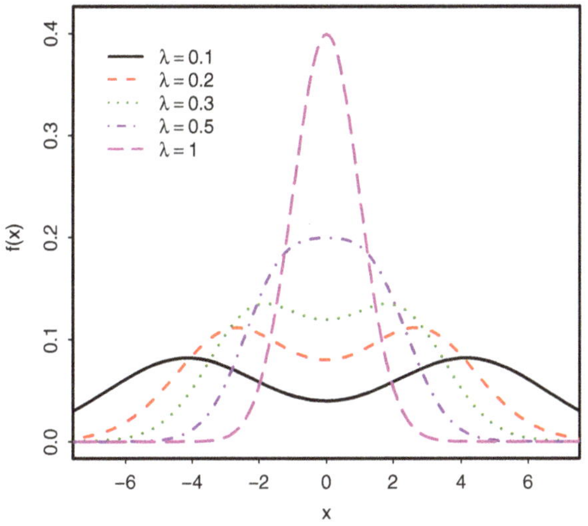

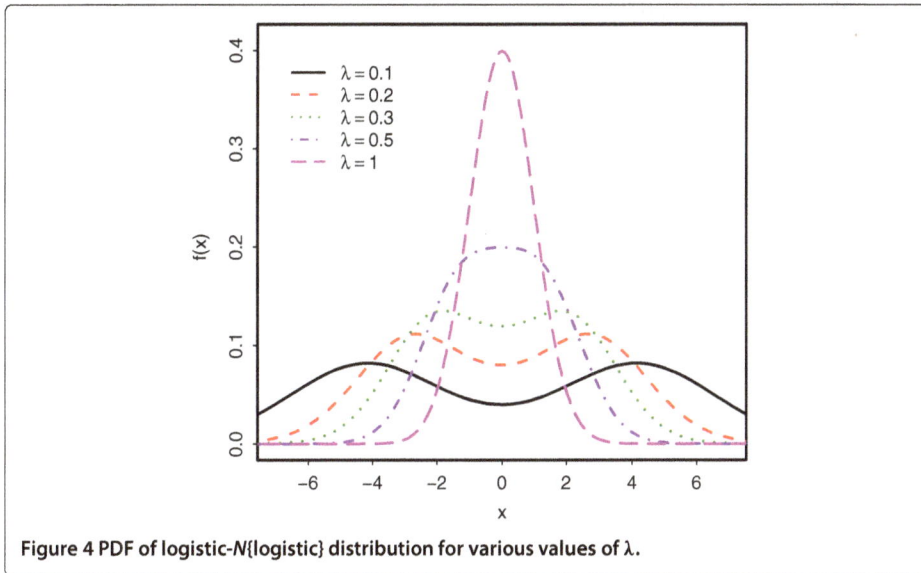

Figure 4 PDF of logistic-*N*{logistic} distribution for various values of λ.

$$f_X(x) = \frac{\lambda h_\phi(x) {H_\phi}^{\lambda-1}(x)}{\left(1 + {H_\phi}^{\lambda}(x)\right)^2}. \tag{4.4}$$

From (2.13), the CDF of (4.4) is given by $F_X(x) = \frac{{H_\phi}^{\lambda}(x)}{1+{H_\phi}^{\lambda}(x)}$.

In Figure 5, various graphs of $f_X(x)$ when $\mu = 0$, $\sigma = 1$ and for various values of λ are provided. These graphs indicate that the distribution is always right skewed. Also, the skewness increases as λ decreases.

5 Applications

To illustrate the flexibility of the GN distributions, we fit some GN distributions to a unimodal data set and a bimodal data set. The unimodal data with $n = 66$ in Table 2

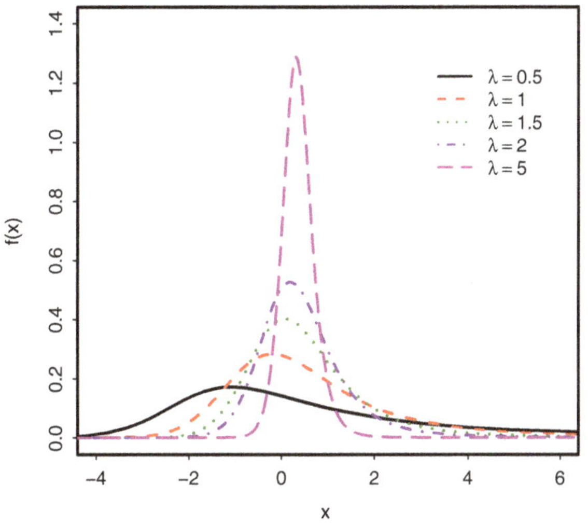

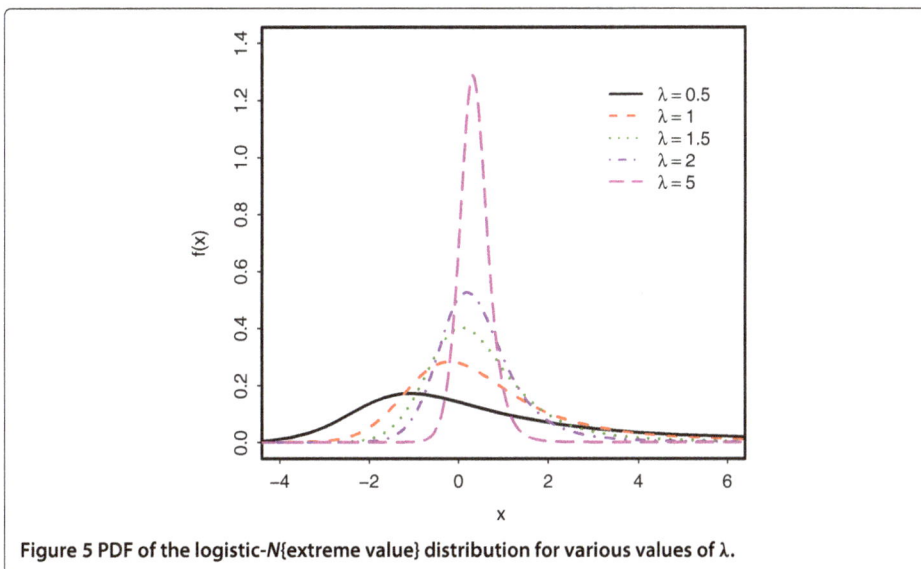

Figure 5 PDF of the logistic-*N*{extreme value} distribution for various values of λ.

Table 2 Breaking stress of carbon fibers data

3.70	2.74	2.73	2.50	3.60	3.11	3.27	2.87	1.47	3.11	3.56
4.42	2.41	3.19	3.22	1.69	3.28	3.09	1.87	3.15	4.90	1.57
2.67	2.93	3.22	3.39	2.81	4.20	3.33	2.55	3.31	3.31	2.85
1.25	4.38	1.84	0.39	3.68	2.48	0.85	1.61	2.79	4.70	2.03
1.89	2.88	2.82	2.05	3.65	3.75	2.43	2.95	2.97	3.39	2.96
2.35	2.55	2.59	2.03	1.61	2.12	3.15	1.08	2.56	1.80	2.53

is obtained from Nichols and Padgett (2006) on the breaking stress of carbon fibers of 50 mm in length. Alzaatreh at el. (2013) fitted the data set to the gamma-normal distribution. They showed that the standard gamma-normal distribution with $\mu = 0$ and $\sigma = 1$ provides a good fit to the data set. The standard form of exponential-N\{exponential\}, exponentiated exponential-N\{exponential\} and Weibull-N\{exponential\} distributions with $\mu = 0$ and $\sigma = 1$ are applied to fit the data set and the results compared with the results from standard gamma-normal distribution. The maximum likelihood estimates, the log-likelihood value, the AIC (Akaike Information Criterion), the Kolmogorov-Smirnov (K-S) test statistic, and the p-value for the K-S statistic for the fitted distributions are reported in Table 3. The results in Table 3 show that all the generalized normal distributions give an adequate fit to the data. However, the K-S values indicate that the gamma-N\{exponential\} distribution provides the best fit among the distributions. Figure 6 displays the histogram and the fitted density functions for the data.

The second application is on a bimodal data set obtained from Emlet et al. (1987) on the asteroid and echinoid egg size. The data is available from the first author. The data consists of 88 asteroid species divided into three types; 35 planktotrophic larvae, 36 lecithotrophic larvae, and 17 brooding larvae. Since the logarithm of the egg diameters of the asteroids data has a bimodal shape, Famoye et al. (2004) applied the beta-normal distribution to the logarithm of the data set. We apply the logistic-N\{logistic\} distribution, which can be bimodal, to fit the same data. The results of the maximum likelihood estimates, the log-likelihood value, the AIC (Akaike Information Criterion), the Kolmogorov-Smirnov (K-S) test statistic, and the p-value for the K-S statistic for the fitted distributions are reported in Table 4. The results in Table 4 show

Table 3 Parameter estimates for the carbon fibers data

Distribution	**Exponential-N\{exponential\}**	**Exponentiated exponential-N\{exponential\}**	**Weibull-N\{exponential\}**	**gamma-N\{exponential\}**
Parameter	$\hat{\theta} = 0.1612$	$\hat{\alpha} = 6.0389$	$\hat{c} = 2.4062$	$\hat{\alpha} = 4.7966$
Estimates	(0.0198)*	(1.3675)	(0.2226)	(0.8076)
		$\hat{\lambda} = 0.3919$	$\hat{\gamma} = 6.9991$	$\hat{\beta} = 1.2932$
		(0.0434)	(0.3780)	(0.2296)
Log-likelihood	-114.8292	-87.0385	-86.0629	-85.9070
AIC	231.6584	178.0770	176.1258	175.8140
K-S	0.2768	0.0854	0.0894	0.0693
K-S p-value	0.0001	0.7215	0.6676	0.9090

*standard error.

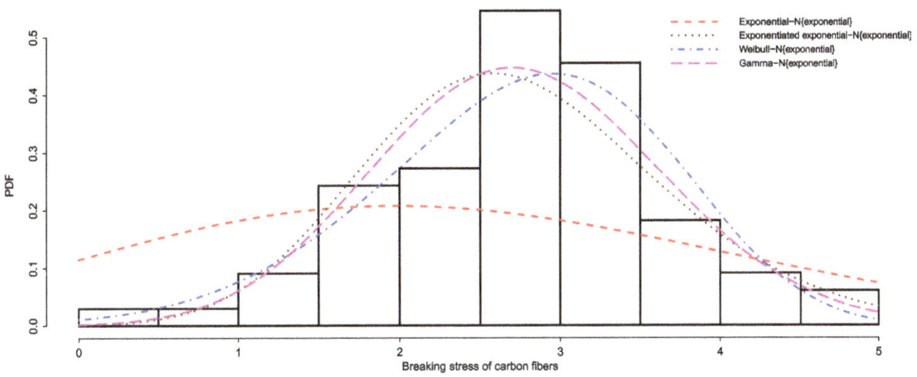

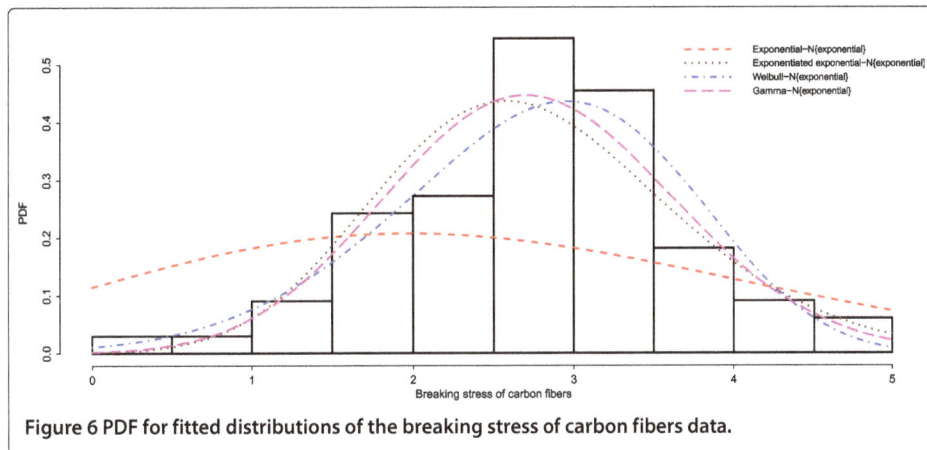

Figure 6 PDF for fitted distributions of the breaking stress of carbon fibers data.

that both the beta-normal and logistic-N\{logistic\} distributions give an adequate fit to the data. However, the K-S values indicate that the logistic-N\{logistic\} distribution provides a better fit. Figure 7 displays the histogram and the fitted density functions for the data.

6 Summary and conclusions

The normal distribution is the most commonly used distribution in both statistical theory and applications. The generalization of the normal distribution is studied using the T-X framework proposed by Alzaatreh et al. (2013). Four types of generalized normal families from the quantile functions of the (i) exponential, (ii) log-logistic, (iii) logistic, and (iv) extreme value distributions are proposed. Some general properties are studied. Four generalized normal distributions are described and some of their properties investigated. It is noticed that the shapes of GN distributions can be symmetric, skewed to the right, skewed to the left or bimodal. This gives the families some flexibility in fitting real world data. Because the GN distributions include the normal distribution as a special case, using the GN distributions to fit data enables one to check if the additional parameters characterize the deviation from the normal distribution. Many types of generalizations of the normal distribution can be derived using the methodology described in this paper. Due to the fact that GN distributions are

Table 4 Parameter estimates for the asteroids data

Distribution	Beta-normal*	Logistic-N\{logistic\}
Parameter	$\hat{\alpha} = 0.0129$	$\hat{\lambda} = 0.1498(0.0185)$
Estimates	$\hat{\beta} = 0.0070$	$\hat{\mu} = 6.0348(0.0685)$
	$\hat{\mu} = 5.7466$	$\hat{\sigma} = 0.2604(0.010)$
	$\hat{\sigma} = 0.0675$	
Log-likelihood	-109.4800	-111.4287
AIC	226.9600	228.4974
K-S	0.1233	0.0988
K-S p-value	0.1377	0.3572

*From Famoye et al. (2004) and the MLE standard errors were not provided.

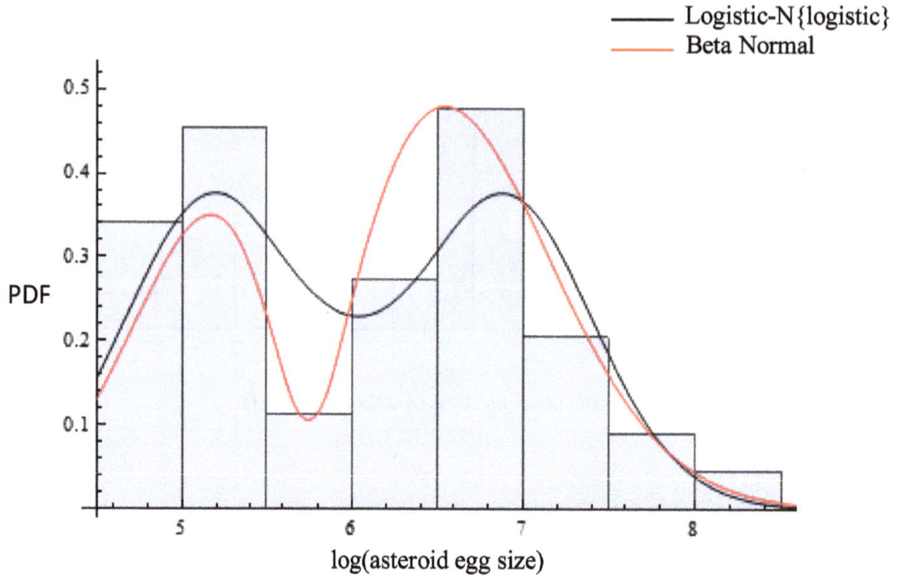

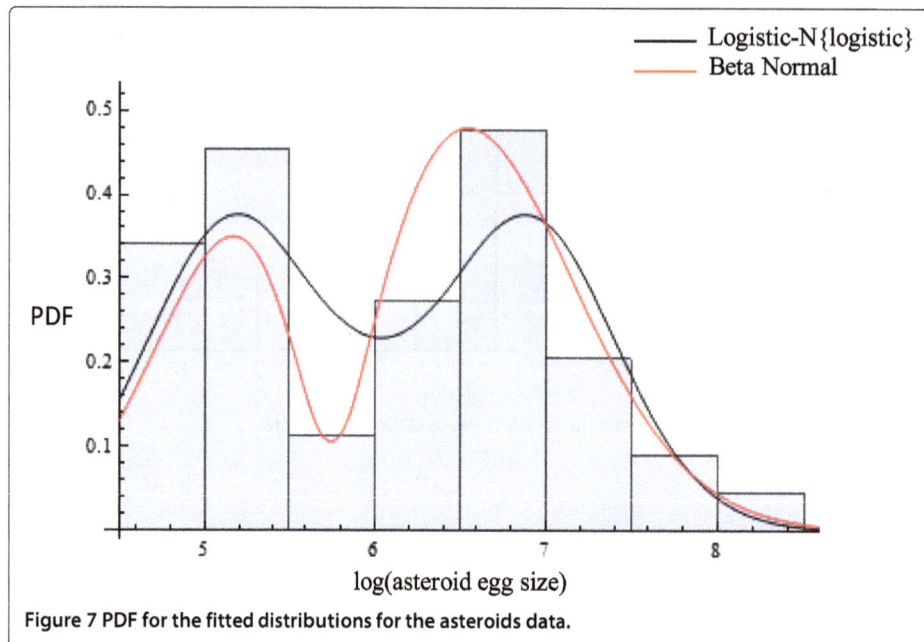

Figure 7 PDF for the fitted distributions for the asteroids data.

natural extensions from the normal distribution, statistical modeling by assuming the error term follows some form of GN distribution will be an interesting topic for future research.

Competing interests

The authors declare that they have no competing interests.

Authors' contributions

The authors, viz AA, CL and FF with the consultation of each other carried out this work and drafted the manuscript together. All authors read and approved the final manuscript.

Acknowledgments

The authors are very grateful to the Associate Editor and the three anonymous reviewers for various constructive comments and suggestions that have greatly improved the presentation of the paper. The authors particularly thank one of the reviewers for suggesting the unified notation to define the *T-R{Y}* family of distributions.

Author details

[1]Department of Mathematics and Statistics, Austin Peay State University, Clarksville, TN 37044, USA. [2]Department of Mathematics, Central Michigan University, Mount Pleasant, MI 48859, USA.

References

Alexander, C, Cordeiro, GM, Ortega, EMM, Sarabia, JM: Generalized beta-generated distributions. Comput. Stat. Data Anal. **56**(6), 1880–1897 (2012)

Aljarrah, MA, Lee, C, Famoye, F: A method of generating *T-X* family of distributions using quantile functions. J. Stat. Distributions Appl. **1**(2), 17 (2014)

Alzaatreh, A, Famoye, F, Lee, C: The gamma-normal distribution: Properties and applications. Comput. Stat. Data Anal. **69**(1), 67–80 (2014)

Alzaatreh, A, Lee, C, Famoye, F: A new method for generating families of continuous distributions. Metron. **71**(1), 63–79 (2013)

Arellano-Valle, RB, Gómez, HW, Quintanan, FA: A new class of skew-normal distributions. Commun. Stat.-Theory Methods. **33**, 1465–1480 (2004)

Arnold, BC, Beaver, RJ: Skewed multivariate models related to hidden truncation and/or selective reporting (with discussion). Test. **11**, 7–35 (2002)

Arnold, BC, Castillo, E, Jose, JM: Distributions with generalized skewed conditionals and mixtures of such distributions. Commun. Stat.-Theory Methods. **36**, 1493–1503 (2007)

Azzalini, A: A class of distributions which includes the normal ones. Scand. J. Stat. **12**, 171–178 (1985)

Balakrishnan, N: Discussion of skewed multivariate models related to hidden truncation and/or selective reporting by Arnold & Beaver. Test. **11**, 37–39 (2002)
Choudhury, K, Abdul, MM: Extended skew generalized normal distribution. Metron. **69**, 265–278 (2011)
Cordeiro, GM, de Castro, M: A new family of generalized distributions. J. Stat. Comput. Simulat. **81**(7), 883–898 (2011)
de Moivre, A: Approximatio ad summam ferminorum binomii $(a + b)^n$ in seriem expansi. Self-published pamphlet (1733)
Emlet, RB, McEdward, LR, Strathmann, RR: Echinoderm larval ecology viewed from the egg. In: Jangoux, M, Lawrence, JM (eds.) *Echinoderm Studies, volume 2*, pp. 55–136. AA Balkema, Rotterdam (1987)
Eugene, N, Lee, C, Famoye, F: The beta-normal distribution and its applications. Commun. Stat.-Theory Methods. **31**(4), 497–512 (2002)
Famoye, F, Lee, C, Eugene, N: Beta-normal distribution: bimodality properties and applications. J. Mod. Appl. Stat. Meth. **3**(1), 85–103 (2004)
Fernández, C, Steel, MFJ: On Bayesian modeling of fat tails and skewness. J. Am. Stat. Assoc. **93**, 359–371 (1998)
Ferreira, JTAS, Steel, MFJ: A constructive representation of univariate skewed distributions. J. Am. Stat. Assoc. **101**, 823–829 (2006)
Gauss, CF: Theoria Motus Corporum Coelestium, pp. 205–224. Perthes u. Besser, Hamburg (1809). Lib. 2, Sec. III
Gauss, CF: Bestimmung der genauigkeit der beobachtungen. Zeitschrift Astronomi. **1**, 185–197 (1816)
Gupta, RC, Gupta, RD: Generalized skew normal model. Test. **13**, 501–520 (2004)
Johnson, NL, Kotz, S, Balakrishnan, N: Continuous Univariate Distributions, volume 1. second edition. John Wiley & Sons, New York (1994)
Kotz, S, Vicari, D: Survey of developments in the theory of continuous skewed distributions. Metron. **63**, 225–261 (2005)
Lee, C, Famoye, F, Alzaatreh, A: Methods for generating families of univariate continuous distributions in the recent decades. WIREs Comput. Stat. **5**, 219–238 (2013)
Nichols, MD, Padgett, WJ: A bootstrap control for Weibull percentiles. Qual. Reliab. Eng. Int. **22**, 141–151 (2006)
Patel, JK, Read, CB, Handbook of the Normal Distribution. second edition. Marcel Dekker, New York (1996)
Pourahmadi, M: Construction of skew-normal random variables: are they linear combination of normals and half-normals? J. Stat. Theory Appl. **3**, 314–328 (2007)
Rényi, A: On measures of entropy and information. In: *Proceedings of the Fourth Berkeley Symposium on Mathematical Statistics and Probability*, pp. 547–561. University of California Press, Berkeley, CA (1961)
Sharafi, M, Behboodian, J: The Balakrishnan skew-normal density. Stat. Pap. **49**, 769–778 (2008)
Song, KS: Rényi information, loglikelihood and an intrinsic distribution measure. J. Stat. Plann. Infer. **93**, 51–69 (2001)
Wolfram. http://functions.wolfram.com/GammaBetaErf/InverseErf/06/01/02/0004/. Retrieved on May 29, 2014
Yadegari, I, Gerami, A, Khaledi, MJ: A generalization of the Balakrishnan skew-normal distribution. Stat. Probability Lett. **78**, 1165–1167 (2008)

9

On the distribution theory of over-dispersion

Evdokia Xekalaki

Correspondence: exek@aueb.gr
Department of Statistics, Athens University of Economics and Business, 76 Patision St., Athens, Greece

Abstract

An overview of the evolution of probability models for over-dispersion is given looking at their origins, motivation, first main contributions, important milestones and applications. A specific class of models called the Waring and generalized Waring models will be a focal point. Their advantages relative to other classes of models and how they can be adapted to handle multivariate data and temporally evolving data will be highlighted.

Keywords: Heterogeneity; Contagion; Clustering; Spells model; Accident proneness; Mixtures; Zero-adjusted models; Biased sampling; Generalized Waring distribution; Generalized Waring process

1. Introduction

Data analysts have often to deal with data that exhibit a variability that differs from what they expect on the basis of the hypothesized model. The phenomenon is known as ***overdispersion*** if the observed variability exceeds the expected variability or ***under-dispersion*** if it is lower than expected.

Such differences between observed and nominal variances can be interpreted as brought about by failures of some of the basic assumptions of the model. These can be classified by the mechanism leading to them. As summarized by Xekalaki (2006), in traditional experimental contexts, they may be caused by deviations from the hypothesized structure of the population, due to lack of independence between individual item responses, contagion, clustering, and heterogeneity. In observational study contexts, on the other hand, they are the result of the method of ascertainment, which can lead to partial distortion of the observations. In both contexts, the observed value x no longer represents an observation on the original variable X, but constitutes an observation on a random variable Y whose distribution (*the observed distribution*) is a distorted version of the distribution of X (*original distribution*).

Such practical situations have been noticed since over a century ago (e.g. Lexis 1879; Student 1919). The *Lexis ratio* appears to be the first statistic suggested for testing for the presence of over- or under-dispersion relative to a binomial hypothesized model in populations structured in clusters. Also, for count data, Fisher (1950) considered using the sample index of dispersion for testing the appropriateness of a Poisson distribution for an observed variable Y.

The paper is structured as follows. Section 2 introduces the reader to the various approaches to modelling overdispersion in the case of traditional experimental contexts. Section 3 highlights approaches in the case of observational study contexts. Section 4 focuses on the case of heterogeneous populations followed by Sections 5 and 6, which look into a particular type of distribution, the generalized Waring distribution, and its relevance in the context of applications under the various scenaria leading to over-dispersion mentioned above. Through the prism of these scenaria, a bivariate version of it is also presented, and its use in applied contexts is discussed in Section 7. A multivariate version of it is also given, and its application potential is outlined in Section 8. Finally, Sections 9 and 10 present a model for temporally evolving data, the multivariate generalized Waring process, and an application illustrating its practical potential.

As the field of accident studies has received much attention, and various theories have been developed for the interpretation of factors underlying an accident situation, most of the models will be presented in accident or actuarial data analysis contexts. Of course the results can be adapted in a great variety of situations with appropriate parameter interpretations so that they can be applied in several other fields ranging from economics, inventory control and insurance through to demometry, biometry, psychometry and web access modeling, as the case is with the application discussed in Section 10.

2. Modelling over - or under - dispersion in traditional experimental contexts

One important, but often ignored by data analysts, implication of using single parameter distributions such as the Poisson distribution to analyse data is that the variance can be determined by the mean, a relation that collapses by the presence of overdispersion. If this is ignored in practice, any form of statistical inference may induce low efficiency, although, for modest amounts of overdispersion this may not be the case (Cox 1983). So, insight into the mechanisms that induce over (or under) dispersion is required when dealing with such data. Such insight can be gained by looking at the above-mentioned potential triggering sources as classified by Xekalaki (2006).

2.1 Lack of independence between individual responses

In accident study related contexts, where one is interested in the total number of reported accidents $Y = \sum_{i=1}^{n} Y_i$ in a total number of accidents, n, that actually occurred, when accidents are reported with equal probabilities $p = P(Y_i = 1) = 1 - P(Y_i = 0)$, but not independently $(Cor(Y_i, Y_j) = \rho \neq 0)$, the mean of Y will still be $E(Y) = np$, but its variance will be $V(Y) = V\left(\sum_{i=1}^{n} Y_i\right) = np(1-p) + 2\binom{n}{2}\rho p(1-p) = np(1-p)(1+\rho(n-1))$, which exceeds that anticipated under a hypothesized independent trial binomial model if $\rho > 0$ (over-dispersion) and is exceeded by it if $\rho < 0$ (under-dispersion).

2.2 Contagion

Another common reason for a variance differing from what is anticipated, is that when the assumption that the probability of the occurrence of an event in a very short interval is constant fails. This framework is the classical *contagion* model (Greenwood and Yule 1920; Xekalaki 1983a).

In data modelling problems faced by actuaries, for example, this model postulates that initially all individuals have the same probability of incurring an accident, but later this probability changes by each accident sustained. It is assumed, specifically, that none of the individuals has had an accident (e.g. new drivers or persons who are just beginning a new type of work), but later the probability with which a person with $Y = y$ accidents by time t will have another accident in the time period from t to $t + dt$ is of the form $(k + my)dt$. This leads to the negative binomial as the distribution of Y with p.f. $P(Y{=}y){=}\binom{k/m}{y}e^{-kt}(1{-}e^{-mt})^y$ with $\mu = E(Y) = k(e^{mt} - 1)/m$, and $V(Y) = ke^{mt}(e^{mt} - 1)/m = \mu e^{mt}$.

2.3 Clustering

A frequently overlooked clustered structure of the population may also induce over - or under - dispersion.

In an accident context again, an accident is regarded as a *cluster* of injuries:

The number Y of injuries incurred by persons involved in N accidents can naturally be thought of as expressed by the sum $Y = Y_1 + Y_2 + \ldots + Y_N$ of the numbers Y_i of injuries resulting from the i - th accident, assumed to be i.i.d. independently of the total number of accidents N, with mean μ and variance σ^2. In this case, $E(Y) = E\left(\sum_{i=1}^{N} Y_i\right) = \mu E(N)$ and $V(Y) = V\left(\sum_{i=1}^{N} Y_i\right) = \sigma^2 E(N) + \mu^2 V(N)$.

So, when N is a Poisson variable with mean $E(N) = \theta = V(N)$, the last relationship leads to overdispersion or underdispersion according as $\sigma^2 + \mu^2$ is greater or less than 1.

The first such model was introduced by Cresswell and Froggatt (1963) in a different accident context whereby each person is liable to spells of weak performance during which all of the person s accidents occur. So, if the number N of spells in a unit time period is Poisson distributed with mean θ, and within spells a person can have 0 accidents with probability $1 - m\log p$, $m > 1/\log p$, $0 < p < 1$ and n accidents $(n \geq 1)$ with probability $m(1 - p)^n/n$, $m, n > 0$ the observed distribution of accidents is the negative binomial distribution with probability function $P(Y = y) = \binom{\theta m + y - 1}{y} p^{\theta m}(1-p)^y$.

This model, known in the literature as the *spells model*, can also lead to other forms of overdispersed distributions (e.g. Xekalaki 1983a, 1984a).

2.4 Heterogeneity

Assuming a homogeneous population when in fact the population is heterogeneous, i.e., when its individuals have constant, but unequal probabilities of sustaining an event can also lead to overdispersion. In this case, each member of the population has its own value of the parameter θ and probability density function $f(\cdot\ ; \theta)$.

So, with θ regarded as the inhomogeneity parameter and varying from individual to individual according to any continuous, discrete, or finite step distribution $G(\cdot)$ of mean μ and variance σ^2, one is led to an observed distribution for Y with probability density function $f_Y(y) = E_G(f(y; \theta)) = \int_\Theta f(y; \theta)dG(\theta)$, where Θ is the parameter space. Models of this type are known as *mixtures*. (For details on their application in the statistical literature see e.g. Karlis and Xekalaki 2003; McLachlan and Peel 2001; Titterington 1990). Under such models, the variance of Y consists of two additive components, one representing the variance part due to the variability of θ and one due to the inherent variability of Y

if θ did not vary, *i.e.*, $V(Y) = V(E(Y|\theta)) + E(V(Y|\theta))$. This offers an explanation as to why mixture models are often referred to as *overdispersion models.*

It should be noted that a similar idea forms the basis for *analysis-of-variance (ANOVA)* models, where the total variability can be split into additive components, the *between groups* and the *within groups* components. In the case of the Poisson (θ) distribution, we have in particular that $V(Y) = E(\theta) + V(\theta)$. Based on the fact that in this case, the factorial moments of Y coincide with the moments of θ about the origin, Carriere (1993) proposed a test of the hypothesis that a Poisson mixture fits a data set.

Mixed Poisson distributions were first introduced by Greenwood and Woods (1919) in the context of accident studies. Assuming that an individuals accident experience $Y|\theta$ is Poisson distributed with parameter θ that was varying from individual to individual according to a gamma distribution with mean μ and index parameter μ/γ, they obtained a negative binomial distribution for Y with probability function $P(Y = y) = \binom{\mu/\gamma + y-1}{y}$ $\{\gamma/(1+\gamma)\}^y(1+\gamma)^{-\mu/\gamma}$ and with mean and variance given respectively by $E(Y) = \mu$ and $V(Y) = \mu(1 + \gamma)$, where γ represents the over-dispersion parameter.

The mixed Poisson process has been popularised in the actuarial literature by Dubourdieu (1938) gamma mixed case was treated by Thyrion (1969).

Numerous other mixtures have since then been proposed in the literature for interpreting overdispersion in data, such as binomial mixtures (e.g. Tripathi et al. 1994), negative binomial mixtures (e.g., Xekalaki 1983a, c, 1984a; Irwin 1975), normal mixtures (e.g. Andrews and Mallows 1974) and exponential mixtures (e.g. Jewell 1982). Discrete Poisson mixtures with finite step distributions for the Poisson parameter θ have also been proposed, the interest being on creating clusters of data by grouping the observations on Y according to some criterion *(cluster analysis)*. The number of clusters can be decided on the basis of a testing procedure for the number of components in the finite mixture (Karlis and Xekalaki 1999).

2.4.1 Heterogeneity in mixture models treating the parameter θ as the dependent variable in a regression model

Heterogeneity in models with explanatory variables can be modelled, by assuming that Y has a parameter θ varying from individual to individual according to some regression model $\theta = \eta(x; \beta) + \varepsilon$, where x is a vector of explanatory variables, β is a vector of regression coefficients, η is a function of a known form and ε has some known distribution. Such models are known in the literature as *random effect models* and have been extensively studied within the broad family of Generalized Linear Models. As a simple example in the case of a single covariate, say X, consider data Y_i , $i = 1, 2, \ldots, n$ coming from a Poisson population with mean θ determined by $\log \theta = \alpha + \beta x + \varepsilon$ for some constants α, β and with ε having a distribution with mean 0 and variance say ϕ. In this case, the marginal distribution of Y is no longer the Poisson distribution. It is a mixed Poisson distribution, with some mixing distribution $g(\cdot)$ clearly depending on the distribution of ε. In particular, $Y \sim Poisson\left(te^{(\alpha+\beta x)}\right) \hat{}_t\, g(t)$ where $t = e^{\varepsilon}$.

Negative Binomial and Poisson Inverse Gaussian regression models have also been proposed as overdispersed alternatives to the Poisson regression model (e.g. Lawless 1987; Dean et al. 1989; Xue and Deddens 1992). The case of a two finite step distribution, the finite Poison mixture regression model of Wang et al. s (1996) results. The similarity of the mixture representation and the random effects one is discussed in Hinde and Demetrio (1998).

In meta-analysis contexts, overdispersion (or underdispersion) refers to variance inflation (or deflation) relative to that anticipated by the fixed effects model. Two possible causes of such phenomena are a population structure in clusters or mixing resulting in a compound distribution. Kulinskaya and Olkin (2014) proposed approaching the problem of specification of a random effects model in meta-analysis in terms of a multiplicative model for the distribution of the effect size parameters that allows inflation or deflation. The model considered was motivated by overdispersion induced by intra-class correlation in the model assumed for the distribution of the i-th effect size estimate. In particular, the variance of the estimator $\hat{\theta}_i$ of the effect size parameter θ_i in the i-th study is assumed to be of the form $\sigma^2_{\hat{\theta}_i} = (1 + \alpha(n_i)\gamma)\sigma_i^2$, where $\alpha(n_i)$ are some known functions of the sample sizes n_i, σ_i^2 is the within the i-th study variance, $i = 1, 2, \ldots, k$ and γ is interpreted as an intra class correlation parameter.

2.4.2 Estimation and testing for overdispersion under mixture models

The structure of mixture models, including random effect models, entails different forms of variance-to-mean relationships. So, viewing the mean and variance of Y as represented by $E(Y) = \mu(\beta)$, and $V(Y) = \sigma^2(\mu(\beta), \lambda)$ respectively for some parameters β, λ a number of estimation approaches have been proposed in the literature based on moment methods (e.g. Breslow 1990; Lawless 1987; Moore 1986) and quasi or pseudo likelihood methods (e.g. Davidian and Carroll 1988; McCullagh and Nelder 1989; Nelder and Pregibon 1987). The above representation for the mean and variance of Y allows also estimation in the case of multiplicative overdispersion as in McCullagh and Nelder (1989).

Testing for the presence of overdispersion or underdispersion, on the other hand, can be done by means of asymptotic arguments. Let $f(y; \theta)$ denote the density function of a random variable Y in the initial model. Cox (1983) showed that, under regularity conditions, the density of y in the overdispersed model, $f_Y(y)$, admits a representation of the form $f_Y(y) = E_\Theta(f(y;\theta)) = f(y;\mu_\theta) + \frac{1}{2}\sigma_\theta^2 \frac{\partial^2 f(y;\mu_\theta)}{\partial \mu_\theta^2} + O(1/n)$, with $\mu_\theta = E(\theta)$, $\sigma_\theta^2 = V(\theta)$ and Θ is the parameter space. This in turn implies that $f_Y(y)$ can be put in the form $f(y; \mu_\theta)(1 + \varepsilon h(y, \phi_\theta))$, where $h(y, \phi_\theta) = \left[\frac{\partial \log f(y;\mu_\theta)}{\partial \mu_\theta}\right]^2 + \frac{\partial^2 \log f(y;\mu_\theta)}{\partial \mu_\theta^2}$.

This representation entails overdispersion if $\varepsilon > 0$, underdispersion if $\varepsilon < 0$ and, of course, none of these complications if $\varepsilon = 0$. Cox (1983) suggested a testing procedure for the hypothesis $\varepsilon = 0$, which can be regarded as a general version of standard dispersion tests.

2.5 Zero adjusted models

It would be interesting to note that another aspect of the population structure that is often responsible for the phenomenon of over-dispersion or under-dispersion is the presence of an excess or a scant number of zeros. Though the models discussed in Sections 2.3 and 2.4 may capture over-dispersion or under-dispersion rather well, they cannot capture excess or scarcity of zeros. In the literature, this question has been addressed by two types of models known as zero-inflated (or zero-deflated) models, and hurdle models. A unified representation of the models is provided by $f(y; \omega) = \omega I_{\{0\}}(y) + (1 - \omega)f_Y(y)$, where Y is the count variable, $I_{\{0\}}(\cdot)$ is the indicator function and ω is a constant, whose values, if in (0,1) render a hurdle model for $f_Y(0) = 0$, a zero-inflated model for $f_Y(0) \neq 0$, while negative values of it render a zero-deflated model.

Obviously, ω can be interpreted as the proportion of excess zeros in the case of the first two models and the above representation explains why there can be regarded as having a dual nature. They are (finite) mixtures, which account for heterogeneity, while at the same time, they are capturing a population structure in two clusters. However, in the case $\omega < 0$ (zero-deflation), the model ceases to admit a mixture interpretation.

Zero-inflated and hurdle models have mostly been used for Poisson, generalized Poisson or negative binomial count distributions in various contexts (e.g. Ridout et al. 2001; Gupta et al. 2004; Famoye and Singh 2006). Gupta et al. (1996) proposed a zero-adjusted generalized Poisson distribution and studied the effect of not using an adjusted model for zero-inflation or -deflation when the occurrence of zeroes differs from the anticipated one. Reviews of such models can be found in Ridout et al. (1998), Gschlößl and Czado (2008) and Ngatchou-Wandji and Paris (2011).

3. Over- or under-dispersion in observational study contexts - the effect of the method of ascertainment

Often, in connection with data collection based on observation or on recording values as produced by nature, the original distribution may not be reproduced due to various reasons. These may lead to partial destruction or partial enhancement (augmentation) of observations. The models that have been introduced to deal with such situations are respectively known as *damage models* introduced by Rao (1963) and *generating models* introduced by Panaretos (1983). The distortion mechanism is usually assumed to be manifested through the conditional distribution of the resulting random variable Y given the value of the original random variable X. Hence, the resulting (observed) distribution is a distorted version of the original distribution that can be represented as a mixture of the distortion mechanism. In particular, in the case of damage, $P(Y = r) = \sum_{n=r}^{\infty} P(Y = r|X = n)P(X = n),\ \ r = 0,\ 1,\ 2,\ \ldots,$ while, in the case of enhancement,

$$P(Y = r) = \sum_{n=1}^{r} P(Y = r|X = n)P(X = n),\ \ r = \ 1,\ 2,\ \ldots.$$

Various forms of distributions have been considered for the distortion mechanism in the above two cases. In the case of damage, the most popular forms have been the binomial distribution Rao (1963), mixtures on p of the binomial distribution (e.g. Panaretos 1982; Xekalaki and Panaretos 1983) whenever damage can be regarded as additive ($Y = X - U$, U independent of Y) or in terms of the uniform distribution in $(0, x)$ (e.g. Dimaki and Xekalaki 1990, 1996; Xekalaki 1984b) whenever damage can be regarded as multiplicative ($Y = [RX]$, R independent of X and uniformly distributed in $(0, 1)$). The latter case has also been considered in the context of continuous distributions by Krishnaji (1970). The generating model was introduced and studied by Panaretos (1983).

Both, the generating model and the damage model offer a perceptive approach in actuarial contexts where one is interested in modelling the distributions of the numbers of accidents, of the damage claims, and of the claimed amounts. These models become relevant due to the fact that people have in general a tendency to under report their accidents, so that the reported (observed) number Y is less than or equal to the actual number

X $(Y \le X)$, but tend to over report damages incurred by them, so that the reported damage Y is greater than or equal to the true damage X $(Y \ge X)$.

Another type of distortion is induced by the adoption of a sampling scheme that assigns to the units in the original distribution unequal probabilities of inclusion in the sample. As a result, the value x of X is observed with a frequency that noticeably differs from that anticipated under the original density function $f_X(x; \theta)$. It represents an observation on a random variable Y whose probability distribution is the results of adjusting the probabilities of the anticipated distribution through weighting them with the probability with which the value x of X is included in the sample. So, if this probability is proportional to some *weight function,* $w(x, \beta)$, $\beta \in R$, the recorded value x is a value of Y having density function $f_Y(x; \theta, \beta) = w(x; \beta) f_x(x; \theta)/E(w(X; \beta))$.

Distributions of this type are known as *weighted distributions* (see, e.g. Cox 1962; Fisher 1934; Patil and Ord 1976; Rao 1985). For $w(x; \beta) = x$, these are known as *size biased distributions.* In actuarial data modelling contexts again, the weight function can represent reporting bias. In the context of reporting accidents or placing damage claims, for example, it can have a value that is directly or inversely analogous to the size x of X, the actual number of incurred accidents or the actual size of the incurred damage. The functions $w(x; \beta) = x$ and $w(x; \beta) = \beta^x$ $(\beta > 1$ *or* $\beta < 1)$ are plausible choices. So, for example, in the case of a Poisson (θ) distributed X, these lead to distributions for Y that are of Poisson type. In particular, the weight function $w(x; \beta) = x$ leads to a shifted Poisson distribution with probability function $P(Y = x) = e^{-\theta}\theta^{x-1}/(x-1)!$, $x = 1, 2, \ldots$, while the choice $w(x; \beta) = \beta^x$ leads to a Poisson distribution $P(Y = x) = e^{-\theta\beta}(\theta\beta)^x/x!$, $x = 0, 1, \ldots$. The value of the variance of the observed variable Y under the first assumption for $w(x; \beta)$ is $1 + \theta$ and exceeds that of X (overdispersion), while under the second assumption it is $\theta\beta$ implying overdispersion for $\beta > 1$ or underdispersion for $\beta < 1$.

4. Looking closer into the case of heterogeneity

Assuming a specific form for the distribution of the population that generated a data set implies that the mean to variance relation is given for this distribution, e.g. the Poisson distribution with a mean to variance ratio equal to unity. As has become obvious from the above, this relationship ceases to hold in real data sets however. This being rarely the case, flexible families have been sought in the literature by allowing the parameter θ of the original distribution to vary according to a distribution with probability density function, say $g(\cdot)$.

As mentioned before, a density function $f_X(\cdot)$ is a mixture on the parameter θ of the distribution function $f(\cdot\ ; \theta)$ *with some mixing distribution* $G_\theta(\cdot)$, which can be continuous, discrete or a finite step distribution, if it can be written in the form $f_X(x) = E_G(f(x; \theta)) = \int_\Theta f(x; \theta) dG(\theta)$, where Θ is the parameter space. An appropriate choice of a mixing distribution allows its parameter to vary and acts as a means of loosening" the structure of the initial model, thus offering more realistic interpretations of the mechanisms that generated the data.

A large number of Poisson mixtures have been developed. (For an extensive review, see Karlis and Xekalaki 2003, 2005). The derivation of the negative binomial distribution, as a mixture of the Poisson distribution with a gamma distribution as the mixing distribution, originally obtained by Greenwood and Yule (1920) constitutes a typical example. Mixtures of the negative binomial distribution have also been widely used in connection with applications in a plethora of fields. These include the Yule distribution (Yule 1924; Irwin 1941; Xekalaki 1983c, 1984b) the Waring distribution (Irwin 1963) and the generalized Waring distribution (Irwin 1968, 1975; Xekalaki 1981, 1983a, 1984a), which contains the Yule distribution and the Waring distribution as a special cases.

In what follows, we focus on the generalized Waring distribution and its relevance in accident data modeling contexts.

5. The generalized Waring distribution

This was introduced by Irwin (1968) in connection to biological data and later was shown by him to arise as an accident distribution (Irwin 1975). It is the distribution with probability generating function given by

$$G(s) = \frac{\rho_{(k)}}{(a+\rho)_{(k)}} {}_2F_1(a,k;a+k+\rho;s),\ \alpha,k,\rho > 0$$

with ${}_2F_1(a,b;c;z)$ denoting the Gauss hypergeometric function $\sum_{r=a}^{x} \{a_{(r)}b_{(r)}z^r\}/\{c_{(r)}r!\}$, where $h_{(l)} = \Gamma(h+l)/\Gamma(h),\ h>0,\ l \in R$.

Irwin s starting point was Waring s expansion (hence the distribution s name) given by $\frac{1}{x-a} = \sum_{r=0}^{\infty} \frac{a_{(r)}}{x_{(r+1)}}$, which he then generalized to $\frac{1}{(x-a)_{(k)}} = \sum_{r=0}^{\infty} \frac{a_{(r)}k_{(r)}}{x_{(k+r)}}\frac{1}{r!},\quad \alpha,k>0.$

Hence, by multiplying both sides by $\rho_{(k)}$, where $\rho = x - a > 0$, the successive terms of the resulting series could he regarded as defining a probability function, which he termed the generalized Waring distribution with parameters α,k,ρ. In particular, the probability function of the generalized Waring distribution with parameters α,k,ρ is given by

$$p_r = \frac{\rho_{(k)}}{(a+\rho)_{(k)}}\frac{a_{(r)}k_{(r)}}{(a+k+\rho)_{(r)}}\frac{1}{r!},\ \alpha,k,\rho>0,\ r=0,\ 1,\ 2,\ \ldots$$

where $h_{(l)} = \Gamma(h+l)/\Gamma(h)$.

Notwithstanding the complexity of its structure, this distribution was shown to offer an insightful tool in the interpretation of accident data as will be seen below. Among its aspects that can be of practical value, is that, as shown by Xekalaki (1983b), it is a discrete self-decomposable distribution in Steutel and van Harn s (1979) sense, hence infinitely divisible, implying that its probability generating function can be put in the form $G(s) = \exp\left\{-\lambda\int_s^1 \frac{1-g(u)}{1-u}du\right\}$, where $\lambda = p_1/p_0$ and $g(\cdot)$ denotes the probability generating function of the distribution with probability function satisfying the recurrence relation

$$q_n = \lambda\left\{n\frac{ak+\rho(a+k+\rho)}{ak(a+k+\rho+n)} - \sum_{j=0}^{n-1} q_j\binom{n}{j}\binom{a+k+\rho+n-1}{j}\Big/\left[\binom{a+n-1}{j}\binom{k+n-1}{j}\right]\right\}$$

6. The generalized Waring distribution in relation to accident theory

The hypotheses that have formed the basis of investigations into the occurrence of accidents since almost a century ago are

(i) ***Pure chance,*** giving rise to the Poisson distribution
(ii) ***True contagion,*** i.e. the hypothesis that initially all individuals have the same probability of incurring an accident but that this probability is modified by each accident sustained.
(iii) ***Apparent contagion (heterogeneity),*** i.e. the hypothesis that individuals have constant but unequal probabilities of having an accident - the resultant distribution being a compound Poisson distribution **("accident proneness" model).**
(iv) ***The "Spells" Model,*** i.e each person is liable to periods of time during which the person's performance is weak (spells). All of the person's accidents occur within those spells. The numbers of accidents within different spells are independent and independent of the number of spells.

As already seen, the negative binomial distribution can be given a an accident proneness and a spells" interpretation in the context of accident theory in terms of a gamma mixed Poisson distribution and a Poisson distribution generalized by a logarithmic distribution (Kemp 1967).

Therefore, a good fit of the negative binomial is no help at all in distinguishing among the proneness", contagion" and spells" hypotheses. This is known as the discrimination problem between the compounded, contagion and generalized models for the negative binomial distribution and has been discussed by Arbous and Kerrich (1951); Bates and Neyman (1952); Gurland (1959) and Cane (1974, 1977). For an extensive bibliography on the accident hypotheses mentioned, see Kemp (1970).

6.1 Irwin's "Proneness" model

As evident, in all three of the above models, the data are treated as if the individuals under observation were exposed to equal environmental risk, a fact criticized by Irwin (1968), who suggested a three-parameter distribution, which he called the univariate generalized Waring distribution" (UGWD). He derived this distribution in a framework that allows separately for random factors, differences in the exposure of individuals to external risk of accident, and differences in proneness.

In particular, his model assumes a non homogeneous population with respect to personal and environmental attributes affecting the occurrence of accidents.

Let the distribution of the number, X, of accidents for individuals of equal proneness ν, and of equal exposure to external risk of accident $\lambda|\nu$, i.e. λ for given ν), have probability generating function

$$G_{X|\lambda}(s) = \exp\{(\lambda|\nu)(s-1)\}$$

in a unit time interval (0, 1). If the distributions of $\lambda|\nu$ and ν in the population at risk can be described by the probability density functions (pdf)

$$\{\nu^{-k}\exp(-\lambda/\nu)\lambda^{k-1}\}/\Gamma(k),\ \nu, k > 0$$

and

$$\left\{\Gamma(a+\rho)\nu^{a-1}(1+\nu)^{-(a+\rho)}\right\}/\{\Gamma(\rho)\Gamma(a)\},\ a,\rho>0$$

respectively, the pgf of the resulting distribution of accidents will be $\{\rho_{(k)2}F_1(a,\ k;\ a+k+\rho;\ s\}/(a+\rho)_{(k)}$, i.e. the univariate generalized Waring distribution with parameters a, k and ρ, which will be denoted by $UGWD(a,\ k;\ \rho)$. Here, ${}_2F_1(a,\ b;\ c;\ z)$ denotes the Gauss hypergeometric function $\sum_{r=a}^{x}\{a_{(r)}b_{(r)}z^r\}/\{c_{(r)}r!\}$, where $h_{(l)}=\Gamma(h+l)/\Gamma(h)$, $h>0$, $l\in R$. For more information about the UGWD the reader is referred to the work of Irwin (1963, 1968, 1975); Xekalaki (1981) and the references therein and Xekalaki (1983a).

6.2 The "Contagion" model

Xekalaki (1983a), extended the assumptions of the classical contagion model developed by Greenwood and Yule (1920) by considering a population of individuals exposed to varying accident risk.

In particular, assume that at time $t=0$ none of the individuals has had an accident. This would be true if, for example, with a population of new drivers or of individuals just beginning a new type of work. Suppose that during the time period from t to $t+dt$ a person with x accidents by time t can incur another accident with a probability of $\{(k+x)/(1+\lambda t)\}\lambda dt$ (independent of the times of the previous accidents), where k is a positive constant and λ refers to the individual s risk exposure. At $t=0$, since $x=0$, the probability of an accident is $k\lambda dt$. Hence, what the model basically assumes is that, initially, the probability of having an accident is not the same for each individual, but depends on the external conditions; later, the probability is also affected by the number of preceding accidents. Under these assumptions and if differences in the exposure to accident risk can be thought of as governed by a distribution with probability density function given by $\{\Gamma(a+\rho)\nu^{a-1}(1+\nu)^{-(a+\rho)}\}/\{\Gamma(\rho)\Gamma(a)\}$, the final distribution of accidents over a unit period of time turns out to be $UGWD(a,\ k;\ \rho)$.

The above derivation of the generalized Waring distribution closely relates to a modeling approach whereby the distribution of accident occurrences in a time internal $(0,\ t)$ is regarded as underpinned by a stochastic process and, in particular, by a pure birth process $\{X_t,\ t=0,\ 1,\ 2,\ \ldots\}$ where the probability of a person to incur an accident in $(t,\ t+dt)$, having had x accidents by time t is $P(X_{t+\delta t}=x+1|X_t=x)=f_\lambda(n,t)\delta t+o(\delta t)$.

Irwin (1941), followed later by Arbous and Kerrich (1951), derived the negative binomial distribution on the hypothesis solving the associated Kolmogorov forward differential equations by a method due to McKendrick (1925). Specifically, assuming that individuals can have during the time period from t to dt, individuals can have 0 accidents with probability $1-f_\lambda(x,\ t)dt$, 1 accident with probability $f_\lambda(x,\ t)dt$ and >1 accidents with probability 0, he solved the resulting system of Kolmogorov forward difference-differential equations

$$\frac{\partial}{\partial t}P_\lambda(0,t)=-f_\lambda(0,t)P_\lambda(0,t)$$
$$\frac{\partial}{\partial t}P_\lambda(x,t)=-f_\lambda(x,t)P_\lambda(x,t)+f_\lambda(x-1,t)P_\lambda(x-1,t),\ x\geq 1$$

in terms of a single difference-differential equation involving the probability generating function $G_\lambda(s;t)$ of X_t given by

$$\frac{\partial}{\partial t}G_\lambda(s;t) = (s-1)\sum_{x=0}^{\infty} s^x f_\lambda(x,t)P_\lambda(x,t)$$

where $G_\lambda(s;t) = \sum_{x=0}^{\infty} P_\lambda(x,t)s^x$. (He obtained this equation by multiplying the i-th equation of the system by s^{i-1}, $i = 1, 2, \ldots$ and summing the resulting equations).

Assuming further that $f_\lambda(x,t) = \lambda(k+mx)$, $k, m > 0$ and subject to the initial conditions $G_\lambda(1;t) = G_\lambda(s;0) = 1$, he obtained for the distribution of accidents

$$G_\lambda(s;t) = \left[e^{\lambda mt} - s\left(e^{\lambda mt}-1\right)\right]^{-k/m},$$

i.e. the probability generating function of the negative binomial distribution with parameters k/m and $(1-e^{-\lambda mt})^{-1}$.

Relaxing Irwin s implicit assumption that all individuals were exposed to the same accident risk, Xekalaki (1981) treated the parameter λ as referring to a variable risk exposure according to an exponential distribution with density $ae^{-a\lambda}$, $a > 0$ and obtained the generalized Waring distribution as the accident distribution. In particular,

$$\begin{aligned}
G_{X_t}(s) &= a\int_0^{\infty} e^{-a\lambda}\left[e^{\lambda mt} - s\left(e^{\lambda mt}-1\right)\right]^{-k/m} d\lambda \\
&= \frac{a(1-s)^{-k/m}}{mt}\int_0^{\infty} e^{-\lambda\frac{a+kt}{mt}}\left(1-\frac{s}{s-1}e^{-\lambda}\right)^{-k/m} d\lambda \\
&= \frac{a(1-s)^{-k/m}}{mt}\frac{\Gamma((a+kt)/(mt))}{\Gamma(1+(a+kt)/(mt))}\,{}_2F_1\left(\frac{k}{m},\frac{a+kt}{mt};\frac{a+kt}{mt}+1;\frac{s}{s-1}\right) \\
&= \frac{a}{a+mt}\,{}_2F_1\left(\frac{k}{m},1;\frac{a}{mt}+\frac{k}{m}+1;s\right)
\end{aligned}$$

which is the probability generating function of the $UGWD\left(\frac{k}{m},1;\frac{a}{mt}\right)$.

This model was considered by Panaretos (1989) for the description of the evolution of surnames. Faddy (1997) provided a unifying approach to under- and over-dispersion relative to the Poisson distribution within a scheme of a similar nature, which generalizes the simple Poisson process that underpins the Poisson distribution. He demonstrated that any count distribution can be obtained by a suitable choice of $f_\lambda(x,t)$ and provided an expression for the system of Kolmogorov forward differential equations in terms of a matrix-exponential function.

Finally, Winkelmann (1995) looked at under- and over-dispersion using renewal theory by exploring the link between duration dependence and dispersion. He demonstrated that discrepancies between observed and nominal variances are conveyed by a hazard function of the waiting times that is not constant, but instead is a decreasing function of time inducing over-dispersion or an increasing function of time inducing under-dispersion.

6.3 The "Spells" model

Further, Xekalaki (1983a) considered a variant of the spells" model due to Cresswell and Froggatt (1963) that rejects the presence of proneness and contagion.

Assume that every individual is liable to spells and that the number of spells in a given time period $(0, t)$ is a Poisson variable with parameter $\theta t, \theta > 0$. Suppose that no accidents occur outside spells and that the probability of an accident within a spell depends on the risk exposure of the particular individual. In particular, suppose that within a spell a person can have

$$\left.\begin{array}{l} 0 \text{ accidents with probability } 1-m\log(1+\lambda) \\ \text{or } n \text{ accidents } (n\geq 1) \text{ with probability } m\{\lambda/(1+\lambda)\}^n/n \end{array}\right\},$$

$0 < m < 1/\log(1+\lambda)$, $\lambda > 0$, where λ is the external risk parameter for the given individual. Assume further that the numbers of accidents arising out of different spells are independent and independent of the number of spells. Then, if differences in the risk exposure can be described by a beta distribution of the second kind with probability density function, $\{\Gamma(a+\rho)v^{a-1}(1+v)^{-(a+\rho)}\}/\{\Gamma(\rho)\Gamma(a)\}$, $a, \rho > 0$, the resulting accident distribution will have probability generating function given by

$$\left\{\rho_{(a)}\,{}_2F_1(a, \theta mt; a+\theta mt+\rho; s)\right\}/(\rho+\theta mt)_{(a)}.$$

Hence, in a unit time period, the number of accidents follows the $UGWD(a, \theta m; \rho)$.

It is worth noticing that the form of the distribution of λ in the last two models is more general than that considered by the proneness model. It is however, a reasonable choice as it implies a beta distribution of the first kind (Pearson Type I) for the parameter $q = \lambda/(1+\lambda)$ of the negative binomial distribution of $X|\lambda$.

6.4 Deciding about the underlying model

It is evident from the above, that three completely different sets of hypotheses give rise to exactly the same form of distribution and that while the *UGWD* may be a plausible model if accident proneness is a accepted as an established fact, a satisfactory fit of it is not to be taken as evidence for the validity of the proneness hypothesis. How can we then discriminate?

Statisticians have always been excited to look for ways of discriminating among different models that give rise to the same distribution. Most attempts seem to have been concentrated on distinguishing between the proneness and contagion models generating the negative binomial distribution. The papers by Bates and Neyman (1952) and Bates (1955) cover part of the work that has been done on the subject, though they primarily focus on distinguishing between different forms of contagion. Shaw and Sichel s (1971) attempt was on proving or disproving proneness by ranking individual accident performance on a scale based on their average interval between successive accidents. However, the first systematic study on how one can discriminate between the proneness and contagion models of the negative binomial distribution appears to be that by Cane (1974).

She demonstrated, however, that one cannot distinguish between the two models, even with knowledge of the time sequence of accidents. She demonstrated, in particular, that the conditional distribution of the times, t_i, $i = 1, 2, \ldots, n$ at which accidents occurred in a time period $(0, T)$ is the same in both cases, namely that of an ordered sample from a uniform distribution over $(0, T)$ with probability density function $n!\, T^{-n}$. In fact, this is the case for any compound Poisson accident distribution whose compounding distribution has finite moments (Cane 1977), hence also for the $UGWD(a, k; \rho)$.

This implies that the availability of information on the times of the occurrence of accidents is not sufficient to guide one s choice between the proneness and contagion models.

However, as demonstrated by Xekalaki (1983a), there appears to exist a possibility in the framework of the Spells model. Consider, in particular, the problem of finding the joint distribution of times t_i, $i = 1, 2, \ldots, n$ of accidents by individuals with n accidents in a unit period of time under the spells model. For fixed λ, accidents occur as events in a generalized Poisson process: $X(t) = \sum_{i=1}^{N(t)} Y_i$, $N(t) \sim Poisson(\theta t)$, where $\theta > 0$, $t \geq 0$ and Y_i are identically and independently distributed with probability density function given by $\{\Gamma(a+\rho)v^{a-1}(1+v)^{-(a+\rho)}\}/\{\Gamma(\rho)\Gamma(a)\}$, $a, \rho > 0$. Consequently, the required probability function can be written as $\int_0^\infty (1+\lambda)^{-\theta m(1-t_n)} \left[\prod_{i=1}^{n} \left\{\lambda m\theta(1+\lambda)^{-\theta m(t_i - t_{i-1})-1} dt_i\right\}\right] dH(\lambda)$, with $H(\cdot)$ denoting the distribution function of the beta distribution of the second kind defined as above. Hence, the required probability is $\left\{\frac{(\theta m)^n \rho_{(a)} a_{(n)}}{(\theta m+\rho)_{(a+n)}}\right\} dt_1 \ldots\ dt_n$. Therefore, conditional on n accidents during a time period from 0 to 1, the joint pdf of t_i, $i = 1, 2, \ldots, n$, is $n!\,(\theta m)^n/(\theta m)_{(n)}$.

The obtained form differs from that arising under the proneness and contagion models. This fact is itself is very interesting as far as establishing the presence of spells is concerned, as it implies the following: if an observed accident distribution of the $UGWD$ type has arisen from the spells model, the time intervals $(0, t_i)$, $i = 1, 2, \ldots, n$, given a total of n accidents, will be jointly distributed with the above density function. Any departure from this distribution is, then, evidence against the spells model. Of course, if on the available evidence one has to reject this form in favor of that obtained by Cane, then one is faced again with the question: proneness or contagion?" This cannot be answered by studying the distribution of t_i.

6.5 What does Irwin's accident model offer beyond a good fit to the data?

The innovation brought by Irwin s accident proneness model does not merely lie in the better fit it provides to accident data, but in the possibility of partitioning the total variance

(σ^2) into three additive components due to proneness $\left(\sigma_\nu^2\right)$, liability $\left(\sigma_\lambda^2\right)$ and randomness $\left(\sigma_R^2\right)$ thus,

$$\sigma^2 = \sigma_\lambda^2 + k^2\sigma_\nu^2 + \sigma_R^2,$$

Where

$$\begin{aligned}
\sigma_\lambda^2 &= ak(a+1)(\rho-1)^{-1}(\rho-2)^{-1} \\
\sigma_\nu^2 &= a(a+\rho-1)^{-1}(\rho-1)^{-2}(\rho-2)^{-1} \\
\sigma_R^2 &= ak(\rho-1)^{-1} \\
\sigma^2 &= ak(a+\rho-1)(k+\rho-1)(\rho-1)^{-2}(\rho-2)^{-1}.
\end{aligned}$$

There is still, however, a problem due to the fact that the $UGWD(a, k; \rho)$ is symmetrical in a and k ($UGWD(a, k; \rho) \sim UGWD(k, a; \rho)$). Hence, although one may consider that $\sigma_\lambda^2 + k^2\sigma_\nu^2$ represents the variance component due to all non-random factors, the mathematics alone cannot determine whether σ_λ^2 represents the liability component and $k^2\sigma_\nu^2$ the proneness component or vice versa. As a consequence, distinguishable estimates for the non-random variance components σ_λ^2 and σ_ν^2 cannot be obtained unless subjective judgement is made. This problem was addressed by Xekalaki (1984a) with the introduction of her bivariate form of the generalized Waring distribution.

7. The bivariate generalized Waring distribution

Generalizing further Irwin s (1963) generalization of Waring s expansion, we have for $k, m, a > 0$,

$$\begin{aligned}
\frac{1}{(x-a)_{(k+m)}} &= \sum_{\ell=0}^{\infty}\sum_{r=0}^{\ell} \frac{a_{(\ell)}(-1)^\ell}{\ell!}\binom{\ell}{r}\Delta^r\frac{1}{x_{(k)}}\Delta^{\ell-r}\frac{1}{(x+k+r)_{(m)}} \\
&= \sum_{r=0}^{\infty}\sum_{\ell=0}^{\infty} \frac{a_{(r+\ell)}(-1)^{r+\ell}}{r!\ell!}\Delta^r\frac{1}{x_{(k)}}\Delta^{\ell}\frac{1}{(x+k+r)_{(m)}} \\
&= \sum_{r=0}^{\infty}\sum_{\ell=0}^{\infty} \frac{a_{(r+\ell)}k_{(r)}m_{(\ell)}}{x_{(k+m+r+\ell)}}\frac{1}{r!}\frac{1}{\ell!}
\end{aligned}$$

If $x > a$, the above series is convergent. Then, by letting $\rho = x - a > 0$ and multiplying both sides by $\rho_{(k+m)}$, leads to a double series of positive terms converging to unity. The general term of the series therefore can be regarded as defining a bivariate discrete probability distribution with probability function

$$p_{r,\ell} = \frac{\rho_{(k+m)}}{(a+\rho)_{(k+m)}}\frac{a_{(r+\ell)}k_{(r)}m_{(\ell)}}{(a+k+m+\rho)_{(r+\ell)}}\frac{1}{r!}\frac{1}{\ell!},\ a, k, m, \rho > 0,\ r, \ell = 0,\ 1,\ 2,\ \ldots$$

In the remainder of the paper, we refer to this distribution as the *bivariate generalized Waring distribution* with parameters a, k, m and ρ and we denote it by $BGWD(a; k, m; \rho)$.

7.1 The BGWD in relation to accident theory

Assume that individuals of proneness ν and liability $\lambda_i|\nu$ for a period i of observation incur, over two non-overlapping time periods, accidents X, Y according to a double Poisson distribution $G_{(X,Y)|\lambda_1,\lambda_2,\nu}(s,t) = \exp\{(\lambda_1|\nu)(s-1) + (\lambda_2|\nu)(t-1)\},\ \lambda_1,\lambda_2 > 0$. Assume further that the liability parameters $\lambda_1|\nu,\ \lambda_2|\nu$ are independently gamma distributed with densities $\left(\Gamma(\theta_i)\nu^{\theta_i}\right)^{-1} e^{-\lambda_i|\nu} \lambda_i^{\theta_i - 1},\ \theta_1 \equiv k,\ \theta_2 \equiv m,\ \nu > 0$, whence for individuals with the same proneness ν, but varying liabilities, the numbers of occurring accidents over the two periods are jointly distributed as the double negative binomial with probability generating function

$$G_{(X,Y)|\nu}(s,t) = \{1 + \nu(1-s)\}^{-k}\{1 + \nu(1-t)\}^{-m}.$$

Letting now the proneness parameter ν be beta distributed with density function $\{\Gamma(a+\rho)\nu^{a-1}(1+\nu)^{-(a+\rho)}\}/\{\Gamma(\rho)\Gamma(a)\},\ a,\rho > 0$, the probability generating function of the joint distribution of accidents over the two periods takes the form

$$\begin{aligned} G_{(X,Y)}(s,t) &= \frac{\Gamma(\rho+a)}{\Gamma(\rho)\Gamma(a)} \int_0^{+\infty} \nu^{a-1}(1+\nu)^{-(a+\rho)}\{1+\nu(1-s)\}^{-k}\{1+\nu(1-t)\}^{-m} d\nu \\ &= \frac{\rho_{(k+m)}}{(a+\rho)_{(k+m)}} F_1(a;k,m;a+k+m+\rho;s,t) \sim BGWD(a;k,m;\rho), \end{aligned}$$

where $F_1(a;b,c;d;u,v) = \sum_{r,s=0}^{\infty} \{a_{(r+s)} b_{(r)} c_{(s)} u^r v^s\} / \{d_{(r+s)} r! s!\}$ is Appells hypergeometric series and $h(l) = \Gamma(h+l)/\Gamma(h),\ h > 0,\ l \in R$.

Regarding separate estimation of the contribution of proneness, liability and randomness in a given accident situation over a period of observation whenever proneness is accepted as an established fact, Xekalaki (1984a) showed that rearranging the observed distribution in two non-overlapping sub-intervals and fitting the $BGWD(a;k,m;\rho)$ to the resulting bivariate accident distribution does enable separate estimation of the variance components. This is demonstrated in Table 1.

Further models leading to the *BGWD* provided by Xekalaki (1984c), provide the framework within which one can also obtain the *BGWD* as an accident distribution under the contagion and the spells accident theories.

Table 1 Estimators of the components of the variance of the generalized waring distribution

Component due to	Marginal variance of X	Marginal variance of Y	Variance of $X+Y$ entire period
Random factors	$\frac{\hat{a}\hat{k}}{\hat{\rho}-1}$	$\frac{\hat{a}\hat{m}}{\hat{\rho}-1}$	$\frac{\hat{a}\left(\hat{k}+\hat{m}\right)}{\hat{\rho}-1}$
Proneness	$\frac{\hat{k}^2\hat{a}(\hat{a}+\hat{\rho}-1)}{(\hat{\rho}-1)^2(\hat{\rho}-2)}$	$\frac{\hat{m}^2\hat{a}(\hat{a}+\hat{\rho}-1)}{(\hat{\rho}-1)^2(\hat{\rho}-2)}$	$\frac{\left(\hat{k}+\hat{m}\right)^2\hat{a}(\hat{a}+\hat{\rho}-1)}{(\hat{\rho}-1)^2(\hat{\rho}-2)}$
Liability	$\frac{\hat{a}\hat{k}(\hat{a}+1)}{(\hat{\rho}-1)(\hat{\rho}-2)}$	$\frac{\hat{a}\hat{m}(\hat{a}+1)}{(\hat{\rho}-1)(\hat{\rho}-2)}$	$\frac{\hat{a}\left(\hat{k}+\hat{m}\right)(\hat{a}+1)}{(\hat{\rho}-1)(\hat{\rho}-2)}$
Total	$\frac{\hat{a}\hat{k}\left(\hat{\rho}+\hat{k}-1\right)(\hat{\rho}+\hat{a}-1)}{(\hat{\rho}-1)^2(\hat{\rho}-2)}$	$\frac{\hat{a}\hat{m}(\hat{\rho}+\hat{m}-1)(\hat{\rho}+\hat{a}-1)}{(\hat{\rho}-1)^2(\hat{\rho}-2)}$	$\frac{\hat{a}\left(\hat{k}+\hat{m}\right)\left(\hat{\rho}+\hat{k}+\hat{m}-1\right)(\hat{\rho}+\hat{a}-1)}{(\hat{\rho}-1)^2(\hat{\rho}-2)}$

8. The multivariate generalized Waring distribution

The n-variate version of the genaralized Waring distribution introduced and studied by Xekalaki (1986) is also obtained as an inverse factorial distribution. Its probability generating function is given by

$$G(\underline{t}) = \frac{\rho_{\left(\sum k_i\right)}}{(a+\rho)_{\left(\sum k_i\right)}} F_D\left(a; k_1, \ldots, k_n; a + \sum_{i=1}^{n} k_i + \rho; \underline{t}\right)$$

with $F_D(a; \beta_1, \ldots, \beta_n; \gamma; \underline{t})$ denoting Lauricellas hypergeometric function given by

$$F_D(a; \beta_1, \ldots, \beta_n; \gamma; \underline{t}) = \sum_{r_1, \ldots, r_n} \frac{a_{\left(\sum r_i\right)}}{\gamma_{\left(\sum r_i\right)}} \prod_{i=1}^{n} \frac{(\beta_i)_{(r_i)} t_i^{r_i}}{(r_i)!}$$

The probability function of it is given by

$$P_{\underline{r}} \equiv P(\underline{X} = \underline{r}) = \frac{\rho_{\left(\sum k_i\right)}}{(a+\rho)_{\left(\sum k_i\right)}} \frac{a_{\left(\sum r_i\right)} (k_1)_{(r_i)} \ldots (k_n)_{(r_n)}}{\left(a+\rho+\sum k_i\right)_{\left(\sum r_i\right)} (r_1)! \ldots (r_n)!},$$
$$r_i = 0, 1, 2, \ldots; \; i = 1, \ldots, n$$

and its probabilities are related by the following first order recurrences, which facilitate their computation

$$\frac{P_{l_1, l_2, \ldots, l_{h-1}, l_h+1, l_{h+1}, \ldots, l_n}}{P_{l_1, l_2, \ldots, l_n}} = \frac{\left[a + \sum_{i=1}^{n} l_i\right](k_n + l_n)}{\left[a + \sum_{i=1}^{n} k_i + \sum_{i=1}^{n} l_i\right](l_n + 1)}, \; l = 0, \; 1, \; 2, \; \ldots;$$
$$i = 1, \; 2, \; \ldots, n$$

An interesting aspect of the bivariate and multivariate versions of the generalized Waring distribution is that their marginal distributions (conditional and unconditional) as well as their convolution are of the same form (*UGWD's*), properties that exhibit a symmetry analogous to that existing in the case of the multivariate normal distribution. Further, the generalized Waring distribution is self-decomposable (Xekalaki 1983b).

9. The Generalized Waring Process (gWp)

Looking into how temporally evolving data from the wide spectrum of application contexts that can reasonably be viewed from the perspective of the frameworks discussed in Sections 6, 7 and 8 can be treated, Xekalaki and Zografi (2008) defined and studied the generalized Waring process. In establishing its definition, the structural properties of both the bivariate and the multivariate versions of the generalized Waring distribution played a significant role. This process, analogously to the case of Poisson and Pólya processes, which can be obtained as limiting cases of it, was shown to be a Markov process.

Let $\{N(t), t \geq 0\}$ be a counting process. This is said to be a *generalized Waring process* with parameters $a, k, \rho > 0$, denoted by $gWp(a, k; \rho)$, if (i) $N(0) = 0$, (ii) $N(t)$ is a Markov process, and (iii) $N(t+h) - N(t)$ has the generalized Waring distribution with parameters $a, k; \rho$ for $h > 0,\ t \geq 0$. The process starts at 0, it has stationary increments and

$$P(N(t)=n) = \frac{\rho_{(kt)}}{(\rho+a)_{(kt)}} \frac{a_{(n)}(kt)_{(n)}}{(a+\rho+kt)_{(n)}} \frac{1}{n!}$$

i.e., $N(t)$ has a generalized Waring distribution with parameters $a, kt; \rho$.

The transition probabilities of the generalized Waring process are given by

$$p_{m,n}(s, s+t) = P\{N(s+t)=n|N(s)=m\} = \frac{\Gamma(a+n)}{\Gamma(a+m)} \frac{(kt)_{(n-m)}}{(n-m)!} \frac{(\rho+ks)_{(a+m)}}{(\rho+ks+kt)_{(a+n)}}$$

$$p_{0,n}(0,t) = P\{N(t)=n|N(0)=0\} = \frac{\rho_{(kt)}}{(\rho+a)_{(kt)}} \frac{a_{(n)}(kt)_{(n)}}{(a+\rho+kt)_{(n)}} \frac{1}{n!} = P(N(t)=n)$$

with the last equality indicating that the generalized Waring process is a non-homogenous Markov process. Its mean and variance are respectively

$$E[N(t)] = \frac{akt}{\rho-1} \text{ and } Var[N(t)] = \frac{akt(\rho+kt-1)(\rho+\alpha-1)}{(\rho-1)^2(\rho-2)}$$

Note that since the generalized Waring process is a stationary process and its mean is of the form $E[N(t)] = \eta t$, the above formula implies that its *intensity is* $\eta = ak/(\rho - 1)$. Its variance can be split into three additive components, thus

$$Var[N(t)] = \sigma^2_{\Lambda(t)} + (kt)^2\sigma^2_\nu + \sigma^2_R$$

with the liability and random components dependent on time. In particular,

$$\sigma^2_{\Lambda(t)} = akt(a+1)(\rho-1)^{-1}(\rho-2)^{-1};\ \sigma^2_\nu = a(a+\rho-1)(\rho-1)^{-2}(\rho-2)^{-1};$$
$$\sigma^2_R = akt(\rho-1)^{-1}.$$

9.1 The generalized Waring process in an accident proneness context

We consider a population which is inhomogeneous with respect to personal and environmental attributes affecting the occurrence of accidents. The terms accident proneness" and accident liability" are again used to refer respectively to a person s predisposition to accidents, and to a person s exposure to external risk of accident with the conditional distribution of the random variable λ given ν describing differences in external risk factors among individuals. Liability fluctuations over a time interval $(t, t+h)$ depend on the length h of the interval and are described by a distribution for $\lambda|\nu$ with probability density function $\lambda^{kh-1}e^{-\lambda/(\nu h)}(\nu h)^{-kh}/\Gamma(kh)$. Allowing further the parameter ν have a beta distribution of the second kind with parameters a and ρ and density function ϕ given by

$\phi(\nu) = \Gamma(a+\rho)\nu^{a-1}(1+\nu)^{-(a+\rho)}/[\Gamma(a)\Gamma(\rho)],\ a, \rho \geq 0$, we obtain for the distribution of the number of accidents $N(t)$:

$$P(N(t+h)-N(t)=n) = \frac{\rho_{(kh)}}{(a+\rho)_{(kh)}} \frac{a_{(n)}(kh)_{(n)}}{(a+\rho+kh)_{(n)}} \frac{1}{n!}$$

and

$$P(N(t)=n) = P_n(t) = \frac{\rho_{(kt)}}{(a+\rho)_{(kt)}} \frac{a_{(n)}(kt)_{(n)}}{(a+\rho+kt)_{(n)}} \frac{1}{n!},\ n = 0,\ 1,\ \ldots$$

So, the process arising in the context of this model, satisfies the defining conditions of the generalized Waring process.

9.2 The generalized Waring process in the context of a spells model

Xekalaki and Zografi (2008) showed that the generalized Waring process could also be used in modeling temporally evolving data in the context of a *spells* model. Assume again that each person is liable to spells and that no accidents can occur outside spells. Let $S(t)$, $t = 0, 1, 2, \ldots$, the number of spells up to a given moment t, be a homogeneous Poisson process with rate k/m, $k > 0$, the number X_i of accidents within a spell i be a random variable with a logarithmic series distribution with parameters m and v and probability function given by $P(X_i = n) = \frac{m}{n}\left(\frac{v}{1+v}\right)^n$, $n \geq 1$ with $P(X_i = 0) = 1 - m \log(1 + v)$, $v > 0$, $0 < m < 1/\log(1 + v)$, and the numbers of accidents arising out of different spells be independent and independent of the number $S(t)$ of spells. Here v is regarded as the external risk parameter, too, which they assumed varying according to a beta distribution of the second kind with parameters a and ρ and probability density function given by $\Gamma(a + \rho)v^{a-1}(1 + v)^{-(a+\rho)}/[\Gamma(a)\Gamma(\rho)]$, $a, \rho \geq 0$. They then showed that the above framework leads to a process conforming with the postulates of the generalized Waring process, thus demonstrating its potential application in the context of the Spells model.

10. An application: modeling the counting process $\{N(s), s > 0\}$ associated with the access pattern of a web site

As an illustration of the application potential of the generalized Waring process in other fields by appropriately adjusting the concepts and terminology used in this paper so as to have natural interpretations, we outline an example of a model for temporally evolving data on web access patterns provided by Xekalaki and Zografi (2008).

In this context, $\{N(s), s > 0\}$ is the counting process associated with the access pattern of a web site, where, for any $t > 0$, $N(t)$ represents the number of visits that the web pages on this particular site get within the interval $(0, t)$. Note that the generalized Waring distribution was cited in Ajiferuke et al. (2004) as used by them to fit observed website visitation data for a given period, i.e, to model counts $N(t_0)$ of web visits on a given fixed time interval $(0, t_0)$.

Except for chance, visits to a web site can be regarded as affected by the intrinsic appeal of the particular site to web users (corresponding to proneness) as well as by exogenous factors (corresponding to external factors) such as, links provided by other sites to the particular site, how well the site is advertised etc.

Letting v denote the intrinsic factors and $\lambda|v$ the exogenous factors. Then assuming that $N(t)|\lambda$ follows a $Poisson(\lambda(t))$ distribution, where $\lambda(t) = \lambda t$ with $\lambda|v$ following a gamma distribution with density $\lambda^{kt-1}e^{-\lambda/(vt)}(vt)^{-kt}/\Gamma(kt)$, and with v following a beta distribution of the second kind with density $\Gamma(a + \rho)v^{a-1}(1 + v)^{-(a+\rho)}/[\Gamma(a)\Gamma(\rho)]$, $a, \rho \geq 0$, then the unconditional distribution of $N(t)$ is the $GWD(a, kt; \rho)$, i.e. the process $\{N(t), t \geq 0\}$ is a generalized Waring process.

10.1 The data

The log files representing the hits on an e-shop site for the period from March 31, 2006 to April 30, 2006 have been used to fit this model. (A log file typically contains information

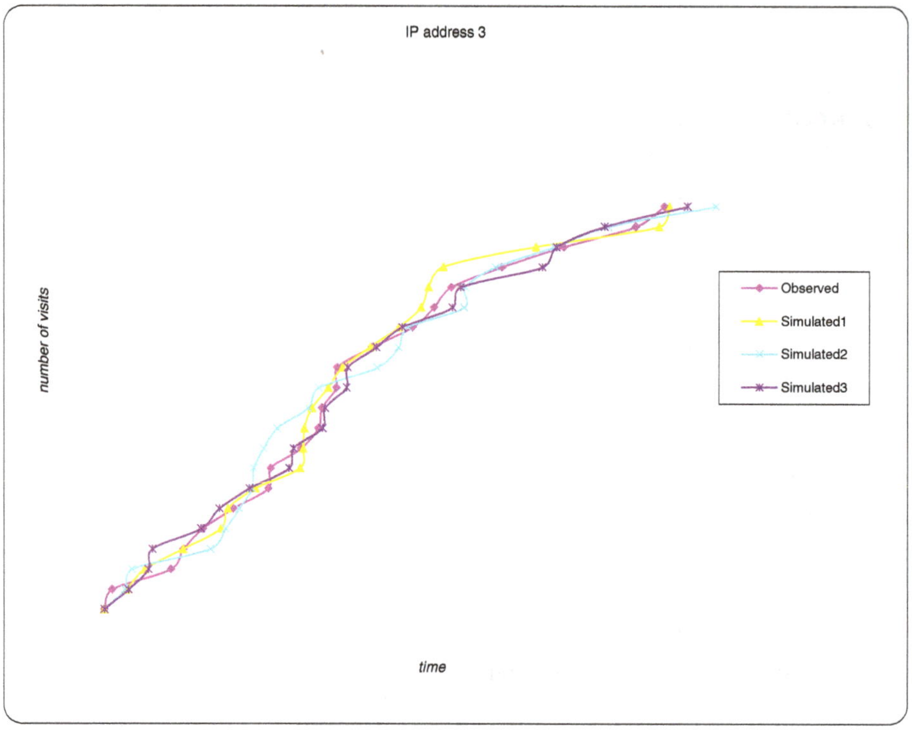

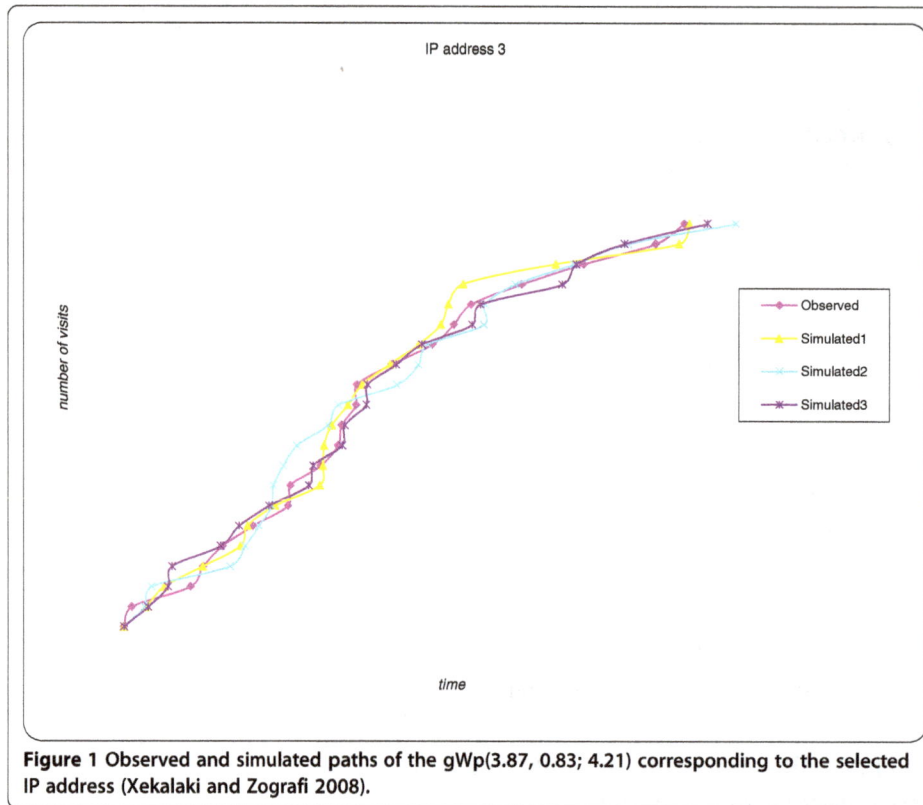

Figure 1 Observed and simulated paths of the gWp(3.87, 0.83; 4.21) corresponding to the selected IP address (Xekalaki and Zografi 2008).

on the times of visits per IP address per day). On the basis of such log files, the visits per day made by each of 468 IP addresses to a web site during the above period were enumerated yielding 468 paths of visits $N_i(t_j)$ made by IP address i up to and including time t_j denoted by $\{N_i(t_j),\ i = 1, 2, \ldots, 468;\ j = 1, 2, \ldots, 31\}$.

Moment estimates of the parameters of the generalized Waring process were obtained employing an estimation procedure for spatial point process data termed in the literature as the centered reduced moment method. The method introduced and

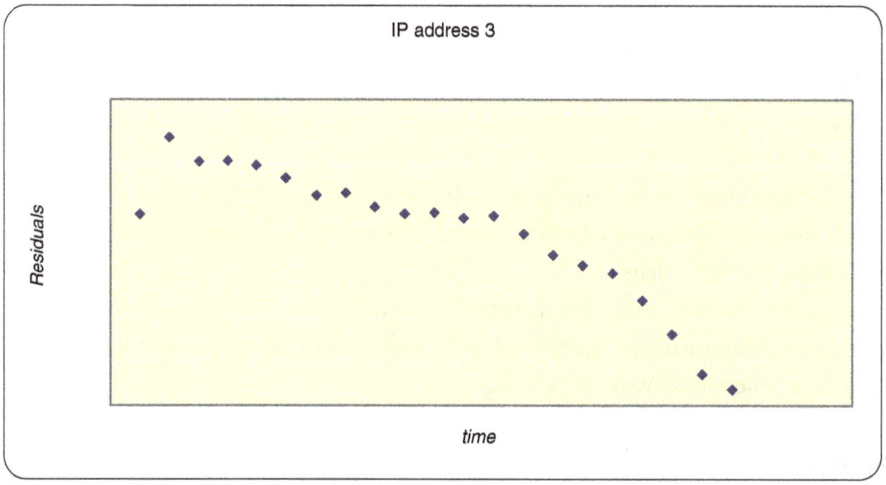

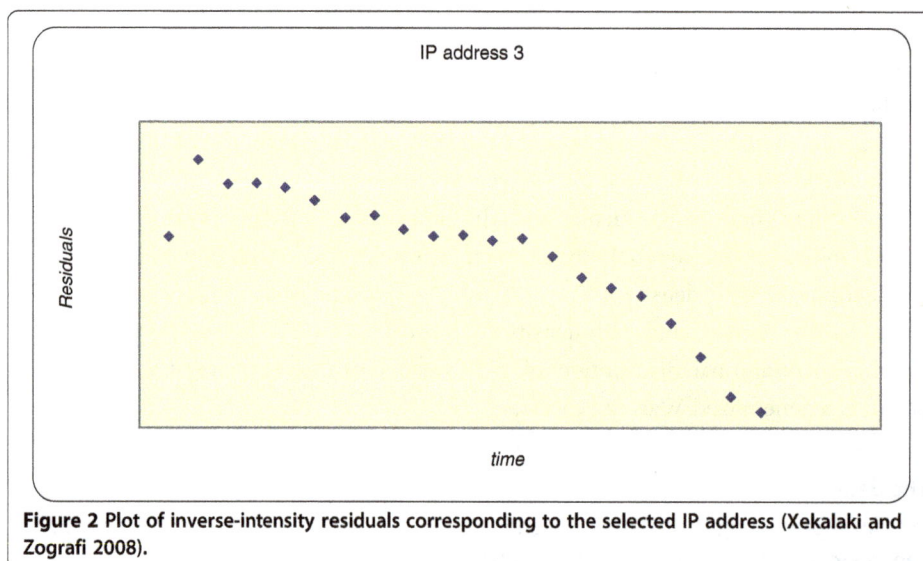

Figure 2 Plot of inverse-intensity residuals corresponding to the selected IP address (Xekalaki and Zografi 2008).

studied by Ripley (1976, 1977) utilizes the intensity of the process and the mean number of further points within distance s of an arbitrary point of the process. In particular, the method utilizes the moment estimators $E(\hat{N}(s)) = \hat{\mu}_1 = \hat{\eta}s = ns/h$, $E(\hat{N}^2(s)) = \hat{\mu}_2 = X/n_{(2)}$, $E(\hat{N}^3(s)) = \hat{\mu}_3 = (Z-X)/n_{(3)}$ with $X = \sum_{i=1}^{n} \sum_{i \neq j} \phi_s^2(x_i, x_j)$, $Z = \sum_{i=1}^{n} \left(\sum_{j \neq i} \phi_s(x_i, x_j) \right) \left(\sum_{k \neq i} \phi_s(x_i, x_k) \right)$, where the quantities involved in the above equations represent weights defined, for each value x_i in the collection of points $\{x_i : i = 1, 2, \ldots, n\}$ of the process within a time interval of length h, as follows: For each x_i in $\{x_i : i = 1, 2, \ldots, n\}$ and a given $s > 0$, consider the interval of center x_i and length s and assign to every point x_j, $j \neq i$ in this interval the weight $\phi_s(x_i, x_j) = \omega(x_i, x_j)^{-1}$, where $\omega(x_i, x_j)$ is the number of other points $\{x_k, k \neq i, k \neq j\}$ of the process that are included in the interval of length $|x_i - x_j|$ and center x_i (see also Diggle and Chetwynd 1991; Chetwynd and Diggle 1998, among others). The standard errors of the thus obtained parameter estimators can in principle be determined by simulation, but the associated computations are formidable. Approximation formulas exist only for the case of homogeneous planar Poisson process, while, for the class of stationary Cox process, there is no obvious way to obtain estimable expressions as noted by Chetwynd and Diggle (1998).

The observed paths were compared to the corresponding time series of simulated realizations of the generalized Waring process over the same time segment. For each IP address, 100 simulated realizations of the $gWp(a, k; \rho)$, were obtained and each of the observed time series paths was compared to the corresponding simulated ones. On average, the realizations of the generalized Waring process exhibited similar structural characteristics, notably recognizable, to those of the paths of the observed time series. For illustration purposes, the path of the observed time series associated with one of the IP addresses considered is presented in Figure 1. In the graph, the path is superimposed by a sample of three of the 100 corresponding simulated realizations of the $gWp(a, k; \rho)$. Inspection of the graph provides a visual appreciation of the degree of similarity in the structural characteristics of the path of the observed and the realized time series.

Following Lewis (1972), Brillinger (1978) and Andersen et al. (1993), the closeness of the observed and realized time series was also checked using diagnostic plots based on the inverse-intensity residuals computed for each value x_j in the collection of points $\{x_j : j = 1, 2, \ldots, n\}$ of the process given by $R_{\hat{\theta}}(B_j, \eta^{-1}) = \sum_{x_i \in B_j} \hat{\eta}(x_i) - \int_{B_j}^{j} I_{R^+}(\hat{\eta}(x))dx$ where $B_j = (0, x_j)$, $\hat{\theta} = (\hat{a}, \hat{k}, \hat{\rho})^{-1}$, $\hat{\eta}(x) = \eta(x, \hat{\theta})$ is the fitted intensity and $I_{R^+}(\cdot)$ is the indicator function. These plots exhibit similar results. The plot corresponding to the data associated to the IP address considered is shown in Figure 2.

Competing interests

The authoress declares that she has no competing interests.

Acknowledgement

The authoress would like to thank the associate editor and the referees for their constructive comments.
This paper is an extended version of the authoress' invited plenary presentation at the International Conference on Statistical Distributions and Applications, Michigan USA, October10-12, 2013.

References

Ajiferuke, I, Wolfram, D, Xie, H: Modelling website visitation and resource usage characteristics by IP address data. In: Julien, H, Thompson, S (eds.). Proceedings of the 32nd Annual Conference of the Canadian Association for Information Science, Manitoba, Canada (2004). Available at: www.cais-acsi.ca/proceedings/2004/ajiferuke_2004.pdf

Andersen, P, Borgan, Ø, Gill, R, Keiding, N: Statistical Models based on Counting Processes. Springer, New York (1993)

Andrews, DF, Mallows, CL: Scale mixtures of normal distributions. J. R. Stat. Soc. B **36**, 99–102 (1974)

Arbous, AG, Kerrich, JE: Accident statistics and the concept of accident proneness. Biometrics **7**, 340–432 (1951)

Bates, GE: Joint distributions of time intervals for the occurrence of successive accidents in a generalized Polya sheme. Ann. Math. Stat. **26**, 705–720 (1955)

Bates, GE, Neyman, J: Contributions to the theory of accident proneness. U. Calif. Publ. Stat. **1**, 215–275 (1952)

Breslow, N: Tests of hypotheses in overdispersed poisson regression and other quasi-likelihood models. J. Am. Stat. Assoc. **85**, 565–571 (1990)

Brillinger, D: Comparative Aspects of the Study of Ordinary Time Series and of Point Processes. In: Krishnaiah, PR (ed.) Developments in Statistics, pp. 33–133. Academic Press, New York (1978)

Cane, VR: The concept of accident proneness. B. Inst. Math. Bulgarian Acad. Sci. **15**, 183–188 (1974)

Cane, VR: A class of non-identifiable stochastic models. J. Appl. Probab. **14**, 475–782 (1977)

Carriere, J: Nonparametric tests for mixed poisson distributions. Insur. Math. Econ. **12**, 3–8 (1993)

Chetwynd, AG, Diggle, PJ: On estimating the reduced second moment measure of a stationary spatial point process. Aust. N.Z. J. Stat. **40**(1), 11–15 (1998)

Cox, DR: Renewal Theory. Barnes & Noble, New York (1962)

Cox, DR: Some remarks on overdispersion. Biometrika **70**, 269–274 (1983)

Cresswell, WL, Froggatt, P: The Causation of Bus Driver Accidents. Oxford University Press, London (1963)

Davidian, M, Carroll, RJ: A note on extended quasi-likelihood. J. R. Stat. Soc. B **50**, 74–82 (1988)

Dean, CB, Lawless, J, Willmot, GE: A mixed poisson-inverse Gaussian regression model. Can. J. Stat. **17**, 171–182 (1989)

Diggle, PJ, Chetwynd, AG: Second- order analysis of spatial clustering for inhomogeneous populations. Biometrics **47**, 1155–1163 (1991)

Dimaki, C, Xekalaki, E: Identifiability of income distributions in the context of damage and generating models. Commun. Stat. A-Theor. **19**(8), 2757–2766 (1990)

Dimaki, C, Xekalaki, E: Additive and multiplicative distortion of observations: some characteristic properties. J. Appl. Stat. Sc. **5**(2/3), 113–127 (1996)

Dubourdieu, J: Remarques Relatives sur la Théorie Mathématique de l'Assurance-Accidents. Bull. Trim. Inst. Actuaries Fran. **44**, 79–146 (1938)

Faddy, MJ: Extending poisson process modelling and analysis of count data. Biometrical J. **39**(4), 431–440 (1997)

Famoye, F, Singh, KP: Zero-inflated generalized poisson regression model with an application to domestic violence. Data. J. Data Sci. **4**, 117–130 (2006)

Fisher, RA: The effects of methods of ascertainment upon the estimation of frequencies. Ann. Eugenic. **6**, 13–25 (1934)

Fisher, RA: The significance of deviations from expectation in a poisson series. Biometrics **6**, 17–24 (1950)

Greenwood, M, Woods, HM: On the Incidence of Industrial Accidents upon Individuals with Special Reference to Multiple Accidents. In: Report of the Industrial Fatigue Research Board, **4**, pp. 1–28. His Majesty's Stationary Office, London (1919)

Greenwood, M, Yule, GU: An inquiry into the nature of frequency distributions representative of multiple happenings with particular reference to the occurrence of multiple attack of disease or repeated accidents. J. R. Stat. Soc. **83**, 255–279 (1920)

Gschlößl, S, Czado, C: Modelling count data with overdispersion and spatial effects. Stat. Pap. **49**, 531–552 (2008)

Gupta, PL, Gupta, RC, Tripathi, RC: Analysis of zero adjusted count data. Comput. Stat. Data An. **23**, 207–218 (1996)

Gupta, PL, Gupta, RC, Tripathi, RC: Score test for zero inflated generalized poisson regression model. Commun. Stat. A-Theor. **33**, 47–64 (2004)

Gurland, J: Some applications of the negative binomial and other contagious distributions. Am. J. Public Health **49**, 1388–1399 (1959)

Hinde, J, Demetrio, CGB: Overdispersion: models and estimation. Comput. Stat. Data An. **27**, 151–170 (1998)

Irwin, JO: Discussion on chambers and Yule's paper. J. R. Stat. Soc. Supplement **7**, 101–109 (1941)

Irwin, JO: The place of mathematics in medical and biological statistics. J. R. Stat. Soc. A **126**, 1–44 (1963)

Irwin, JO: The generalized Waring distribution applied to accident theory. J. R. Stat. Soc. A **131**, 205–225 (1968)

Irwin, JO: The Generalized Waring Distribution. J. R. Stat. Soc. A **138**, 18–31 (1975). Part I), 204-227 (Part II), 374-384 (Part III

Jewell, N: Mixtures of exponential distributions. Ann. Stat. **10**, 479–484 (1982)

Karlis, D, Xekalaki, E: On testing for the number of components in a mixed poisson model. Ann. I. Stat. Math. **51**, 149–162 (1999)

Karlis, D, Xekalaki, E: Mixtures Everywhere. In: Panaretos, J (ed.) Stochastic Musings: Perspectives from the Pioneers of the Late 20th Century, pp. 78-95. Laurence Erlbaum, USA (2003)

Karlis, D, Xekalaki, E: Mixed poisson distributions. Int. Stat. Rev. **73**(1), 35–58 (2005)

Kemp, CD: On a contagious distribution suggested for accident data. Biometrics **23**, 241–255 (1967)

Kemp, CD: "Accident proneness" and discrete distribution theory. In: Patil, GP (ed.) Random Counts in Scientific Work, Vol.2, pp. 41–65. State College: Pennsylvania State University Press, University Park, USA (1970)

Krishnaji, N: Characterization of the Pareto distribution through a model of under-reported incomes. Econometrica **38**, 251–255 (1970)

Kulinskaya, E, Olkin, I: An overdispersion model in meta-analysis. Stat. Model. **14**(1), 49–76 (2014)

Lawless, JF: Negative binomial and mixed poisson regression. Can. J. Stat. **15**, 209–225 (1987)

Lewis, PAW: Recent results in the statistical analysis of univariate point processes. In: Lewis, PAW (ed.) Stochastic Point Processes, pp. 1–54. Wiley, New York (1972)

Lexis, W: Über die Theorie der Statilität Statistischer Reichen. Jahrb. Nationalökon. u. Statist. **32**, 60–98 (1879)

McCullagh, P, Nelder, JA: Generalized Linear Models, 2nd edn. Chapman & Hall, London (1989)

McKendrick, AG: The applications of mathematics to medical problems. Proc. Edinb. Math. Soc. **44**, 98–130 (1925). doi:10.1017/S0013091500034428

McLachlan, JA, Peel, D: Finite Mixture Models. Wiley, New York (2001)

Moore, DF: Asymptotic properties of moment estimators for overdispersed counts and proportions. Biometrika **23**, 583–588 (1986)

Nelder, JA, Pregibon, D: An extended quasi-likelihood function. Biometrika **74**, 221–232 (1987)

Ngatchou-Wandji, J, Paris, C: On the zero-inflated count models with application to modelling annual trends in incidences of some occupational allergic diseases in France. J. Data Sci. **9**, 639–659 (2011)

Panaretos, J: An extension of the damage model. Metrika **29**, 189–194 (1982)

Panaretos, J: A generating model involving Pascal and logarithmic series distributions. Commun. Stat. A-Theor. **12**(7), 841–848 (1983)

Panaretos, J: On the evolution of surnames. Int. Stat. Rev. **57**(2), 161–167 (1989)

Patil, GP, Ord, JK: On size-biased sapling and related form-invariant weighted distributions. Sankhya **38**, 48–61 (1976)

Rao, CR: On discrete distributions arising out of methods of ascertainment. Sankhya **A25**, 311–324 (1963)

Rao, CR: Weighted Distributions Arising Out of Methods of Ascertainment. In: Atkinson, AC, Fienberg, SE (eds.) A Celebration of Statistics, Chapter 24, pp. 543–569. Springer-Verlag, New York (1985)

Ridout, M, Demetrio, CGB, Hinde, J: Models for count data with many zeros, pp. 179–190. Invited paper presented at the Nineteenth International Biometric Conference, Cape Town, South Africa (1998)

Ridout, M, Hinde, J, Demetrio, CGB: A score test for testing zero inflated Poisson regression model against zero inflated negative binomial alternatives. Biometrics **57**, 219–223 (2001)

Ripley, BD: The second-order analysis of stationary point processes. J. Appl. Probab. **13**, 255–266 (1976)

Ripley, BD: Modelling spatial patterns (with Discussion). J. R. Stat. Soc. B **39**, 172–212 (1977)

Shaw, L, Sichel, HS: Accidents Proneness. Pergamon Press, Oxford (1971)

Steutel, FW, van Harn, K: Discrete analogues of self-decomposability and stability. Ann. Prob. **7**, 893–899 (1979)

Student: An explanation of deviations from poisson's law in practice. Biometrika **12**, 211–215 (1919)

Thyrion, P: Extension of the collective risk theory. Skand. Aktuaritidskrift **52**(Supplement), 84–98 (1969)

Titterington, DM: Some recent research in the analysis of mixture distributions. Statistics **21**, 619–641 (1990)

Tripathi, R, Gupta, R, Gurland, J: Estimation of parameters in the beta binomial model. Ann. I. Stat. Math. **46**, 317–331 (1994)

Wang, P, Puterman, M, Cokburn, I, Le, N: Mixed poisson regression models with covariate dependent rates. Biometrics **52**, 381–400 (1996)

Winkelmann, R: Duration dependence and dispersion in count-data models. J. Bus. Econ. Stat. **13**(4), 467–474 (1995)

Xekalaki, E: Chance mechanisms for the univariate generalized Waring distribution and related characterizations. In: Taillie, C, Patil, GP, Baldessari, B (eds.) Statistical Distributions in Scientific Work, Vol. 4 (Models, Structures and Characterizations), pp. 157–171. Reidel, Dordrecht (1981)

Xekalaki, E: The univariate generalized Waring distribution in relation to accident theory: proneness, spells or contagion? Biometrics **39**(3), 887–895 (1983a)

Xekalaki, E: Infinite divisibility, completeness and regression properties of the univariate generalized Waring distribution. Ann. I. Stat. Math. A **35**, 279–289 (1983b)

Xekalaki, E: A property of the Yule distribution and its applications. Commun. Stat. A-Theor. **12**(10), 1181–1189 (1983c)

Xekalaki, E: The bivariate generalized Waring distribution and its application to accident theory. J. R. Stat. Soc. A **147**(3), 488–498 (1984a)

Xekalaki, E: Linear regression and the Yule distribution. J. Econometrics **24**(1), 397–403 (1984b)

Xekalaki, E: Models leading to the bivariate generalized Waring distribution. Utilitas Math. **25**, 263–290 (1984c)

Xekalaki, E: The multivariate generalized Waring distribution. Commun. Stat. A-Theor. **15**(3), 1047–1064 (1986)

Xekalaki, E: Under- and Overdispersion. Enc. Act. Sci. **3**, (2006) [http://onlinelibrary.wiley.com/doi/10.1002/9780470012505.tau003/abstract]

Xekalaki, E, Panaretos, J: Identifiability of compound poisson distributions. Scand. Actuar. J. **66**, 39–45 (1983)

Xekalaki, E, Zografi, M: The generalized Waring process and its application. Commun. Stat. A-Theor. **37**(12), 1835–1854 (2008)

Xue, D, Deddens, J: Overdispersed negative binomial models. Commun. Stat. A-Theor. **21**, 2215–2226 (1992)

Yule, GW: A mathematical theory of evolution based on the conclusions of J.C. Willis, F.R.S. Philos. T. R. Soc. B **213**, 21–87 (1924)

10

The Marshall-Olkin extended Weibull family of distributions

Manoel Santos-Neto[1*], Marcelo Bourguignon[2*], Luz M Zea[3], Abraão DC Nascimento[4] and Gauss M Cordeiro[4]

*Correspondence: mn.neco@gmail.com; m.p.bourguignon@gmail.com
[1] Departamento de Estatística, Universidade Federal de Campina Grande, Bodocongó, 58429-970, Campina Grande, PB, Brazil
[2] Departamento de Estatística, Universidade Federal do Piauí, Ininga, 64049-550, Teresina, PI, Brazil
Full list of author information is available at the end of the article

Abstract
We introduce a new class of models called the Marshall-Olkin extended Weibull family of distributions based on the work by Marshall and Olkin (Biometrika 84:641–652, 1997). The proposed family includes as special cases several models studied in the literature such as the Marshall-Olkin Weibull, Marshall-Olkin Lomax, Marshal-Olkin Fréchet and Marshall-Olkin Burr XII distributions, among others. It defines at least twenty-one special models and thirteen of them are new ones. We study some of its structural properties including moments, generating function, mean deviations and entropy. We obtain the density function of the order statistics and their moments. Special distributions are investigated in some details. We derive two classes of entropy and one class of divergence measures which can be interpreted as new goodness-of-fit quantities. The method of maximum likelihood for estimating the model parameters is discussed for uncensored and multi-censored data. We perform a simulation study using Markov Chain Monte Carlo method in order to establish the accuracy of these estimators. The usefulness of the new family is illustrated by means of two real data sets.

Mathematics Subject Classification (2010): 60E05; 62F03; 62F10; 62P10

Keywords: Extended Weibull distribution; Hazard rate function; Marshall-Olkin distribution; Maximum likelihood estimation; Survival function

1 Introduction

The Weibull distribution has assumed a prominent position as statistical model for data from reliability, engineering and biological studies (McCool 2012). This model has been exaustively used for describing *hazard rates* – an important quantity of survival analysis. In the context of monotone hazard rates, some results from the literature suggest that the Weibull law is a reasonable choice due to its negatively and positively skewed density shapes. However, this distribution is not a good model for describing phenomenon with non-monotone failure rates, which can be found on data from applications in reliability and biological studies. Thus, extended forms of the Weibull model have been sought in many applied areas. As a solution for this issue, the inclusion of additional parameters to a well-defined distribution has been indicated as a good methodology for providing more flexible new classes of distributions.

Marshall and Olkin (1997) derived an important method of including an extra shape parameter to a given baseline model thus defining an extended distribution. The Marshall and Olkin ("$\mathcal{MO}$" for short) transformation furnishes a wide range of behaviors with respect to the baseline distribution. The geometrical and inferential properties associated with the generated distribution depend on the values of the extra parameter. These characteristics provide more flexibility to the $\mathcal{MO}$ generated distributions. Considering the proportional odds model, Sankaran and Jayakumar (2008) presented a detailed discussion about the physical interpretation of the $\mathcal{MO}$ family.

This family has a relationship with the odds ratio associated with the baseline distribution. Let X be a distributed $\mathcal{MO}$ random variable which describes the lifetime relative to each individual in the population with a vector of p-covariates $\boldsymbol{z} = (z_1, \ldots, z_p)^\top$, where $(\cdot)^\top$ denotes the transposition operator. Then, the cumulative distribution function (cdf) of X is given by

$$\overline{F}(x; \boldsymbol{z}) = \frac{k(\boldsymbol{z})\, \overline{G}(x)}{1-[1-k(\boldsymbol{z})]\, \overline{G}(x)}, \tag{1}$$

where $k(\boldsymbol{z}) = \lambda_G(x)/\lambda_F(x; \boldsymbol{z})$ is a non-negative function such that $\boldsymbol{z}$ is independent of the time x, $\lambda_F(x; \boldsymbol{z})$ is the proportional odds model [for a discussion about such modeling, see Sankaran and Jayakumar (2008)] and $\lambda_G(x) = G(x)/\overline{G}(x)$ represents an arbitrary odds for the baseline distribution.

In this paper, we consider $k(\boldsymbol{z}) = \delta$. Before, however, it is important to highlight two important properties of the $\mathcal{MO}$ transformation: (i) the stability and (ii) geometric extreme stability (Marshall and Olkin 1997). In other words, the $\mathcal{MO}$ distribution possesses a stability property in the sense that if the method is applied twice, it returns to the same distribution. In addition, the following stochastic behavior can also be verified: let $\{X_1, \ldots, X_N\}$ be a random sample from the population random variable equipped with the survival function (1) at $k(\boldsymbol{z}) = \delta$. Suppose that N has the geometric distribution with probability p and that this quantity is independent of X_i, for $i = 1, \ldots, N$. Then, $U = min(X_1, \ldots, X_N)$ and $V = max(X_1, \ldots, X_N)$ are random variables having survival functions (1) such that $k(\boldsymbol{z})$ can be equal to p and p^{-1}, respectively, i.e., the $\mathcal{MO}$ transform satisfies the geometric extreme stability property.

Due to these advantages, many papers have employed the $\mathcal{MO}$ transformation. In Marshall and Olkin work, the exponential and Weibull distributions were generalized. Subsequently, the $\mathcal{MO}$ extension was applied to several well-known distributions: Weibull (Ghitany *et al.* 2005, Zhang and Xie 2007), Pareto (Ghitany 2005), gamma (Ristić *et al.* 2007), Lomax (Ghitany *et al.* 2007) and linear failure-rate (Ghitany and Kotz 2007) distributions. More recently, general results have been addressed by Barreto-Souza *et al.* (2013) and Cordeiro and Lemonte (2013). In this paper, we aim to apply the $\mathcal{MO}$ generator to the extended Weibull ($\mathcal{EW}$) class of distributions to obtain a new more flexible family to describe reliability data. The proposed family can also be applied to other fields including business, environment, informatics and medicine in the same way as it was originally done with the Birnbaum-Saunders and other lifetime distributions.

Let $\overline{G}(x) = 1-G(x)$ and $g(x) = dG(x)/dx$ be the survival and density functions of a continuous random variable Y with baseline cdf $G(x)$. Then, the $\mathcal{MO}$ extended distribution has survival function given by

$$\overline{F}(x;\delta) = \frac{\delta\overline{G}(x)}{1-\overline{\delta}\,\overline{G}(x)} = \frac{\delta\overline{G}(x)}{G(x)+\delta\overline{G}(x)}, \quad x \in \mathcal{X} \subseteq \mathbb{R},\ \delta > 0, \tag{2}$$

where $\overline{\delta} = 1-\delta$.

Clearly, $\delta = 1$ implies $\overline{F}(x) = \overline{G}(x)$. The family (2) has probability density function (pdf) given by

$$f(x;\delta) = \frac{\delta g(x)}{[1-\bar{\delta}\,\overline{G}(x)]^2}, \quad x \in \mathcal{X} \subseteq \mathbb{R},\ \delta > 0.$$

Its hazard rate function (hrf) becomes

$$\tau(x;\delta) = \frac{g(x)}{\overline{G}(x)[1-\bar{\delta}\,\overline{G}(x)]}, \quad x \in \mathcal{X} \subseteq \mathbb{R},\ \delta > 0.$$

Further, the class of extended Weibull ($\mathcal{EW}$) distributions pioneered by Gurvich *et al.* (1997) has achieved a prominent position in lifetime models. Its cdf is given by

$$G(x;\alpha,\boldsymbol{\xi}) = 1 - \exp[-\alpha H(x;\boldsymbol{\xi})], \quad x \in \mathcal{D} \subseteq \mathbb{R}_+,\ \alpha > 0, \tag{3}$$

where $H(x;\boldsymbol{\xi})$ is a non-negative monotonically increasing function which depends on the parameter vector $\boldsymbol{\xi}$. The corresponding pdf is given by

$$g(x;\alpha,\boldsymbol{\xi}) = \alpha \exp[-\alpha H(x;\boldsymbol{\xi})]\ h(x;\boldsymbol{\xi}), \tag{4}$$

where $h(x;\boldsymbol{\xi})$ is the derivative of $H(x;\boldsymbol{\xi})$.

Different expressions for $H(x;\boldsymbol{\xi})$ in Equation (3) define important models such as:

(i) $H(x;\boldsymbol{\xi}) = x$ gives the exponential distribution;
(ii) $H(x;\boldsymbol{\xi}) = x^2$ leads to the Rayleigh (Burr type-X) distribution;
(iii) $H(x;\boldsymbol{\xi}) = \log(x/k)$ leads to the Pareto distribution;
(iv) $H(x;\boldsymbol{\xi}) = \beta^{-1}[\exp(\beta x) - 1]$ gives the Gompertz distribution.

In this paper, we derive a new family of distributions by compounding the $\mathcal{MO}$ and $\mathcal{EW}$ classes. We define a new generated family in order to provide a "better fit" in certain practical situations. The compounding procedure follows by taking the $\mathcal{EW}$ class (3) as the baseline model in Equation (2). The *Marshall-Olkin extended Weibull ($\mathcal{MOEW}$) family* of distributions contains some special models as those listed in Table 1 with the corresponding $H(\cdot;\cdot)$ and $h(\cdot;\cdot)$ functions and the parameter vectors.

The paper unfolds as follows. Section 2 presents the cdf and pdf of the proposed distribution and some expansions for the density function. The main statistical properties of the new family are derived in Section 3 including the moments, moment generating function (mgf) and incomplete moments, quantile function (qf), random number generator, skewness and kurtosis measures, order statistics, mean deviations and average lifetime functions. In Section 4, we derive four measures of information theory: Shannon and Rényi entropies, cross entropy and Kullback-Leibler divergence. The maximum likelihood method to estimate the model parameters is adopted in Section 5. Two special models are studied in some details in Section 6. We perform a simulation study using Monte Carlo's experiments in order to assess the accuracy of the maximum likelihood estimators (MLEs) in Section 7.1 and two applications to real data in Section 7.2. Conclusions and some future lines of research are addressed in Section 8.

2 The $\mathcal{MOEW}$ family

The cdf of the new family of distributions is given by

$$F(x;\delta,\alpha,\boldsymbol{\xi}) = \frac{1-\exp[-\alpha H(x;\boldsymbol{\xi})]}{1-\bar{\delta}\exp[-\alpha H(x;\boldsymbol{\xi})]}, \quad x \in \mathcal{D}, \tag{5}$$

Table 1 Special models and the corresponding functions $H(x;\ \xi)$ and $h(x;\ \xi)$

Distribution	$H(x;\xi)$	$h(x;\xi)$	α	ξ	References
Exponential ($x \geq 0$)	x	1	α	$\emptyset$	Johnson *et al.* (1994)
Pareto ($x \geq k$)	$\log(x/k)$	$1/x$	α	k	Johnson *et al.* (1994)
Burr XII ($x \geq 0$)	$\log(1+x^c)$	$cx^{c-1}/(1+x^c)$	α	c	Rodriguez (1977)
Lomax ($x \geq 0$)	$\log(1+x)$	$1/(1+x)$	α	$\emptyset$	Lomax (1954)
Log-logistic ($x \geq 0$)	$\log(1+x^c)$	$cx^{c-1}/(1+x^c)$	1	c	Fisk (1961)
Rayleigh ($x \geq 0$)	x^2	$2x$	α	$\emptyset$	Rayleigh (1880)
Weibull ($x \geq 0$)	x^γ	$\gamma x^{\gamma-1}$	α	γ	Johnson *et al.* (1994)
Fréchet ($x \geq 0$)	$x^{-\gamma}$	$-\gamma x^{-(\gamma+1)}$	α	γ	Fréchet (1927)
Linear failure rate($x \geq 0$)	$ax+bx^2/2$	$a+bx$	1	$[a,b]$	Bain (1974)
Modified Weibull ($x \geq 0$)	$x^\gamma \exp(\lambda x)$	$x^{\gamma-1}\exp(\lambda x)(\gamma+\lambda x)$	α	$[\gamma,\lambda]$	Lai *et al.* (2003)
Weibull extension ($x \geq 0$)	$\lambda[\exp(x/\lambda)^\beta - 1]$	$\beta\exp(x/\lambda)^\beta(x/\lambda)^{\beta-1}$	α	$[\gamma,\lambda,\beta]$	Xie *et al.* (2002)
Phani ($0<\mu<x<\sigma<\infty$)	$[(x-\mu)/(\sigma-x)]^\beta$	$\beta[(x-\mu)/(\sigma-x)]^{\beta-1}[(\sigma-\mu)/(\sigma-t)^2]$	α	$[\mu,\sigma,\beta]$	Phani (1987)
Weibull Kies ($0<\mu<x<\sigma<\infty$)	$(x-\mu)^{\beta_1}/(\sigma-x)^{\beta_2}$	$(x-\mu)^{\beta_1-1}(\sigma-x)^{-\beta_2-1}[\beta_1(\sigma-x)+\beta_2(x-\mu)]$	α	$[\mu,\sigma,\beta_1,\beta_2]$	Kies (1958)
Additive Weibull ($x \geq 0$)	$(x/\beta_1)^{\alpha_1}+(x/\beta_2)^{\alpha_2}$	$(\alpha_1/\beta_1)(x/\beta_1)^{\alpha_1-1}+(\alpha_2/\beta_2)(x/\beta_2)^{\alpha_2-1}$	1	$[\alpha_1,\alpha_2,\beta_1,\beta_2]$	Xie and Lai (1995)
Traditional Weibull ($x \geq 0$)	$x^b[\exp(cx^d-1)]$	$bx^{b-1}[\exp(cx^d)-1]+cdx^{b+d-1}\exp(cx^d)$	α	$[b,c,d]$	Nadarajah and Kotz (2005)
Gen. power Weibull ($x \geq 0$)	$[1+(x/\beta)^{\alpha_1}]^\theta - 1$	$(\theta\alpha_1/\beta)[1+(x/\beta)^{\alpha_1}]^{\theta-1}(x/\beta)^{\alpha_1}$	1	$[\alpha_1,\beta,\theta]$	Nikulin and Haghighi (2006)
Flexible Weibull extension($x \geq 0$)	$\exp(\gamma x-\beta/x)$	$\exp(\gamma x-\beta/x)(\gamma+\beta/x^2)$	1	$[\gamma,\beta]$	Bebbington *et al.* (2007)
Gompertz ($x \geq 0$)	$\beta^{-1}[\exp(\beta x)-1]$	$\exp(\beta x)$	α	β	Gompertz (1825)
Exponential power ($x \geq 0$)	$\exp[(\lambda x)^\beta]-1$	$\beta\lambda\exp[(\lambda x)^\beta](\lambda x)^{\beta-1}$	1	$[\lambda,\beta]$	Smith and Bain (1975)
Chen ($x \geq 0$)	$\exp(x^b)-1$	$bx^{b-1}\exp(x^b)$	α	b	Chen (2000)
Pham ($x \geq 0$)	$(a^x)^\beta - 1$	$\beta(a^x)^\beta\log(a)$	1	$[a,\beta]$	Pham (2002)

where $\alpha > 0$ and $\delta > 0$. Using (5), we can express its survival function as

$$\overline{F}(x;\delta,\alpha,\boldsymbol{\xi}) = \frac{\delta \exp[-\alpha H(x;\boldsymbol{\xi})]}{1-\overline{\delta}\exp[-\alpha H(x;\boldsymbol{\xi})]}, \quad x \in \mathcal{D} \tag{6}$$

and the associated hrf reduces to

$$\tau(x;\delta,\alpha,\boldsymbol{\xi}) = \frac{\alpha\, h(x;\boldsymbol{\xi})}{1-\overline{\delta}\exp[-\alpha H(x;\boldsymbol{\xi})]}, \quad x \in \mathcal{D}. \tag{7}$$

The corresponding pdf is given by

$$f(x;\delta,\alpha,\boldsymbol{\xi}) = \frac{\delta\,\alpha\, h(x;\boldsymbol{\xi}) \exp[-\alpha H(x;\boldsymbol{\xi})]}{\{1-\overline{\delta}\exp[-\alpha H(x;\boldsymbol{\xi})]\}^2}, \tag{8}$$

where $H(x;\boldsymbol{\xi})$ can be any special distribution listed in Table 1.

Hereafter, let X be a random variable having the $\mathcal{MOEW}$ pdf (8) with parameters δ,α and $\boldsymbol{\xi}$, say $X \sim \mathcal{MOEW}(\delta,\alpha,\boldsymbol{\xi})$. Equation (8) extends several distributions which have been studied in the literature.

The $\mathcal{MO}$ Pareto (Ghitany 2005) is obtained by taking $H(x;\,\boldsymbol{\xi}) = \log(x/k)$ $(x \geq k)$. Further, for $H(x;\,\boldsymbol{\xi}) = x^{\gamma}$ we obtain the $\mathcal{MO}$ Weibull (Ghitany *et al.* 2005, Zhang and Xie 2007). The $\mathcal{MO}$ Lomax (Ghitany *et al.* 2007) and $\mathcal{MO}$ log-logistic are derived from (8) by taking $H(x;\,\boldsymbol{\xi}) = \log(1+x^c)$ with $c = 1$ and $H(x;\,\boldsymbol{\xi}) = \log(1+x^c)$ with $\alpha = 1$, respectively. For $H(x;\,\boldsymbol{\xi}) = a\,x + b\,x^2/2$ and $\alpha = 1$, Equation (8) reduces to the $\mathcal{MO}$ linear failure rate (Ghitany and Kotz 2007). In the same way, for $H(x;\,\boldsymbol{\xi}) = \log(1+x^c)$, we have the $\mathcal{MO}$ Burr XII (Jayakumar and Mathew 2008). Finally, we obtain the $\mathcal{MO}$ Fréchet (Krishna *et al.* 2013) from Equation (8) by setting $H(x;\,\boldsymbol{\xi}) = x^{-\gamma}$. Table 1 displays some useful quantities and corresponding parameter vectors for special distributions.

A general approximate goodness-of-fit test for the null hypothesis $H_0 : X_1,\ldots,X_n$ with X_i following $F(x;\boldsymbol{\theta})$, where the form of F is known but the p-vector $\boldsymbol{\theta} = (\delta,\alpha,\boldsymbol{\xi})^{\top}$ is unknown, was proposed by Chen and Balakrishnan (1995). This method is based on the Cramér-von Mises (CM) and Anderson-Darling (AD) statistics and, in general, the smaller the values of these statistics, the better the fit. In this paper, such methodology is applied to provide goodness-of-fit tests for the distributions under study.

Some results in the following sections can be obtained numerically in any software such as MAPLE (Garvan 2002), MATLAB (Sigmon and Davis 2002), MATHEMATICA (Wolfram 2003), Ox (Doornik 2007) and R (R Development Core Team 2009). The Ox (for academic purposes) and R are freely available at `http://www.doornik.com` and `http://www.r-project.org`, respectively. The results can be computed by taking in the sums a large positive integer value in place of ∞.

2.1 Expansions for the density function

For any positive real number a, and for $|z| < 1$, we have the generalized binomial expansion

$$(1-z)^{-a} = \sum_{k=0}^{\infty} \frac{(a)_k}{k!}\, z^k, \tag{9}$$

where $(a)_k = \Gamma(a+k)/\Gamma(a) = a(a+1)\ldots(a+k-1)$ is the ascending factorial and $\Gamma(\cdot)$ is the gamma function. Applying (9) to (8), for $0 < \delta < 1$, gives

$$f(x;\delta,\alpha,\boldsymbol{\xi}) = \sum_{j=0}^{\infty} \eta_j\, g(x;(j+1)\alpha,\boldsymbol{\xi}), \tag{10}$$

where $\eta_j = \delta\bar{\delta}^j$ and $g(x; (j+1)\alpha, \boldsymbol{\xi})$ denotes the $\mathcal{EW}$ density function with parameters $(j+1)\alpha$ and $\boldsymbol{\xi}$. Otherwise, for $\delta > 1$, after some algebra, we can express (8) as

$$f(x; \delta, \alpha, \boldsymbol{\xi}) = \frac{g(x; \alpha, \boldsymbol{\xi})}{\delta\left\{1 - (1 - 1/\delta)\left[1 - \exp(-\alpha H(x; \boldsymbol{\xi}))\right]\right\}^2}. \tag{11}$$

In this case, we can verify that $|(1 - 1/\delta)\left[1 - \exp(-\alpha H(x; \boldsymbol{\xi}))\right]| < 1$. Then, applying twice the expansion (9) in Equation (11), we obtain

$$f(x; \delta, \alpha, \boldsymbol{\xi}) = \sum_{j=0}^{\infty} \nu_j\, g(x; (j+1)\alpha, \boldsymbol{\xi}), \tag{12}$$

where

$$\nu_j = \nu_j(\delta) = \frac{(-1)^j}{\delta (j+1)!} \sum_{k=j}^{\infty} (k+1)!\,(1 - 1/\delta)^k.$$

We can verify that $\sum_{j=0}^{\infty} \eta_j = \sum_{j=0}^{\infty} \nu_j = 1$. Then, the $\mathcal{MOEW}$ density function can be expressed as an infinite linear combination of $\mathcal{EW}$ densities. Equations (10) and (12) have the same form except for the coefficients $\eta_j's$ in (10) and $\nu_j's$ in (12). They depend only on the generator parameter δ. For simplicity, we can write

$$f(x; \delta, \alpha, \boldsymbol{\xi}) = \sum_{j=0}^{\infty} w_j\, g(x; (j+1)\alpha, \boldsymbol{\xi}), \tag{13}$$

where

$$w_j = \begin{cases} \eta_j, & \text{if } 0 < \delta < 1, \\ \nu_j, & \text{if } \delta > 1, \end{cases}$$

and η_j and ν_j are given by (10) and (12), respectively. Thus, some mathematical properties of (13) can be obtained directly from those $\mathcal{EW}$ properties. For example, the ordinary, incomplete, inverse and factorial moments and the mgf of X follow immediately from those quantities of the $\mathcal{EW}$ distribution.

3 General properties

3.1 Moments, generating function and incomplete moments

The nth ordinary moment of X can be obtained from (13) as

$$\mathrm{E}(X^n) = \sum_{j=0}^{\infty} w_j\, E(Y_j^n),$$

where from now on $Y_j \sim \mathcal{EW}((j+1)\alpha, \boldsymbol{\xi})$ denotes a random variable having the $\mathcal{EW}$ density function $g(y; (j+1)\alpha, \boldsymbol{\xi})$.

The mgf and the kth incomplete moment of X follow from (13) as

$$M_X(t) = E\left(e^{tX}\right) = \sum_{j=0}^{\infty} w_j\, M_j(t)$$

and

$$T_k(z) = \sum_{j=0}^{\infty} w_j\, T_k^{(j)}(z), \tag{14}$$

where $M_j(t)$ is the mgf of Y_j and $T_k^{(j)}(z) = \int_{-\infty}^{z} x^k\, g(x; (j+1)\alpha, \boldsymbol{\xi})\mathrm{d}x$ comes directly from the $\mathcal{EW}$ model.

3.2 Quantile function and random number generator

The qf of X follows by inverting (5) and it can be expressed in terms of $H^{-1}(\cdot)$ as

$$Q(u) = H^{-1}\left(\frac{1}{\alpha}\log\left(\frac{1-\bar{\delta}\,u}{1-u}\right), \xi\right). \tag{15}$$

In Table 2, we provide the function $H^{-1}(x;\xi)$ for some special models.

Hence, the generator for X can be given by the algorithm:

Algorithm 1 Random number generator for the $\mathcal{MOEW}$ distribution

1: Generate $U \sim U(0,1)$.

2: Specify a function $H^{-1}(\cdot;\cdot)$ such as anyone in Table 2 and use (15).

3: Obtain an outcome of X by $X = Q(U)$.

The $\mathcal{MOEW}$ distributions can be very useful in modeling lifetime data and practitioners may be interested in fitting one of these models. We provide a script using the R language to generate the density, distribution function, hrf, qf, random numbers, Anderson-Darling test, Cramer-von Mises test and likelihood ratio (LR) tests. This script can be be obtained from the authors upon requested.

3.3 Mean deviations

The mean deviations of X about the mean and the median are given by

$$\delta_1 = \int_{\mathcal{D}} |x-\mu| f(x;\delta,\alpha,\xi)\,\mathrm{d}x \quad \text{and} \quad \delta_2 = \int_{\mathcal{D}} |x-M| f(x;\delta,\alpha,\xi)\,\mathrm{d}x,$$

respectively, where $\mu = E(X)$ denotes the mean and $M = Median(X)$ the median. The median follows from the nonlinear equation $F(M;\delta,\alpha,\xi) = 1/2$. So, these quantities reduce to

$$\delta_1 = 2\,\mu\,F(\mu;\delta,\alpha,\xi) - 2\,T_1(\mu) \quad \text{and} \quad \delta_2 = \mu - 2\,T_1(M),$$

where $T_1(z)$ is the first incomplete moment of X obtained from (14) as

$$T_1(z) = \sum_{j=0}^{\infty} w_j\, T_1^{(j)}(z),$$

and $T_1^{(j)}(z) = \int_{-\infty}^{z} x\, g(x;(j+1)\alpha,\xi)\,\mathrm{d}x$ is the first incomplete moment of Y_j.

An important application of the mean deviations is related to the Bonferroni and Lorenz curves. These curves are useful in economics, reliability, demography, medicine

Table 2 The $H^{-1}(x;\xi)$ function

Distribution	$H^{-1}(x;\xi)$
Exponential power	$\frac{[\log(x+1)]^{1/\beta}}{\lambda}$
Chen	$[\log(x+1)]^{1/\beta}$
Weibull extension	$\lambda\left[\log\left(\frac{x}{\lambda}+1\right)\right]^{1/\beta}$
Log-Weibull	$\sigma\log(x)+\mu$
Kies	$\frac{x^{1/\beta}\sigma+\mu}{x^{1/\beta}+1}$
Gen. Power Weibull	$\beta\left[(x+1)^{1/\theta}-1\right]^{1/\alpha_1}$
Gompertz	$\frac{\log(\beta x+1)}{\beta}$
Pham	$\left[\frac{\log(x+1)}{\log(a)}\right]^{1/\beta}$

and other fields. For a given probability p, they are defined by $B(p) = T_1(q)/(p\,\mu)$ and $L(p) = T_1(q)/\mu$, respectively, where $q = Q(p)$ is the qf of X given by (15) at $u = p$.

3.4 Average lifetime and mean residual lifetime functions

The average lifetime is given by

$$t_m = \int_0^\infty [1 - F(x;\delta,\alpha,\boldsymbol{\xi})]\,\mathrm{d}x = \sum_{j=0}^{\infty} w_j \int_0^\infty \overline{G}(x;(j+1)\alpha,\boldsymbol{\xi})\mathrm{d}x.$$

In fields such as actuarial sciences, survival studies and reliability theory, the mean residual lifetime has been of much interest; see, for a survey, Guess and Proschan (1988). Given that there was no failure prior to x_0, the residual life is the period from time x_0 until the time of failure. The mean residual lifetime is given by

$$\begin{aligned} m(x_0;\delta,\alpha,\boldsymbol{\xi}) &= \mathrm{E}\,(X - x_0 | X \geq x_0;\delta,\alpha,\boldsymbol{\xi}) = \int\limits_{\{x:x>x_0\}} \frac{(x-x_0)f(x;\delta,\alpha,\boldsymbol{\xi})}{\Pr(X > x_0)}\mathrm{d}x \\ &= [\Pr(X > x_0)]^{-1}\int_0^\infty y f(x_0 + y;\delta,\alpha,\boldsymbol{\xi})\mathrm{d}y \\ &= \left[\overline{F}(x_0;\delta,\alpha,\boldsymbol{\xi})\right]^{-1}\sum_{j=0}^{\infty} w_j \int_0^\infty y\, g(x_0 + y;(j+1)\alpha,\boldsymbol{\xi})\mathrm{d}y. \end{aligned}$$

The last integral can be computed from the baseline $\mathcal{EW}$ distribution. Further, $m(x_0;\delta,\alpha,\boldsymbol{\xi}) \to E(X)$ as $x_0 \to 0$.

4 Information theory measures

The seminal idea about information theory was pioneered by Hartley (1928), who defined a logarithmic measure of information for communication. Subsequently, Shannon (1948) formalized this idea by defining the entropy and mutual information concepts. The relative entropy notion (which would later be called *divergence*) was proposed by Kullback and Leibler (1951). The Kullback-Leibler's measure can be understood like a comparison criterion between two distributions. In this section, we derive two classes of entropy measures and one class of divergence measures which can be understood as new goodness-of-fit quantities such those discussed by Seghouane and Amari (2007). All these measures are defined for one element or between two elements in the $\mathcal{MOEW}$ family.

4.1 Rényi entropy

The Rényi entropy of X with pdf (8) is given by

$$H_{\mathrm{R}}^{s}(X) = \frac{1}{1-s}\log\left(\int_{\mathcal{D}} f(x;\delta,\alpha,\boldsymbol{\xi})^s \mathrm{d}x\right),$$

where $s \in (0,1) \cup (1,\infty)$.

It is a difficult problem to obtain $H_{\mathrm{R}}^{s}(X)$ in closed-form for the $\mathcal{MOEW}$ family. So, we derive an expansion for this quantity.

By using (9), $f(x;\delta,\alpha,\boldsymbol{\xi})^s$ can be expanded as

$$f(x;\delta,\alpha,\boldsymbol{\xi})^s = \sum_{j=0}^{\infty} w_j' \exp[-(j+s)\alpha H(x;\boldsymbol{\xi})]\; h(x;\boldsymbol{\xi})^s, \tag{16}$$

where

$$w_j' = \begin{cases} \eta_j'(\alpha,\delta) = \frac{\alpha^s \delta^s (2s)_j \bar{\delta}^j}{j!}, & \text{for } 0 < \delta < 1, \\ \nu_j'(\alpha,\delta) = \frac{\alpha^s \delta^{-s}}{j!} \sum_{k=0}^{\infty} \frac{(2s)_k (k)_j}{k!} (1-1/\delta)^k, & \text{for } \delta > 1. \end{cases}$$

The proof of this expansion is given in Appendix 8.

Finally, based on Equation (16), the Rényi entropy can be expressed as

$$H_R^s(X) = \frac{1}{1-s} \log \left\{ \sum_{j=0}^{\infty} w_j' \int_{\mathcal{D}} \exp[-(j+s)\alpha H(x;\boldsymbol{\xi})]\, h(x;\boldsymbol{\xi})^s dx \right\}.$$

An advantage of this expansion is its dependence of an integral which has closed-form for some $\mathcal{EW}$ distributions.

4.2 Shannon entropy

The Shannon entropy of X is given by

$$H_S(X) = E_X \left\{ -\log[f(X;\delta,\alpha,\boldsymbol{\xi})] \right\},$$

where the log-likelihood function corresponding to one observation follows from (8) as

$$\log[f(x;\delta,\alpha,\boldsymbol{\xi})] = \log(\delta\alpha) + \log[\, h(x;\boldsymbol{\xi})] - \alpha H(x;\boldsymbol{\xi}) - 2\log\left\{1 - \bar{\delta}\exp[-\alpha H(x;\boldsymbol{\xi})]\right\}.$$

Thus, it can be reduced to

$$H_S(X) = -\log(\alpha\delta) + 2E\left\{\log\left[1 - \bar{\delta}\bar{G}(X;\boldsymbol{\xi})\right]\right\} - E\left\{\log\left[h(X;\boldsymbol{\xi})\right]\right\} + \alpha E\left[H(X;\boldsymbol{\xi})\right].$$

4.3 Cross entropy and Kullback-Leibler divergence and distance

Let X and Y be two random variables with common support $\mathbb{R}_+$ whose densities are $f_X(x;\boldsymbol{\theta}_1)$ and $f_Y(y;\boldsymbol{\theta}_2)$, respectively. Cover and Thomas (1991) defined the *cross entropy* as

$$C_X(Y) = E_X\left\{-\log\left[f_Y(X;\boldsymbol{\theta}_2)\right]\right\} = -\int_0^{\infty} f_X(z;\boldsymbol{\theta}_1)\log\left[f_Y(z;\boldsymbol{\theta}_2)\right] dz.$$

We consider that $X \sim \mathcal{MOEW}(\delta_x, \alpha_x, \boldsymbol{\xi}_x)$ and $Y \sim \mathcal{MOEW}(\delta_y, \alpha_y, \boldsymbol{\xi}_y)$. After some algebraic manipulations, we obtain

$$\begin{aligned} C_X(Y) &= -\int_{\mathcal{D}} f_X(z;\delta_x,\alpha_x,\boldsymbol{\xi}_x)\log\left[f_Y(z;\delta_y,\alpha_y,\boldsymbol{\xi}_y)\right] dz \\ &= -\log\left(\delta_y\alpha_y\right) - E_X\left\{\log\left[h(X;\boldsymbol{\xi}_y)\right]\right\} + \alpha_y E_X\left[H(X;\boldsymbol{\xi}_y)\right] \\ &\quad + 2E_X\left\{\log\left[1 - \bar{\delta}\bar{G}(X;\boldsymbol{\xi}_y)\right]\right\}. \end{aligned} \tag{17}$$

An important measure in information theory is the Kullback-Leibler divergence given by

$$D(X||Y) = C_X(Y) - H_S(X) = E_X\left\{\log\left[\frac{f_X(X;\delta_x,\alpha_x,\boldsymbol{\xi}_x)}{f_Y(X;\delta_y,\alpha_y,\boldsymbol{\xi}_y)}\right]\right\}. \tag{18}$$

Applying (4.2) and (17) in Equation (18) gives

$$\begin{aligned} D(X||Y) = &\log\left(\frac{\delta_x\alpha_x}{\delta_y\alpha_y}\right) + E_X\left\{\log\left[\frac{h(X;\boldsymbol{\xi}_x)}{h(X;\boldsymbol{\xi}_y)}\right]\right\} + 2E_X\left\{\log\left[\frac{1-\bar{\delta}\bar{G}(X;\boldsymbol{\xi}_y)}{1-\bar{\delta}\bar{G}(X;\boldsymbol{\xi}_x)}\right]\right\} \\ &+ \alpha_y E_X[H(X;\boldsymbol{\xi}_y)] - \alpha_x E_X[H(X;\boldsymbol{\xi}_x)]. \end{aligned} \tag{19}$$

According to Cover and Thomas (1991), the Kullback-Leibler measure $D(X||Y)$ is the quantification of the error considering that the Y model is true when the data follow the X distribution. For example, this measure has been proposed as essential parts of test statistics, which has seen strongly applied to contexts of radar synthetic aperture image processing in both univariate (Nascimento et al. 2010) and polarimetric (or multivariate) (Nascimento et al. 2014) perspectives.

In order to work with measures that satisfied the non-negativity, symmetry and definiteness properties, Nascimento et al. (2010) considered the symmetrization of (19)

$$
\begin{aligned}
d_{\mathrm{KL}}(X,Y) &= \frac{1}{2}\,[D(X\|Y) + D(Y\|X)] \\
&= \int_{\mathcal{D}} \underbrace{\left(f_X\left(x;\delta_x,\alpha_x,\boldsymbol{\xi}_x\right) - f_Y\left(x;\delta_y,\alpha_y,\boldsymbol{\xi}_y\right)\right) \log\left(\frac{f_X(x;\delta_x,\alpha_x,\boldsymbol{\xi}_x)}{f_Y(x;\delta_x,\alpha_x,\boldsymbol{\xi}_x)}\right)}_{\equiv \mathit{IntegrandKL}(x,y)} \mathrm{d}x,
\end{aligned}
$$

which is given by

$$
\begin{aligned}
2\,d_{\mathrm{KL}}(X,Y) = {} & \alpha_y \left\{\mathrm{E}_X\left[H(X;\boldsymbol{\xi}_y)\right] - \mathrm{E}_Y\left[H(Y;\boldsymbol{\xi}_y)\right]\right\} + \alpha_x \left\{\mathrm{E}_Y\left[H(Y;\boldsymbol{\xi}_x)\right] - \mathrm{E}_X[H(X;\boldsymbol{\xi}_x)]\right\} \\
& + \mathrm{E}_X\left\{\log\left[\frac{h\left(X;\boldsymbol{\xi}_x\right)}{h\left(X;\boldsymbol{\xi}_y\right)}\right]\right\} + \mathrm{E}_Y\left\{\log\left[\frac{h\left(Y;\boldsymbol{\xi}_y\right)}{h\left(Y;\boldsymbol{\xi}_x\right)}\right]\right\} \\
& + 2\,\mathrm{E}_X\left\{\log\left[\frac{1-\bar{\delta}\bar{G}\left(X;\boldsymbol{\xi}_y\right)}{1-\bar{\delta}\bar{G}\left(X;\boldsymbol{\xi}_x\right)}\right]\right\} + 2\,\mathrm{E}_Y\left\{\log\left[\frac{1-\bar{\delta}\bar{G}\left(Y;\boldsymbol{\xi}_x\right)}{1-\bar{\delta}\bar{G}\left(Y;\boldsymbol{\xi}_y\right)}\right]\right\}.
\end{aligned}
\tag{20}
$$

Although this measure does not satisfy the triangle inequality, it is usually called the *Kullback-Leibler* distance (*Jensen-Shannon* divergence). The new measure can be used to answer questions like "how could one quantify the difference in selecting the Phani model with three parameters as the baseline distribution instead of the Weibull Kies distribution which has four parameters?".

As an illustration for (20), we initially consider two distinct elements of the generated special model from the specifications: $H(x;\beta) = \beta^{-1}[\exp(\beta x) - 1]$ and $h(x;\beta) = \exp(\beta x)$ in (8). This model will be presented with more details in future sections and its parametric space is represented by the vector (δ,α,β). Suppose that we are interested in quantifying the influence of a nuisance degree ϵ in the parameter α over the distance between two distinct elements, $(2,1,3)$ and $(2,1+\epsilon,3)$, at such parametric space. Figure 1(a) displays

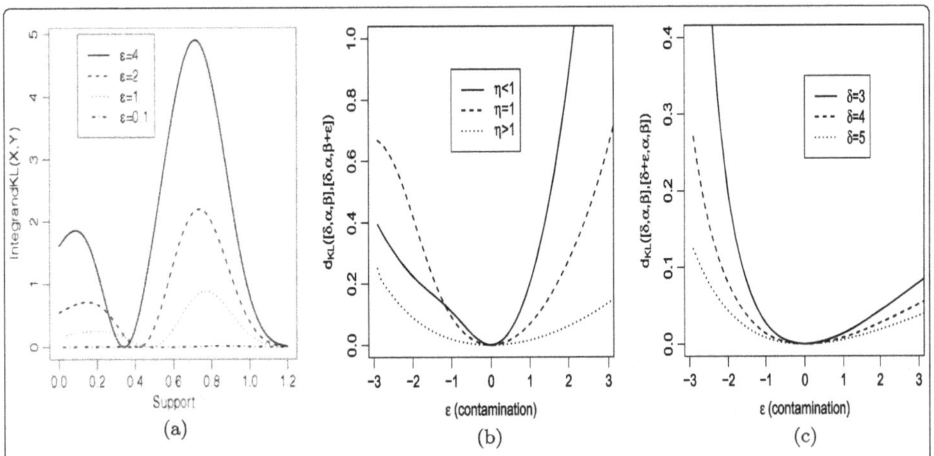

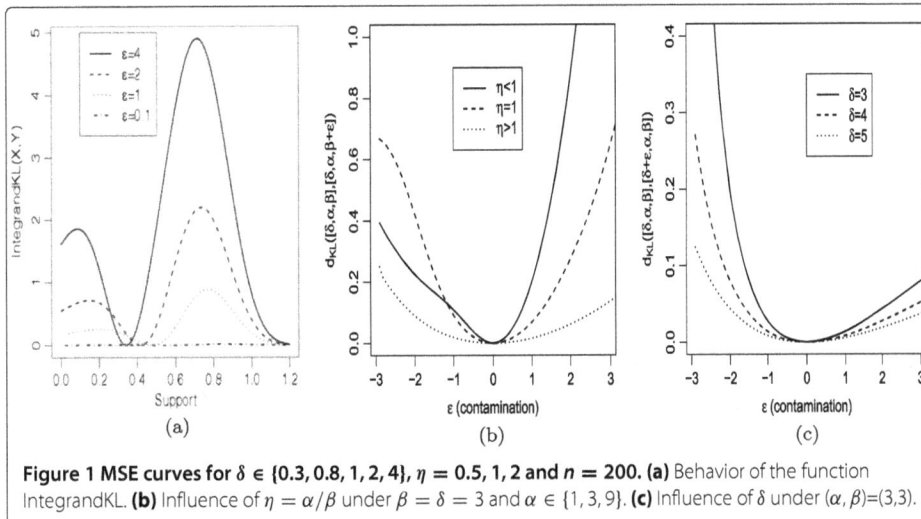

Figure 1 MSE curves for $\delta \in \{0.3, 0.8, 1, 2, 4\}$, $\eta = 0.5, 1, 2$ and $n = 200$. (a) Behavior of the function IntegrandKL. **(b)** Influence of $\eta = \alpha/\beta$ under $\beta = \delta = 3$ and $\alpha \in \{1, 3, 9\}$. **(c)** Influence of δ under (α, β)=(3,3).

the integrand of (20) for $\epsilon = 0.1$, 1, 2 and 4 for which the distances (or areas) associated with $d_{\mathrm{KL}}(X, Y)$ are 6.50×10^{-3}, 3.56×10^{-1}, 9.46×10^{-1} and 2.25, respectively. It is notable that $d_{\mathrm{KL}}(X, Y)$ takes smaller values for more closer points (or, equivalently, for more closer fits) and, therefore, (20) consists of new goodness-of-fit measures. In Figures 1(b) and 1(c), we show the influence of $\eta = \alpha/\beta$ on $d_{\mathrm{KL}}([\delta, \alpha, \beta], [\delta, \alpha, \beta + \epsilon])$ (for $\beta = \delta = 3$ and $\alpha \in \{1, 3, 9\}$) and of δ on $d_{\mathrm{KL}}([\delta, \alpha, \beta], [\delta + \epsilon, \alpha, \beta])$ (for $\beta = \alpha = 3$ and $\delta \in \{3, 4, 5\}$). For all cases, the contamination ϵ takes values in the interval $(-2.9, 2.9)$.

5 Estimation

Here, we present a general procedure for estimating the $\mathcal{MOEW}$ parameters from one observed sample and from multi-censored data. Additionally, we provide a discussion about how one can test the significance of additional parameter at the proposed class. Let $x_1, \ldots, x_n$ be a sample of size n from X. The log-likelihood function for the vector of parameters $\boldsymbol{\theta} = (\delta, \alpha, \boldsymbol{\xi}^\top)^\top$ can be expressed as

$$\ell(\boldsymbol{\theta}) = n \log(\delta\alpha) + \sum_{i=1}^{n} \log\left[h(x_i; \boldsymbol{\xi})\right] - \alpha \sum_{i=1}^{n} H(x_i; \boldsymbol{\xi}) - 2 \sum_{i=1}^{n} \log\left\{1 - \bar{\delta} \exp[-\alpha H(x_i; \boldsymbol{\xi})]\right\}.$$

From the above log-likelihood, the components of the score vector, $\mathbf{U}(\theta) = (U_\delta, U_\alpha, U_{\boldsymbol{\xi}}^\top)^\top$, are given by

$$U_\delta(\boldsymbol{\theta}) = \frac{\partial \ell(\theta)}{\partial \delta} = \frac{n}{\delta} - 2 \sum_{i=1}^{n} \frac{\exp[-\alpha H(x_i; \boldsymbol{\xi})]}{1 - \bar{\delta} \exp[-\alpha H(x_i; \boldsymbol{\xi})]},$$

$$U_\alpha(\boldsymbol{\theta}) = \frac{\partial \ell(\theta)}{\partial \alpha} = \frac{n}{\alpha} - \sum_{i=1}^{n} H(x_i; \boldsymbol{\xi}) - 2\bar{\delta} \sum_{i=1}^{n} \frac{H(x_i; \boldsymbol{\xi}) \exp[-\alpha H(x_i; \boldsymbol{\xi})]}{1 - \bar{\delta} \exp[-\alpha H(x_i; \boldsymbol{\xi})]} \quad \text{and}$$

$$U_{\xi_k}(\boldsymbol{\theta}) = \frac{\partial \ell(\theta)}{\partial \xi_k} = \sum_{i=1}^{n} \frac{1}{h(x_i; \boldsymbol{\xi})} \frac{\partial h(x_i; \boldsymbol{\xi})}{\partial \xi_k} - \alpha \sum_{i=1}^{n} \frac{\partial H(x_i; \boldsymbol{\xi})}{\partial \xi_k}$$
$$- 2\bar{\delta}\alpha \sum_{i=1}^{n} \frac{\partial H(x_i; \boldsymbol{\xi})}{\partial \xi_k} \frac{\exp[-\alpha H(x_i; \boldsymbol{\xi})]}{1 - \bar{\delta} \exp[-\alpha H(x_i; \boldsymbol{\xi})]}.$$

Finally, the partitioned observed information matrix for the $\mathcal{MOEW}$ family is

$$J(\boldsymbol{\theta}) = -\begin{pmatrix} U_{\delta\delta} & U_{\delta\alpha} & | & U_{\delta\boldsymbol{\xi}}^\top \\ U_{\alpha\delta} & U_{\alpha\alpha} & | & U_{\alpha\boldsymbol{\xi}}^\top \\ -- & -- & -- & -- \\ U_{\delta\boldsymbol{\xi}} & U_{\alpha\boldsymbol{\xi}} & | & U_{\boldsymbol{\xi}\boldsymbol{\xi}} \end{pmatrix},$$

whose elements are

$$U_{\delta\delta}(\boldsymbol{\theta}) = -n\delta^{-2}, U_{\delta\alpha}(\boldsymbol{\theta}) = 2 \sum_{i=1}^{n} \frac{H(x_i; \boldsymbol{\xi}) \exp[-\alpha H(x_i; \boldsymbol{\xi})]}{\left\{1 - \bar{\delta} \exp[-\alpha H(x_i; \boldsymbol{\xi})]\right\}^2},$$

$$U_{\delta\xi_k}(\boldsymbol{\theta}) = 2\alpha \sum_{i=1}^{n} \frac{\partial H(x_i; \boldsymbol{\xi})}{\partial \xi_k} \frac{\exp[-\alpha H(x_i; \boldsymbol{\xi})]}{\left\{1 - \bar{\delta} \exp[-\alpha H(x_i; \boldsymbol{\xi})]\right\}^2},$$

$$U_{\alpha\alpha}(\boldsymbol{\theta}) = -\frac{n}{\alpha^2} + 2\bar{\delta} \sum_{i=1}^{n} \frac{H(x_i; \boldsymbol{\xi})^2 \exp[-\alpha H(x_i; \boldsymbol{\xi})]}{\left\{1 - \bar{\delta} \exp[-\alpha H(x_i; \boldsymbol{\xi})]\right\}^2},$$

$$U_{\alpha\xi_k}(\boldsymbol{\theta}) = -2\bar{\delta}\sum_{i=1}^{n}\frac{\partial H(x_i;\boldsymbol{\xi})}{\partial \xi_k}\frac{\exp[-\alpha H(x_i;\boldsymbol{\xi})]}{1-\bar{\delta}\exp[-\alpha H(x_i;\boldsymbol{\xi})]}\left[1-\frac{\alpha H(x_i;\boldsymbol{\xi})}{1-\bar{\delta}\exp[-\alpha H(x_i;\boldsymbol{\xi})]}\right] + \sum_{i=1}^{n}\frac{\partial H(x_i;\boldsymbol{\xi})}{\partial \xi_k} \quad \text{and}$$

$$U_{\xi_k\xi_j}(\boldsymbol{\theta}) = \sum_{i=1}^{n}\frac{1}{h(x_i;\boldsymbol{\xi})}\left[\frac{\partial^2 h(x_i;\boldsymbol{\xi})}{\partial \xi_k\xi_j} - \frac{1}{h(x_i;\boldsymbol{\xi})}\frac{\partial h(x_i;\boldsymbol{\xi})}{\partial \xi_k}\frac{\partial h(x_i;\boldsymbol{\xi})}{\partial \xi_j}\right] - \alpha\sum_{i=1}^{n}\frac{\partial^2 H(x_i;\boldsymbol{\xi})}{\partial \xi_k\xi_j}$$

$$-2\alpha\bar{\delta}\sum_{i=1}^{n}\frac{\exp[-\alpha H(x_i;\boldsymbol{\xi})]}{1-\bar{\delta}\exp[-\alpha H(x_i;\boldsymbol{\xi})]}\left[\frac{\partial^2 H(x_i;\boldsymbol{\xi})}{\partial \xi_k\xi_j} - \frac{\partial H(x_i;\boldsymbol{\xi})}{\partial \xi_k}\frac{\alpha H(x_i;\boldsymbol{\xi})}{1-\bar{\delta}\exp[-\alpha H(x_i;\boldsymbol{\xi})]}\right].$$

When some standard regularity conditions are satisfied (Cox and Hinkley 1974), one can verify that $\sqrt{n}\left(\left[\widehat{\alpha},\widehat{\delta},\widehat{\boldsymbol{\xi}}\right]^\top - [\alpha,\delta,\boldsymbol{\xi}]^\top\right)$ converges *in distribution* to the multivariate $N_{p+2}\left(\mathbf{0},\mathcal{K}([\alpha,\delta,\boldsymbol{\xi}])^{-1}\right)$ distribution, where p denotes the dimension of $\boldsymbol{\xi}$ and $\mathcal{K}([\alpha,\delta,\boldsymbol{\xi}])$ is the expected information matrix for which the limit identity $\lim_{n\to\infty} J_n([\alpha,\delta,\boldsymbol{\xi}]) = \mathcal{K}([\alpha,\delta,\boldsymbol{\xi}])$ is satisfied. Based on this result, one can compute confidence regions for the $\mathcal{MOEW}$ parameters. Such regions can be used as decision criteria in several practical situations.

For checking if δ is statistically different from one, i.e. for testing the null hypothesis $H_0 : \delta = 1$ against $H_1 : \delta \neq 1$, we use the LR statistic given by $\text{LR} = 2\left\{\ell(\widehat{\boldsymbol{\theta}}) - \ell(\widetilde{\boldsymbol{\theta}})\right\}$, where $\widehat{\boldsymbol{\theta}}$ is the vector of unrestricted MLEs under H_1 and $\widetilde{\boldsymbol{\theta}}$ is the vector of restricted MLEs under H_0. Under the null hypothesis, the limiting distribution of LR is a χ_1^2 distribution. If the test statistic exceeds the upper $100(1-\alpha)\%$ quantile of the χ_1^2 distribution, then we reject the null hypothesis.

Censored data occur very frequently in lifetime data analysis. Some mechanisms of censoring are identified in the literature as, for example, types I and II censoring (Lawless 2003). Here, we consider the general case of multi-censored data: there are $n = n_0 + n_1 + n_2$ subjects of which n_0 is known to have failed at the times $x_1, \ldots, x_{n_0}$, n_1 is known to have failed in the interval $[s_{i-1}, s_i]$, $i = 1, \ldots, n_1$, and n_2 survived to a time r_i , $i = 1, \ldots, n_2$, but not observed any longer. Note that type I censoring and type II censoring are contained as particular cases of multi-censoring. The log-likelihood function of $\boldsymbol{\theta} = (\delta, \alpha, \boldsymbol{\xi}^\top)^\top$ for this multi-censoring data reduces to

$$\begin{aligned}\ell(\boldsymbol{\theta}) &= n_0\log(\delta\alpha) + \sum_{i=1}^{n_0}\log\left[h(x_i;\boldsymbol{\xi})\right] - \alpha\sum_{i=1}^{n_0}H(x_i;\boldsymbol{\xi}) - 2\sum_{i=1}^{n_0}\log\left\{1-\bar{\delta}\exp[-\alpha H(x_i;\boldsymbol{\xi})]\right\}\\ &+\sum_{i=1}^{n_1}\log\left\{\frac{1-\exp[-\alpha H(s_i;\boldsymbol{\xi})]}{1-\bar{\delta}\exp[-\alpha H(s_i;\boldsymbol{\xi})]} - \frac{1-\exp[-\alpha H(s_{i-1};\boldsymbol{\xi})]}{1-\bar{\delta}\exp[-\alpha H(s_{i-1};\boldsymbol{\xi})]}\right\}\\ &+n_2\log(\delta) - \alpha\sum_{i=1}^{n_2}H(r_i;\boldsymbol{\xi}) - 2\sum_{i=1}^{n_2}\log\left\{1-\bar{\delta}\exp[-\alpha H(r_i;\boldsymbol{\xi})]\right\}.\end{aligned} \tag{21}$$

The score functions and the observed information matrix corresponding to (21) is too complicated to be presented here.

6 Two special models

In this section, we study two special $\mathcal{MOEW}$ models, namely the Marshall-Olkin modified Weibull ($\mathcal{MOMW}$) and Marshall-Olkin Gompertz ($\mathcal{MOG}$) distributions. We provide plots of the density and hazard rate functions for some parameters to illustrate the flexibility of these distributions.

6.1 The $\mathcal{MOMW}$ model

For $H(x;\lambda,\gamma) = x^{\gamma}\exp(\lambda x)$ and $h(x;\lambda,\gamma) = x^{\gamma-1}\exp(\lambda x)(\gamma+\lambda x)$, we obtain the $\mathcal{MOMW}$ distribution. Its density function is given by

$$f(x;\alpha,\delta,\lambda,\gamma) = \delta\alpha(\gamma+\lambda x)x^{\gamma-1}\frac{\exp[\lambda x - \alpha x^{\gamma}\exp(\lambda x)]}{\left\{1-\bar{\delta}\exp[-\alpha x^{\gamma}\exp(\lambda x)]\right\}^{2}}, \quad x>0,$$

where $\lambda,\gamma \geq 0$. If $\delta = 1$, it leads to the special case of the modified Weibull ($\mathcal{MW}$) distribution (Lai *et al.* 2003). In addition, when $\lambda = 0$, it gives the Weibull distribution. Its cdf and hrf are given by

$$F(x;\alpha,\delta,\lambda,\gamma) = \frac{1-\exp[-\alpha x^{\gamma}\exp(\lambda x)]}{1-\bar{\delta}\exp[-\alpha x^{\gamma}\exp(\lambda x)]}$$

and

$$\tau(x;\alpha,\delta,\lambda,\gamma) = \frac{\alpha x^{\gamma-1}\exp(\lambda x)(\gamma+\lambda x)}{1-\bar{\delta}\exp[-\alpha x^{\gamma}\exp(\lambda x)]},$$

respectively. In Figures 2(a), 2(b), 2(c) and 2(d), we note some different shapes of the $\mathcal{MOMW}$ pdf. Further, Figures 3(a), 3(b), 3(c) and 3(d) display plots of the $\mathcal{MOMW}$ hrf, which can have increasing, decreasing, non-monotone and bathtub forms.

The rth raw moment of the $\mathcal{MOMW}$ distribution comes from (13) as

$$\mathrm{E}\left(X^{r}\right) = \sum_{j=1}^{\infty} w_j\,\mu_r(j), \tag{22}$$

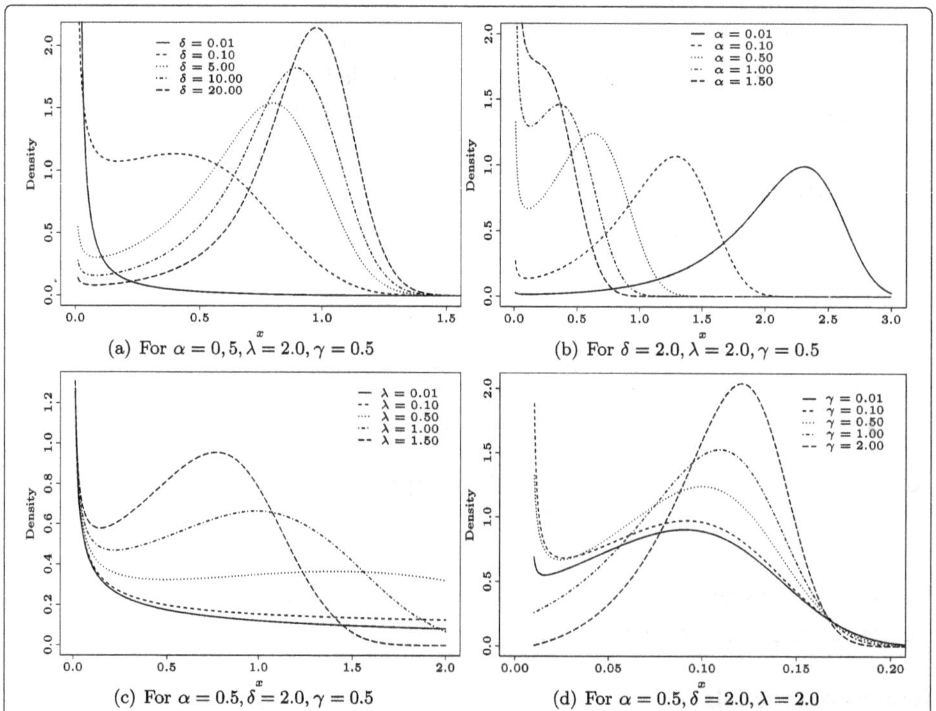

(a) For $\alpha = 0,5, \lambda = 2.0, \gamma = 0.5$ (b) For $\delta = 2.0, \lambda = 2.0, \gamma = 0.5$

(c) For $\alpha = 0.5, \delta = 2.0, \gamma = 0.5$ (d) For $\alpha = 0.5, \delta = 2.0, \lambda = 2.0$

Figure 2 The $\mathcal{MOMW}$ density functions. (a) For $\alpha = 0.5, \lambda = 2.0, \gamma = 0.5$. **(b)** For $\delta = 2.0, \lambda = 2.0, \gamma = 0.5$. **(c)** For $\delta = 5.0, \delta = 2.0, \gamma = 0.5$. **(d)** For $\alpha = 0.5, \delta = 2.0, \lambda = 2.0$.

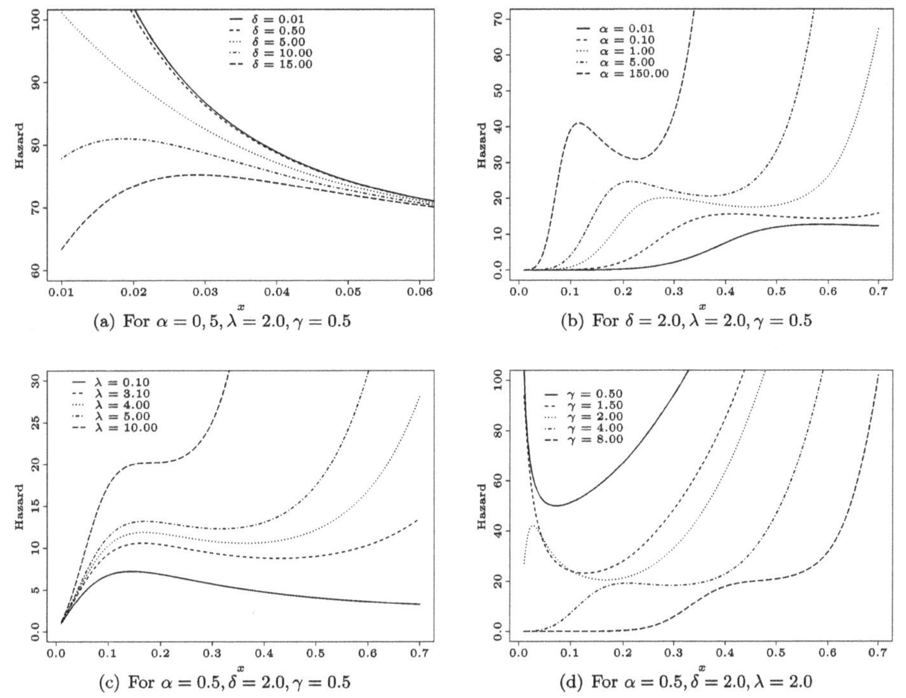

Figure 3 The $\mathcal{MOWN}$ hrfs. (a) For $\alpha = 0.5, \lambda = 2.0, \gamma = 0.5$. **(b)** For $\delta = 2.0, \lambda = 2.0, \gamma = 0.5$. **(c)** For $\delta = 5.0, \delta = 2.0, \gamma = 0.5$. **(d)** For $\alpha = 0.5, \delta = 2.0, \lambda = 2.0$.

where $\mu_r(j) = \int_0^\infty x^r g(x;\, (j+1)\alpha, \gamma, \lambda))dx$ denotes the rth raw moment of the $\mathcal{MW}$ distribution with parameters $(j+1)\alpha, \gamma$ and λ. Carrasco *et al.* (2008) determined an infinite representation for $\mu_r(j)$ given by

$$\mu_r(j) = \sum_{i_1,\ldots,i_r=1}^{\infty} \frac{A_{i_1,\ldots,i_r}\,\Gamma(s_r/\gamma + 1)}{[(j+1)\alpha]^{s_r/\gamma}}, \tag{23}$$

where

$$A_{i_1,\ldots,i_r} = a_{i_1}, \ldots, a_{i_r} \quad \text{and} \quad s_r = i_1, \ldots, i_r,$$

and

$$a_i = \frac{(-1)^{i+1} i^{i-2}}{(i-1)!} \left(\frac{\lambda}{\gamma}\right)^{i-1}.$$

Hence, the $\mathcal{MOMW}$ moments can be obtained directly from (22) and (23).

Let $x_1, \ldots, x_n$ be a sample of size n from $X \sim \mathcal{MOMW}(\alpha, \delta, \lambda, \gamma)$. The log-likelihood function for the vector of parameters $\boldsymbol{\theta} = (\alpha, \delta, \lambda, \gamma)^\top$ can be expressed as

$$\ell(\boldsymbol{\theta}) = n\log(\delta\alpha) + \sum_{i=1}^{n} \log(\gamma + \lambda\, x_i) + (\gamma - 1)\sum_{i=1}^{n} \log(x_i) + \lambda \sum_{i=1}^{n} x_i - \alpha \sum_{i=1}^{n} x_i^{\lambda} \exp(\lambda\, x_i)$$
$$-2\sum_{i=1}^{n} \log\left(1 - \bar{\delta} \exp\left[-\alpha x_i^{\gamma} \exp(\lambda\, x_i)\right]\right).$$

6.2 The $\mathcal{MOG}$ model

For $H(x;\beta) = \beta^{-1}[\exp(\beta x)-1]$ and $h(x;\beta) = \exp(\beta x)$, we obtain the $\mathcal{MOG}$ distribution. Its pdf is given by

$$f(x;\alpha,\delta,\beta) = \frac{\delta\alpha \exp\{\beta x - \alpha/\beta[\exp(\beta x)-1]\}}{\{1-\bar{\delta}\exp\{-\alpha/\beta[\exp(\beta x)-1]\}\}^2}, \quad x > 0,$$

where $-\infty < \beta < \infty$. For $\delta = 1$, it follows the Gompertz distribution as a special case. The $\mathcal{MOG}$ model is a special case of the Marshall-Olkin Makeham distribution (EL-Bassiouny and Abdo 2009). The cdf and hrf of the $\mathcal{MOG}$ distribution are given by

$$F(x;\alpha,\delta,\beta) = \frac{1-\exp\{-\alpha/\beta[\exp(\beta x)-1]\}}{1-\bar{\delta}\exp\{-\alpha/\beta[\exp(\beta x)-1]\}}$$

and

$$\tau(x;\alpha,\delta,\beta) = \frac{\alpha\exp(\beta x)}{1-\bar{\delta}\exp\{-\alpha/\beta[\exp(\beta x)-1]\}}.$$

Figures 4(a), 4(b) and 4(c) display some plots of the density functions for some values of α, δ and β. The hrf of the Gompertz distribution is increasing ($\beta > 0$) and decreasing ($\beta < 0$). Besides these two forms, Figures 5(a), 5(b) and 5(c) indicate that the $\mathcal{MOG}$ hrf can be bathtub shaped.

From Equation (15), the $\mathcal{MOG}$ qf becomes

$$Q(u) = \beta^{-1}\log\left[\frac{\beta}{\alpha}\log\left(\frac{1-\bar{\delta}u}{1-u}\right)+1\right].$$

Let $x_1,\ldots,x_n$ be a sample of size n from the $\mathcal{MOG}$ model. The log-likelihood function for the vector of parameters $\boldsymbol{\theta} = (\delta,\alpha,\beta)^\top$ can be expressed as

$$\ell(\boldsymbol{\theta}) = n\log(\delta\alpha) + \beta\sum_{i=1}^n x_i - \frac{\alpha}{\beta}\sum_{i=1}^n[\exp(\beta x_i)-1]$$
$$-2\sum_{i=1}^n \log\left(1-\bar{\delta}\exp\{-\alpha[\exp(\beta x_i)-1]/\beta\}\right).$$

7 Simulation and applications

This section is divided in two parts. First, we perform a simulation study in order to assess the performance of the MLEs on some points at the parametric space of one of the special

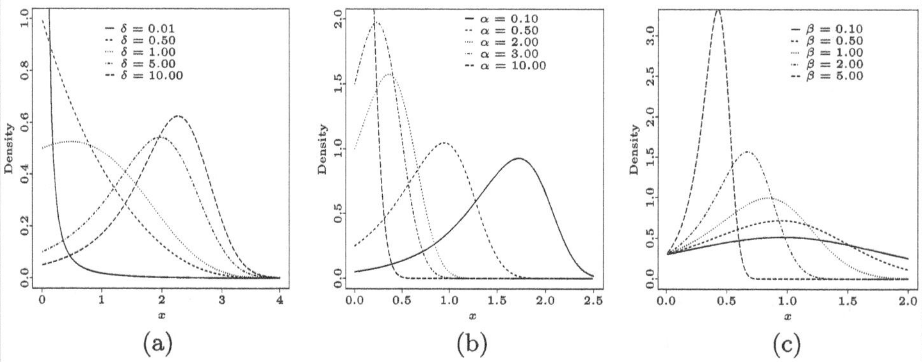

Figure 4 The $\mathcal{MOG}$ density functions. (a) For $\alpha = 0.5, \beta = 0.7$. **(b)** For $\delta = 2.0, \beta = 2.0$. **(c)** For $\delta = 5.0, \alpha = 1.5$.

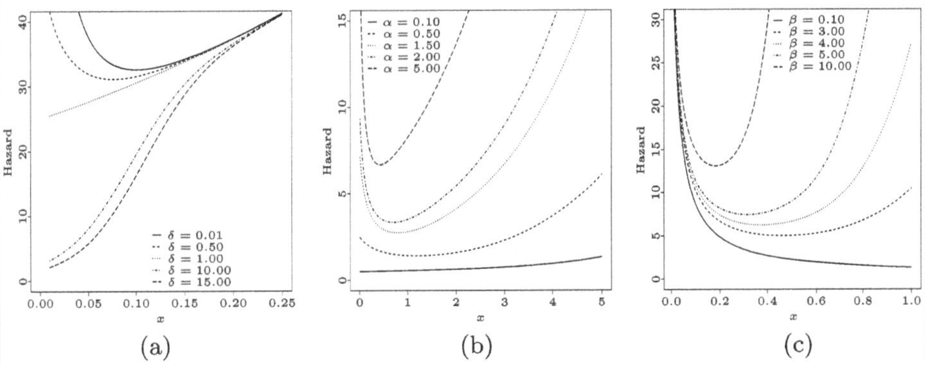

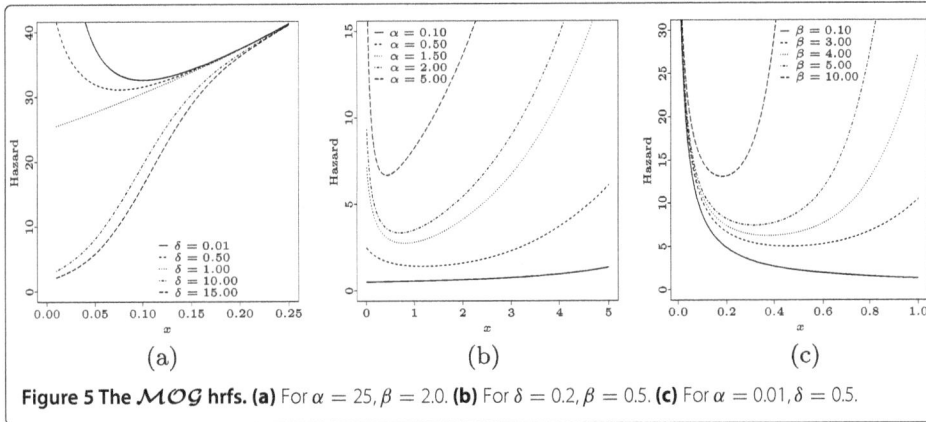

Figure 5 The $\mathcal{MOG}$ hrfs. (a) For $\alpha = 25, \beta = 2.0$. **(b)** For $\delta = 0.2, \beta = 0.5$. **(c)** For $\alpha = 0.01, \delta = 0.5$.

models. Second, an application to real data provides evidence in favor of one distribution in the $\mathcal{MOEW}$ class.

7.1 Simulation study

We present a simulation study by means of Monte Carlo's experiments in order to assess the performance of the MLEs described in Section 5. To that end, we work with the $\mathcal{MOG}$ distribution. One of advantages of this model is that its cdf has tractable analytical form. This fact implies in a simple random number generation (RNG) determined by the $\mathcal{MOG}$ qf given in Section 6.2. The $\mathcal{MOG}$ generator is illustrated in Figure 6.

The simulation study is conducted in order to quantify the influence of $\eta = \alpha/\beta$ over the estimation of the extra parameter δ. It is known that $\eta > 1$ gives the Gompertz distribution which presents mode at zero or, for $\eta < 1$, having their modes at $x^* = \beta^{-1}[1 - \log(\eta)]$. An initial discussion using the Kullback-Leibler distance derived in Section 4.3 points out that increasing the contamination (or the bias of the estimates) can affect the quality of fit.

In this study, the following scenarios are taken into account. For the sample size $n = 50, 100, 150, 200$, we adopt as the true parameters the following cases:

(i) Scenario $\eta < 1$: $(\alpha, \beta) = (1, 2)$ and $\delta \in \{0.3, 1, 4\}$;
(ii) Scenario $\eta = 1$: $(\alpha, \beta) = (2, 2)$ and $\delta \in \{0.3, 1, 4\}$;
(iii) Scenario $\eta > 1$: $(\alpha, \beta) = (4, 2)$ and $\delta \in \{0.3, 1, 4\}$.

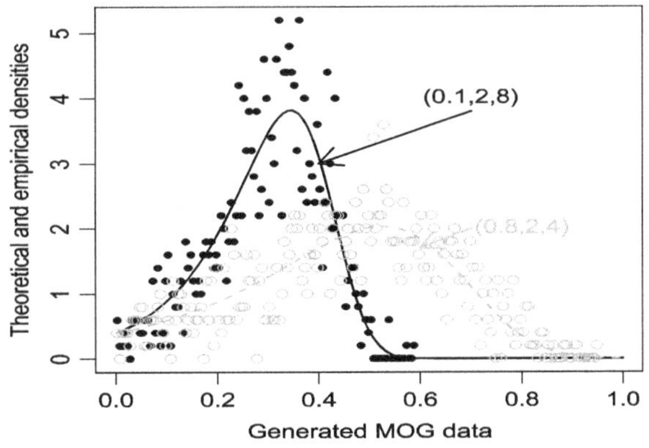

Figure 6 Illustration of the $\mathcal{MOG}$ generator for two points at the parametric space.

Also, we use 10,000 Monte Carlo's replications and, at each one of them, we quantify (i) the average of the MLEs and (ii) the mean square error (MSEs).

Table 3 gives the results of the simulation study. In general, the MLEs present smaller values of the biases and MSEs when the sample size increases. It is important to highlight the following atypical case: for the MLEs of α at the scenarios $(\alpha, \delta, \beta) \in \{(1,4,2), (2,1,2), (4,0.3,2), (4,1,2)\}$ and of δ at $(4,0.3,2)$, the associated biases do not have an inverse monotonic relationship with sample sizes, as expected.However, based on the fact

Table 3 Performance of the MLEs for the $\mathcal{MOG}$ distribution

(α,δ,β)	n	$\overline{\widehat{\theta}}_i(MSE(\widehat{\theta}_i))$					
		$\overline{\widehat{\alpha}}$ (MSE ($\widehat{\alpha}$))		$\overline{\widehat{\delta}}$ (MSE ($\widehat{\delta}$))		$\overline{\widehat{\beta}}$ (MSE ($\widehat{\beta}$))	
				For $\eta < 1$			
$(1,0.3,2)$	50	1.201	(2.837)	0.478	(0.883)	2.502	(1.698)
·	100	1.181	(1.745)	0.406	(0.290)	2.320	(1.238)
·	150	1.156	(1.299)	0.385	(0.195)	2.249	(1.015)
·	200	1.103	(1.008)	0.358	(0.134)	2.244	(0.899)
$(1,1,2)$	50	1.202	(1.965)	1.620	(5.938)	2.425	(1.630)
·	100	1.134	(1.199)	1.361	(2.690)	2.305	(1.145)
·	150	1.079	(0.884)	1.231	(1.638)	2.288	(0.979)
·	200	1.063	(0.735)	1.180	(1.244)	2.250	(0.845)
$(1,4,2)$	50	0.965	(0.810)	4.764	(26.798)	2.544	(1.561)
·	100	0.958	(0.544)	4.398	(14.813)	2.390	(1.025)
·	150	0.959	(0.443)	4.283	(11.454)	2.328	(0.831)
·	200	0.970	(0.369)	4.246	(8.953)	2.262	(0.653)
				For $\eta = 1$			
$(2,0.3,2)$	50	2.246	(7.571)	0.426	(0.473)	2.787	(3.543)
·	100	2.137	(4.502)	0.361	(0.172)	2.561	(2.473)
·	150	2.073	(3.279)	0.341	(0.116)	2.471	(1.981)
·	200	2.011	(2.596)	0.324	(0.083)	2.434	(1.698)
$(2,1,2)$	50	2.161	(5.462)	1.481	(4.886)	2.687	(3.051)
·	100	2.012	(3.115)	1.199	(1.798)	2.543	(2.157)
·	150	1.947	(2.277)	1.100	(1.062)	2.483	(1.763)
·	200	1.923	(1.874)	1.056	(0.787)	2.430	(1.507)
$(2,4,2)$	50	1.805	(2.404)	4.534	(21.279)	2.785	(2.783)
·	100	1.817	(1.681)	4.202	(12.456)	2.572	(1.869)
·	150	1.828	(1.390)	4.097	(9.474)	2.487	(1.527)
·	200	1.861	(1.153)	4.075	(7.495)	2.388	(1.184)
				For $\eta > 1$			
$(4,0.3,2)$	50	3.770	(13.137)	0.336	(0.191)	3.400	(6.701)
·	100	3.737	(8.129)	0.304	(0.072)	2.951	(4.152)
·	150	3.731	(6.119)	0.298	(0.051)	2.764	(3.184)
·	200	3.685	(4.865)	0.289	(0.038)	2.676	(2.613)
$(4,1,2)$	50	3.845	(13.615)	1.272	(3.153)	3.149	(6.239)
·	100	3.735	(7.757)	1.076	(1.043)	2.833	(4.060)
·	150	3.717	(5.760)	1.024	(0.634)	2.689	(3.150)
·	200	3.721	(4.759)	1.000	(0.472)	2.588	(2.601)
$(4,4,2)$	50	3.608	(8.172)	4.605	(21.140)	3.036	(5.150)
·	100	3.677	(5.234)	4.262	(11.467)	2.668	(2.989)
·	150	3.737	(4.039)	4.172	(8.228)	2.510	(2.169)
·	200	3.796	(3.247)	4.138	(6.370)	2.389	(1.588)

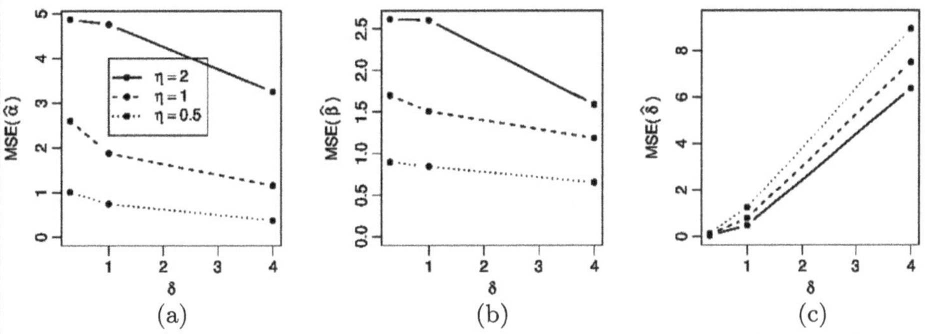

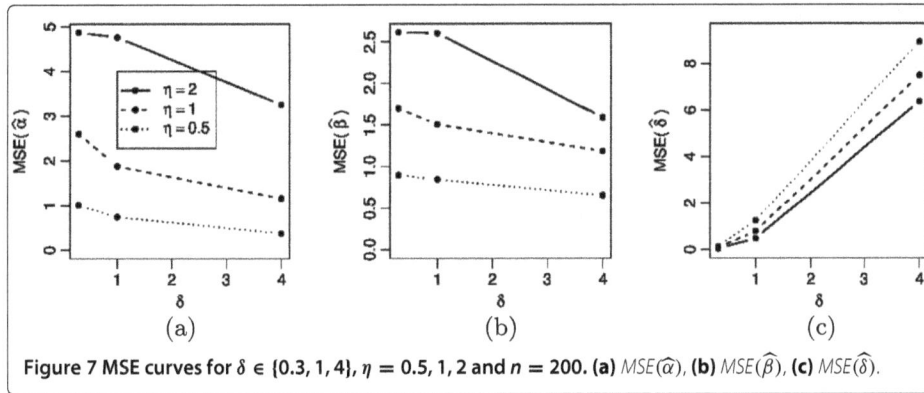

Figure 7 MSE curves for $\delta \in \{0.3, 1, 4\}$, $\eta = 0.5, 1, 2$ and $n = 200$. (a) $MSE(\widehat{\alpha})$, (b) $MSE(\widehat{\beta})$, (c) $MSE(\widehat{\delta})$.

that their MSEs tend to zero, we can expect that there exists a sample size n_0 such that biases of the MLEs decrease when the sample sizes increase from n_0.

The results provide evidence that the scenarios under the condition $\eta > 1$ yield a hard estimation (having larger variation ranges of the MSEs than those obtained for the cases when $\eta < 1$) for α and β parameters, and that the MLEs present smaller values of the MSEs under such conditions. Figure 7 illustrates the above behavior for the cases $\delta \in \{0.3, 0.8, 1, 2, 4\}$ and $n = 200$. In summary, the scenario with less numerical problems is $(\eta, \delta) = (2, 0.1)$, whereas that one which requires more attention for estimating the $\mathcal{MOG}$ parameters is $(\eta, \delta) = (0.5, 4)$.

7.2 Applications

Here, the usefulness of the $\mathcal{MOEW}$ distribution is illustrated by means of two real data sets.

7.2.1 Uncensored data

Here, we compare the fits of some special models of the $\mathcal{MOEW}$ family using a real data set. The estimation of the model parameters is performed by the maximum likelihood method discussed in Section 5. We use the `maxLik` function of the `maxLik` package in `R` language. In this function, if the argument "method" is not specified, a suitable method is selected automatically. For this application, we use the Newton-Raphson method. The data represent the percentage of body fat determined by underwater weighing for 250 men. For more details about the data see http://lib.stat.cmu.edu/datasets/bodyfat.

Table 4 provides some descriptive measures. They suggest an empirical distribution which is slightly asymmetric and platykurtic.

We compare the classical models and generalized models within the $\mathcal{MO}$ family. The null hypothesis $H_0 : \delta = 1$ is tested against $H_1 : \delta \neq 1$ using the LR statistic. The comparisons are presented in Table 5. For the $\mathcal{MOW}$ and $\mathcal{MOEP}$ models, one cannot say that the parameter δ is statistically different from one at the 10% significance level. Based on this result, we fit the $\mathcal{W}$, exponential power ($\mathcal{EP}$), $\mathcal{MOG}$ and Marshall-Olkin flexible Weibull extension ($\mathcal{MOFWE}$) models to the current data (see Table 1). These models are compared with two other three-parameter models, namely: the modified Weibull ($\mathcal{MW}$)

Table 4 Descriptive statistics

Mean	Mode	Median	Std. Desv.	Skewness	Kurtosis	Min	Max
19.30	20.40	19.25	8.23	0.19	2.62	3.00	47.50

Table 5 Comparison of fitted models using the LR test

Null hypothesis	Models	LR statistic	*p*-value
$\delta = 1$	$\mathcal{G} \times \mathcal{MOG}$	11.2963	0.0008
	$\mathcal{W} \times \mathcal{MOW}$	0.7638	0.3822
	$\mathcal{EP} \times \mathcal{MOEP}$	2.1959	0.1384
	$\mathcal{FWE} \times \mathcal{MOFWE}$	12.3659	0.0004

and generalized Birnbaum-Saunders ($\mathcal{GBS}$) (Owen 2006) distributions. The $\mathcal{GBS}$ density is given by

$$f(x;\phi,\eta,\kappa) = \frac{1}{\phi\sqrt{2\pi}\,\eta x^{\kappa}}\left(1-\kappa+\frac{\eta\kappa}{x}\right)\exp\left[-\frac{1}{2\phi^2}\frac{(x-\eta)^2}{\eta x^{2\kappa}}\right], \quad x > 0.$$

In Table 6, we present the MLEs (standard errors in parentheses) of the parameters of the fitted $\mathcal{MOFWE}$, $\mathcal{MOG}$, $\mathcal{EP}$, $\mathcal{W}$, $\mathcal{MW}$ and $\mathcal{GBS}$ distributions. Also, we provide the goodness-of-fit measures (p-values in parentheses). Thus, these values indicate that the null models are strongly rejected for the $\mathcal{MOFWE}$ and $\mathcal{MOG}$ distributions, since the associated p-values are much lower than 0.001.

Table 7 gives the values of the Akaike information criterion (AIC), Bayesian information criterion (BIC), consistent Akaike information criterion (CAIC) and Hannan-Quinn information criterion (HQIC). Since the values of the AIC, CAIC and HQIC are smaller for the $\mathcal{MOFWE}$ distribution compared to those values of the other fitted models. Thus, this new distribution seems to be a very competitive model to explain the current data.

Figures 8(a) and 8(b) display the estimated density and survival functions of the $\mathcal{MOFWE}$ distribution. The plots confirm the excellent fit of this distribution to the data. Figure 8(c) shows that the estimated $\mathcal{MOFWE}$ hrf is an increasing curve.

7.2.2 Censored data

Now, we consider a set of remission times from 137 cancer patients [Lee and Wang (2003), pag. 231]. Lee and Wang (2003) showed that the log-logistic ($\mathcal{LL}$) model provides a good fit to the data. Ghitany *et al.* (2005) compared the fits of the $\mathcal{MOW}$ and $\mathcal{W}$ models to these data. Now, we present a more detailed study by comparing the fitted $\mathcal{W}$, $\mathcal{LL}$, $\mathcal{EP}$, $\mathcal{MOW}$, Marshall-Olkin log-logistic ($\mathcal{MOLL}$), $\mathcal{MOEP}$ and $\mathcal{GBS}$ models to these data.

Table 6 MLEs and goodness-of-fit statistics

Model	Estimates (standard errors)				Goodness-of-fit (*p*-value)	
	$\widehat{\alpha}$(or $\widehat{\phi}$)	$\widehat{\delta}$(or $\widehat{\eta}$)	$\widehat{\gamma}$ (or $\widehat{\kappa}$)	$\widehat{\beta}$ (or $\widehat{\lambda}$)	AD	CM
$\mathcal{MOFWE}$	1	2.9136	0.0552	14.3666	**0.1082**	**0.0115**
	×	(1.1321)	(0.0022)	(3.7615)	**(0.9939)**	**(0.9987)**
$\mathcal{MOG}$	0.1289	18.8183	–	0.0183	0.6825	0.0938
	(0.0151)	(2.9308)	×	(0.0063)	(0.0739)	(0.1361)
$\mathcal{EP}$	1	1	0.0359	1.7778	0.2537	0.0273
	×	×	(0.0008)	(0.0870)	(0.7301)	(0.8800)
$\mathcal{W}$	0.0004	1	2.5373	–	0.4344	0.0667
	(0.0002)	×	(0.1434)	×	(0.2985)	(0.3079)
$\mathcal{MW}$	0.0007	1	2.2292	0.0149	0.2761	0.0384
	(0.0007)	×	(0.4384)	(0.0191)	(0.6546)	(0.7094)
$\mathcal{GBS}$	1.3189	18.7623	0.1328	×	0.5672	0.0876
	(0.1847)	(0.5784)	(0.0513)	×	(0.1404)	(0.1642)

Table 7 Statistics AIC, BIC, CAIC and HQIC

Models	AIC	BIC	CAIC	HQIC
$\mathcal{MOFWE}$	**1753.989**	1764.553	**1754.087**	**1758.241**
$\mathcal{MOG}$	1767.305	1777.870	1767.403	1771.557
$\mathcal{EP}$	1764.136	1771.178	1764.184	1766.970
$\mathcal{W}$	1756.843	**1763.886**	1756.892	1759.678
$\mathcal{MW}$	1757.997	1768.561	1758.094	1762.248
$\mathcal{GBS}$	1761.136	1771.701	1761.234	1765.388

The functions $H(x;\gamma,c) = \log(1+\gamma\, x^c)$ and $h(x;\gamma,c) = \gamma\, c\, x^{c-1}/(1+\gamma\, x^c)$ are associated with the $\mathcal{LL}$ model.

The hypothesis that the underlying distribution is $\mathcal{W}$ (or $\mathcal{EP}$) versus the alternative hypothesis that the distribution is the $\mathcal{MOW}$ (or $\mathcal{MOEP}$) is rejected with p-value = 0.0055 (or p-value = <0.0001). Further, the hypothesis test that the underlying distribution is $\mathcal{LL}$ versus the $\mathcal{MOLL}$ distribution yields the p-value $= 1.0000$. Thus, we compare the $\mathcal{MOW}$, $\mathcal{MOEP}$, $\mathcal{LL}$ and $\mathcal{GBS}$ models to determine which model gives the best fit to the current data.

Table 8 lists the MLEs (and corresponding standard errors in parentheses) of the parameters and the values of the AD and CM statistics (their p-values in parentheses). The figures in this table, specially the p-values, suggest that the $\mathcal{MOW}$ distribution yields a better fit to these data than the other three distributions.

Table 9 lists the values of the AIC, BIC, CAIC and HQIC statistics. The figures in this table indicate that there is a competitiveness among the $\mathcal{MOW}$, $\mathcal{MOEP}$ and $\mathcal{LL}$ models. However, if we observe the Figures 9(a), 9(b) and 9(c), we note that the $\mathcal{MOW}$ and $\mathcal{MOEP}$ models present better fits to the current data.

Figure 9(d) really shows that the $\mathcal{MOW}$ and $\mathcal{MOEP}$ distributions present good fits to the current data. We can conclude that the $\mathcal{MOW}$ and $\mathcal{MOEP}$ distributions are excellent alternatives to explain this data set.

8 Conclusion

In this paper, the Marshall-Olkin extended Weibull family of distributions is proposed and some of its mathematical properties are studied. The maximum likelihood procedure is

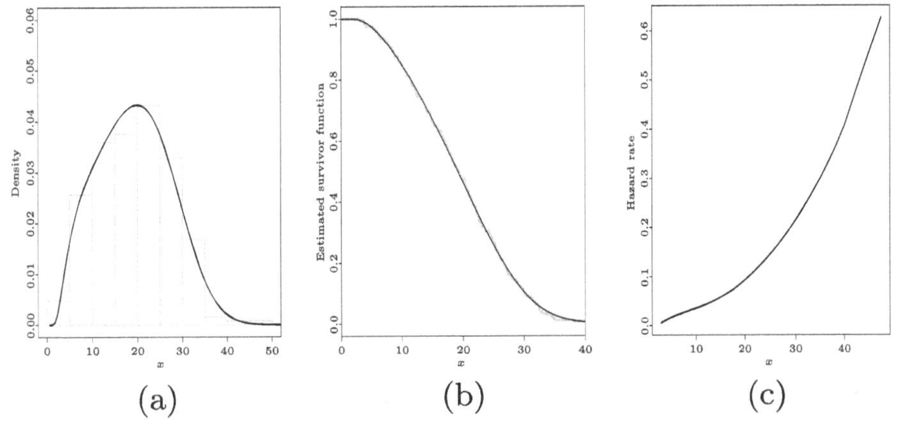

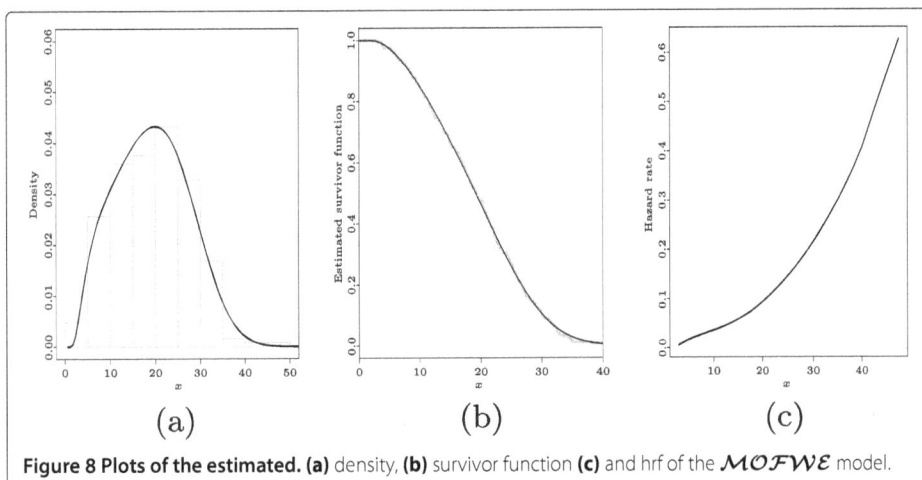

Figure 8 Plots of the estimated. (a) density, **(b)** survivor function **(c)** and hrf of the $\mathcal{MOFWE}$ model.

Table 8 MLEs and goodness-of-fit statistics

Model	Estimates (standard errors)				Goodness-of-fit (p-value)	
	$\widehat{\alpha}$ (or $\widehat{\phi}$)	$\widehat{\delta}$ (or $\widehat{\eta}$)	$\widehat{\gamma}$ (or $\widehat{\kappa}$ or $\widehat{\beta}$)	$\widehat{c}$ (or $\widehat{\lambda}$)	AD	CM
$\mathcal{MOW}$	0.0037	0.0736	1.5719	–	**0.1889**	**0.0264**
	(0.0043)	(0.0727)	(0.1616)	×	**(0.8994)**	**(0.8908)**
$\mathcal{MOEP}$	–	0.0233	0.0144	1.6012	0.2057	0.0279
	×	(0.0165)	(0.1423)	(0.0042)	(0.8686)	(0.8733)
$\mathcal{GBS}$	1.6313	7.1422	0.3356	–	1.2753	0.2116
	(0.1226)	(0.7374)	(0.0314)	×	(0.0025)	(0.0038)
$\mathcal{LL}$	–	–	0.0427	1.6900	0.2891	0.0380
	×	×	(0.0118)	(0.1249)	(0.6101)	(0.7164)

used for estimating the model parameters. Two special models in the family are described with some details. In order to assess the performance of the maximum likelihood estimates, a simulation study is performed by means of Monte Carlo experiments. Special models of the proposed family are compared (through goodness-of-fit measures) with other well-known lifetime models by means of two real data sets. The proposed model outperforms classical lifetime models to these data.

Appendix: An expansion for $f(x;\delta,\alpha,\xi)F(x;\delta,\alpha,\xi)^c$

Here, we obtain an expansion for the quantity $f(x;\delta,\alpha,\xi)F(x;\delta,\alpha,\xi)^c$. First, we consider an expansion for $F(x;\delta,\alpha,\xi)^c$. Based on (5), the power of the cdf can be expressed as

$$F(x;\delta,\alpha,\xi)^c = \underbrace{\{1-\exp[-\alpha H(x;\boldsymbol{\xi})]\}^c}_{\equiv A}\ \underbrace{\{1-\overline{\delta}\exp[-\alpha H(x;\boldsymbol{\xi})]\}^{-c}}_{\equiv B}.$$

Applying expansion (9), we have

$$A = \sum_{k=0}^{\infty}(-1)^k\binom{c}{k}\exp[-k\alpha H(x;\boldsymbol{\xi})].$$

Now, we expand the quantity B. Equation (9) under the restriction $\delta < 1$ (implying that $\overline{\delta}\exp[-\alpha H(x;\boldsymbol{\xi})] < 1$) yields

$$B = \sum_{j=0}^{\infty}\frac{(c)_j}{j!}\,\overline{\delta}^{\,j}\exp[-j\alpha H(x;\boldsymbol{\xi})].$$

Moreover, it is clear that $\delta = 1$ implies $B = 1$. Finally, for $\delta > 1$ (i.e., $\{1-\overline{\delta}\exp[-\alpha H(x;\boldsymbol{\xi})]\} > 1$), the quantity B can be rewritten as

$$B = \left\{1-[1-\{1-\overline{\delta}\exp[-\alpha H(x;\boldsymbol{\xi})]\}^{-1}]\right\}^c.$$

Table 9 Statistics AIC, BIC, CAIC and HQIC

Models	AIC	BIC	CAIC	HQIC
$\mathcal{MOW}$	**843.1171**	851.8770	**843.2975**	846.6769
$\mathcal{MOEP}$	843.1898	851.9498	843.3703	846.7497
$\mathcal{GBS}$	858.3686	867.1285	858.5490	861.9284
$\mathcal{LL}$	843.7586	**849.5986**	843.8481	**846.1318**

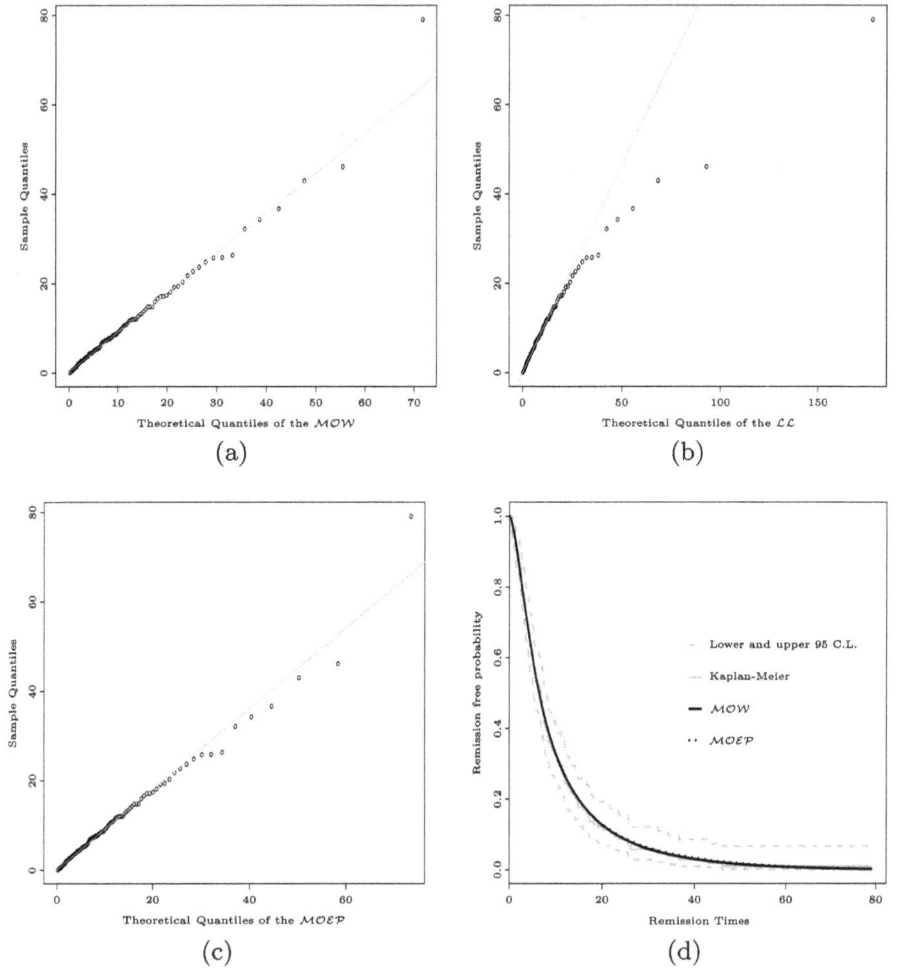

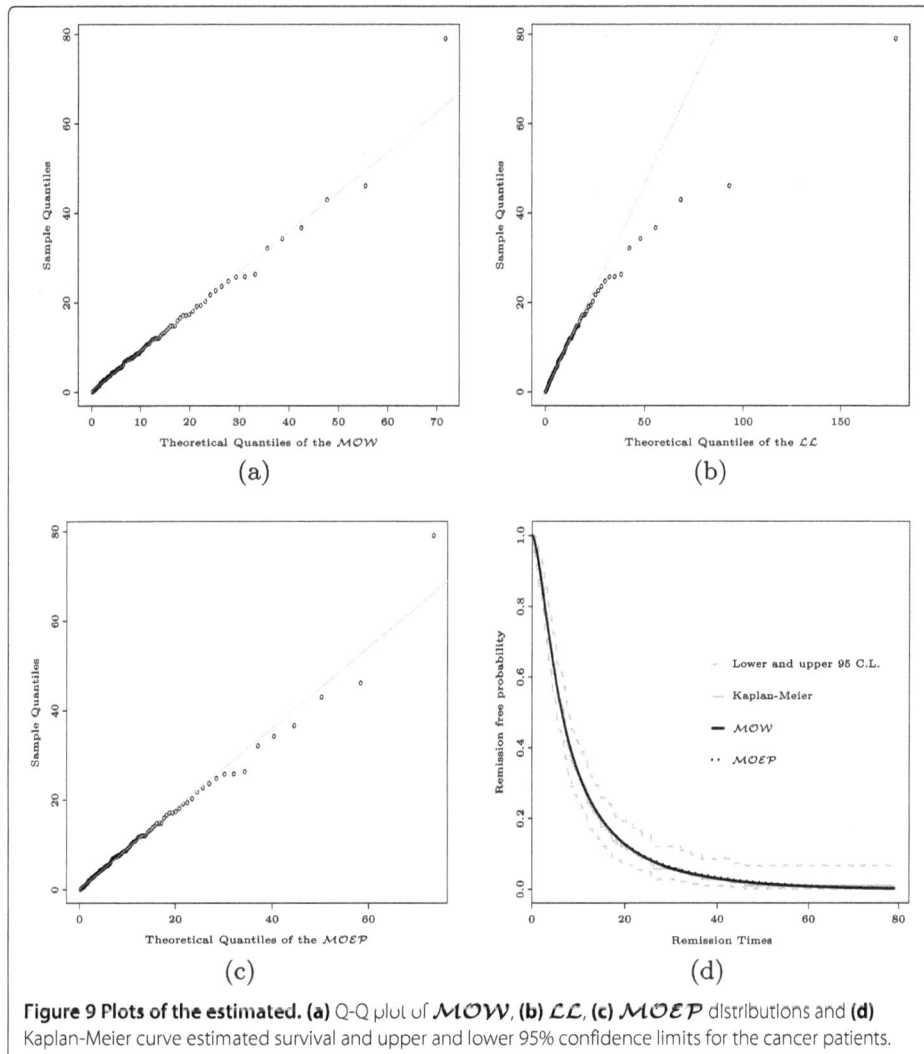

Figure 9 Plots of the estimated. (a) Q-Q plot of $\mathcal{MOW}$, **(b)** $\mathcal{LL}$, **(c)** $\mathcal{MOEP}$ distributions and **(d)** Kaplan-Meier curve estimated survival and upper and lower 95% confidence limits for the cancer patients.

Using the binomial expansion, we have

$$B=\sum_{j=0}^{\infty}(-1)^j\binom{c}{j}\left[1-\left\{1-\bar{\delta}\exp\left[-\alpha H(x;\boldsymbol{\xi})\right]\right\}^{-1}\right]^j.$$

Thus,

$$\begin{aligned}F(x;\delta,\alpha,\xi)^c &= \mathbb{I}_{(\delta<1)}\sum_{j,k=0}^{\infty}(-1)^k\,\frac{(c)_j}{j!}\binom{c}{k}\,\bar{\delta}^j\,\exp[-(j+k)\alpha H(x;\boldsymbol{\xi})]\\ &+\mathbb{I}_{(\delta=1)}\sum_{k=0}^{\infty}(-1)^k\binom{c}{k}\,\exp[-k\alpha H(x;\boldsymbol{\xi})]\\ &+\mathbb{I}_{(\delta>1)}\sum_{j,k=0}^{\infty}(-1)^{j+k}\binom{c}{k}\binom{c}{j}\,\exp[-k\alpha H(x;\boldsymbol{\xi})]\\ &\times[1-\{1-\bar{\delta}\exp[-\alpha H(x;\boldsymbol{\xi})]\}^{-1}]^j.\end{aligned}$$

Hence, based on Equation (13), the following expansion holds

$$f(x;\delta,\alpha,\xi)F(x;\delta,\alpha,\xi)^c = \left(\sum_{\nu=0}^{\infty} w_\nu g\,(x;(\nu+1)\alpha,\boldsymbol{\xi})\right) F(x;\delta,\alpha,\xi)^c = \mathbb{I}_{(\delta<1)} \sum_{j,k,\nu=0}^{\infty} (-1)^k$$

$$\times\, w_\nu \frac{(c)_j}{j!} \binom{c}{k} \bar{\delta}^{\,j} \exp[-(j+k)\alpha H(x;\boldsymbol{\xi})]\; g(x;(\nu+1)\alpha,\boldsymbol{\xi})$$

$$+\, \mathbb{I}_{(\delta=1)} \sum_{k,\nu=0}^{\infty} (-1)^k w_\nu \binom{c}{k} \exp[-k\alpha H(x;\boldsymbol{\xi})]\; g(x;(\nu+1)\alpha,\boldsymbol{\xi})$$

$$+\, \mathbb{I}_{(\delta>1)} \sum_{j,k,\nu=0}^{\infty} (-1)^{j+k} w_\nu \binom{c}{k}\binom{c}{j} \exp[-k\alpha H(x;\boldsymbol{\xi})]$$

$$\times \left[1 - \{1 - \bar{\delta} \exp[-\alpha H(x;\boldsymbol{\xi})]\}^{-1}\right]^j g(x;(\nu+1)\alpha,\boldsymbol{\xi}). \quad (24)$$

Competing interests
The authors declare that they have no competing interests.

Authors' contributions
The authors MS-N, MB, LMZ, ADCN and GMC proposed a new class of models named the Marshall-Olkin extended Weibull distributions and investigated some of its structural properties including ordinary and incomplete moments, generating and quantile functions, mean deviations, information theory measures and some types of entropies. Two special models were discussed and the estimation of the family model parameters was performed by maximum likelihood. They provided a simulation study and two applications to real data. All authors read and approved the final manuscript.

Acknowledgements
The authors gratefully acknowledge financial support from CAPES and CNPq. The authors are also grateful to three referees and an associate editor for helpful comments and suggestions.

Author details
[1]Departamento de Estatística, Universidade Federal de Campina Grande, Bodocongó, 58429-970, Campina Grande, PB, Brazil. [2]Departamento de Estatística, Universidade Federal do Piauí, Ininga, 64049-550, Teresina, PI, Brazil. [3]Departamento de Estatística, Universidade Federal do Rio Grande do Norte, Campus Universitário Lagoa Nova, 59078-970, Natal, RN, Brazil. [4]Departamento de Estatística, Universidade Federal de Pernambuco, Cidade Universitária, 50740-540, Recife, PE, Brazil.

References

Bain, LJ: Analysis for the linear failure-rate life-testing distribution. Technometrics. **16**, 551–559 (1974)
Barreto-Souza, W, Lemonte, AJ, Cordeiro, GM: General results for the Marshall and Olkin's family of distributions. An. Acad. Bras. Cienc. **85**, 3–21 (2013)
Bebbington, M, Lai, CD, Zitikis, R: A flexible Weibull extension. Reliability Eng. Syst. Saf. **92**, 719–726 (2007)
Carrasco, JMF, Ortega, EMM, Cordeiro, GM: A generalized modified Weibull distribution for lifetime modeling. Comput. Stat. Data Anal. **53**, 450–462 (2008)
Chen, Z: A new two-parameter lifetime distribution with bathtub shape or increasing failure rate function. Stat. Probability Lett. **49**, 155–161 (2000)
Chen, G, Balakrishnan, N: A general purpose approximate goodness-of-fit test. J. Qual. Technol. **27**, 154–161 (1995)
Cordeiro, GM, Lemonte, AJ: On the Marshall-Olkin extended Weibull distribution. Stat. Paper. **54**, 333–353 (2013)
Cover, TM, Thomas, JA: Elements of Information Theory. John Wiley & Sons, New York (1991)
Cox, DR, Hinkley, DV: Theoretical Statistics. Chapman and Hall, London (1974)
Doornik, J: Ox 5: object-oriented matrix programming language. 5th ed. Timberlake Consultants, London (2007)
EL-Bassiouny, AH, Abdo, NF: Reliability properties of extended makeham distributions. Comput. Methods Sci. Technol. **15**, 143–149 (2009)
Fisk, PR: The graduation of income distributions. Econometrica. **29**, 171–185 (1961)
Fréchet, M: Sur la loi de probabilité. Ann. Soc. Polon. Math. **6**, 93–93 (1927)
Garvan, F: The Maple Book. Chapman and Hall/CRC, London (2002)
Gompertz, B: On the nature of the function expressive of the law of human mortality and on the new model of determining the value of life contingencies. Philos. Trans. R. Soc. Lond. **115**, 513–585 (1825)
Guess, F, Proschan, F: Mean residual life: Theory and applications. In: Krishnaiah, PR, Rao, CR (eds.) *Handbook of Statistics, vol. 7*, pp. 215–224. Elsevier, (1988). http://dx.doi.org/10.1016/S0169-7161(88)07014-2
Ghitany, ME: Marshall-Olkin extended Pareto distribution and its application. Int. J. Appl. Math. **18**, 17–31 (2005)

Ghitany, ME, Kotz, S: Reliability properties of extended linear failure-rate distributions. Probability Eng. Informational Sci. **21**, 441–450 (2007)

Ghitany, ME, AL-Hussaini, EK, AL-Jarallah: Marshall-Olkin extended Weibull distribution and its application to Censored data. J. Appl. Stat. **32**, 1025–1034 (2005)

Ghitany, ME, AL-Awadhi, FA, Alkhalfan, LA: Marshall-Olkin extended Lomax distribution and its applications to censored data. Comm. Stat. Theor. Meth. **36**, 1855–1866 (2007)

Gurvich, M, DiBenedetto, A, Ranade, S: A new statistical distribution for characterizing the random strength of brittle materials. J. Mater. Sci. **32**, 2559–2564 (1997)

Hartley, RVLL: Transmission of information. Bell Syst. Techn. J. **7**, 535–563 (1928)

Jayakumar, K, Mathew, T: On a generalization to Marshall-Olkin scheme and its application to Burr type XII distribution. Stat. Paper. **49**, 421–439 (2008)

Johnson, NL, Kotz, S, Balakrishnan, N: Continuous Univariate Distributions, Vol. 1. Wiley, New York (1994)

Kies, JA: The Strength of Glass, NRL Report 5093. Naval Research Lab., Washington, DC (1958)

Krishna, E, Jose, KK, Ristić, M: Applications of Marshal-Olkin Fréchet distribution. Comm. Stat. Simulat. Comput. **42**, 76–89 (2013)

Kullback, S, Leibler, RA: On information and sufficiency. Ann. Math. Stat. **22**, 79–86 (1951)

Lai, CD, Xie, M, Murthy, DNP: A modified Weibull distribution. Trans. Reliab. **52**, 33–37 (2003)

Lawless, JF: Statistical Models and Methods for Lifetime Data. 2nd ed. Wiley, New York (2003)

Lee, ET, Wang, JW: Statistical Methods for Survival Data Analysis. 3rd ed. Wiley, New York (2003)

Lomax, KS: Business failures; another example of the analysis of failure data. J. Am. Stat. Assoc. **49**, 847–852 (1954)

Marshall, A, Olkin, I: A new method for adding a parameter to a family of distributions with application to the exponential and Weibull families. Biometrika. **84**, 641–652 (1997)

McCool, JI: Using the Weibull Distribution: Reliability, Modeling and Inference. John Wiley & Sons, New Jersey (2012)

Nadarajah, S, Kotz, S: On some recent modifications of Weibull distribution. IEEE Trans. Reliab. **54**, 561–562 (2005)

Nascimento, ADC, Cintra, RJ, Frery, AC: Hypothesis testing in speckled data with stochastic distances. IEEE Trans. Geosci. Remote Sensing. **48**, 373–385 (2010)

Nascimento, ADC, Horta, MM, Frery, AC, Cintra, RJ: Comparing edge detection methods based on stochastic entropies and distances for PolSAR imagery. IEEE J. Selected Topics Appl. Earth Observations Remote Sensing. **7**, 648–663 (2014)

Nikulin, M, Haghighi, F: A chi-squared test for the generalized power Weibull family for the head-and-neck cancer censored data. J. Math. Sci. **133**, 1333–1341 (2006)

Owen, WJ: A new three-parameter extension to the Birnbaum-Saunders distribution. IEEE Trans. Reliab. **55**, 475–479 (2006)

Pham, H: A vtub-shaped hazard rate function with applications to system safety. Int. J. Reliab. Appl. **3**, 1–16 (2002)

Phani, KK: A new modified Weibull distribution function. Commun. Am. Ceramic Soc. **70**, 182–184 (1987)

R Development Core Team: R: A Language and Environment for Statistical Computing. R Foundation for Statistical Computing, Vienna (2009)

Rayleigh, JWS: On the resultant of a large number of vibrations of the same pitch and of arbitrary phase. Phil. Mag. **10**, 73–78 (1880)

Ristić, MM, Jose, KK, Ancy, J: A Marshall-Olkin gamma distribution and minification process. STARS: Stress Anxiety Res. Soc. **11**, 107–117 (2007)

Rodriguez, N: A guide to the Burr type XII distributions. Biometrika. **64**, 129–134 (1977)

Sankaran, PG, Jayakumar, K: On proportional odds model. Stat. Paper. **49**, 779–789 (2008)

Seghouane, A-K, Amari, S-I: The AIC criterion and symmetrizing the Kullback–Leibler divergence. IEEE Trans. Neural Netw. **18**, 97–106 (2007)

Shannon, CE: A mathematical theory of communication. Bell Syst. Techn. J. **27**, 379–423 (1948)

Sigmon, K, Davis, TA: MATLAB Primer. 6th ed. Chapman and Hall/CRC, London (2002)

Smith, RM, Bain, LJ: An exponential power life testing distribution. Comm. Stat. Theor. Meth. **4**, 469–481 (1975)

Wolfram, S: The Mathematica Book. 5th ed. Wolfram Media, Cambridge (2003)

Xie, M, Lai, D: Reliability analysis using additive Weibull model with bathtub-shaped failure rate function. Reliab. Eng. Syst. Saf. **52**, 87–93 (1995)

Xie, M, Tang, Y, Goh, TN: A modified Weibull extension with bathtub-shaped failure rate function. Reliab. Eng. Syst. Saf. **76**, 279–285 (2002)

Zhang, T, Xie, M: Failure data analysis with extended Weibull distribution. Comm. Stat. Simulat. Comput. **36**, 579–592 (2007)

11

Chi-p distribution: characterization of the goodness of the fitting using L^p norms

George Livadiotis

Correspondence: glivadiotis@swri.edu
Southwest Research Institute, San Antonio, TX, USA

Abstract

This paper derives (1) the Chi-p distribution, i.e., the analog of Chi-square distribution but for datasets that follow the General Gaussian distribution of shape p, and (2) develops the statistical test for characterizing the goodness of the fitting with L^p norms. It is shown that the statistical test has double role when the fitting method is induced by the L^p norms: For given the shape parameter p, the test is rated based on the estimated p-value. Then, a convenient characterization of the fitting rate is developed. In addition, for an unknown shape parameter and if the fitting is expected to be good, then those L^p norms that correspond to unlikely p-values are rejected with a preference to the norms that maximized the p-value. The statistical test methodology is followed by an illuminating application.

1. Introduction

The fitting of a given dataset $\{f_i \pm \sigma_{f_i}\}_{i=1}^{N}$ to the values $\{V_i\}_{i=1}^{N}$ of a statistical model $V(X;\alpha)$ in the domain $X \in D_x \subseteq \mathfrak{R}$ (McCullagh 2002; Adèr 2008), involves finding the optimal parameter value $\alpha = \alpha^*$ in $\alpha \in D_\alpha \subseteq \mathfrak{R}$ that minimizes the total square deviations (TSD) between model and data,

$$TSD(\alpha)^2 = \sum_{i=1}^{N} \sigma_{f_i}^{-2} [f_i - V(x_i;\alpha)]^2, \tag{1}$$

where the inverse of the variance of the data measurements $\left\{ w_i = \sigma_{f_i}^{-2} \right\}_{i=1}^{N}$ is weighting the summation. The deviations may be also defined using the total absolute deviations (TAD),

$$TAD(\alpha) = \sum_{i=1}^{N} \sigma_{f_i}^{-2} |f_i - V(x_i;\alpha)|. \tag{2}$$

A class of generalized fitting methods has been considered by Livadiotis (2007), using the metric induced by the p-norms L^p, $p \geq 1$, that denotes a complete normalized vector space with finite Lebesgue integral. The total deviations (TD) are now defined by

$$TD(\alpha)^p = \sum_{i=1}^{N} \sigma_{f_i}^{-p} |f_i - V(x_i;\alpha)|^p. \tag{3}$$

The least square method based on the Euclidean norm, $p = 2$, and the least absolute deviations method based on the "Taxicab" norm, $p = 1$, are some cases of the general fitting methods based on the L^p-norms (see Burden and Faires 1993; for more applications of the fitting methods based on L^p norms, see: Sengupta 1984; Livadiotis and Moussas 2007; Livadiotis 2008; 2012; for fitting methods based on other effect sizes e.g., correlation, see: Livadiotis and McComas 2013a).

The goodness of the least square fitting is typically measured using the estimated Chi-square value, that is the least squared value, $\chi^2_{\text{est}} = TSD(\alpha^*)^2$. Then, this χ^2_{est} is compared with the Chi-square distribution, to examine whether such a value is frequent or not (see next sections). However, this test can apply only to datasets $\{f_i \pm \sigma_{f_i}\}_{i=1}^{N}$ that follow the normal distribution $f_i \sim N\left(\mu_{f_i}, \sigma_{f_i}\right)$. There is no similar test for cases where the dataset follows the General Gaussian distribution of shape p, $f_i \sim GG\left(\mu_{f_i}, \sigma_{f_i}, p\right)$ (see Section 2 and Appendix A). Livadiotis (2012) showed the connection between the fitting with L^p norms, as in Eq. (3), and datasets that follow the General Gaussian distributions, $f_i \sim GG\left(\mu_{f_i}, \sigma_{f_i}, p\right)$.

The purpose of this paper is to (1) construct the formulation of the Chi-p distribution, the analog of Chi-square distribution but for datasets that follow the General Gaussian distribution of shape p, and (2) develop the statistical test for characterizing the goodness of the fitting with L^p norms, which corresponds to datasets that follow the General Gaussian distribution of shape p. Therefore, in Section 2, we revisit the Chi-square derivation, and following similar steps, we construct the Chi-p distribution. In Section 3, we develop the statistical test for characterizing the goodness of the fitting with L^p norms, using the Chi-p distribution and the p-value. In Section 4, we provide an application of the statistical test. Finally, in Section 5, we summarize the conclusions. Appendix A briefly describes the General Gaussian distribution, while Appendix B shows the mathematical derivation of the surface of the sphere of higher dimensions in L^p space.

2. Chi-*p* distribution

We first revisit the derivation of Chi-square distribution. This distribution is necessary to test the goodness of fitting of measurements that follow the Gaussian distribution. This test applies to datasets $\{x_i \pm \sigma_{x_i}\}_{i=1}^{N}$ that follow the normal distribution $x_i \sim N\left(\mu_{x_i}, \sigma_{xi}\right)$. The Chi-square is given by

$$\chi^2 = \sum_{i=1}^{N} \left(\frac{x_i - \mu_{x_i}}{\sigma_{x_i}}\right)^2, \tag{4}$$

that is the sum of squares of N independent random variables. The distribution of this sum is given by

$$P(X;N)dX = \frac{2^{-\frac{N}{2}}}{\Gamma\left(\frac{N}{2}\right)} e^{-\frac{1}{2}X} X^{\frac{N}{2}-1} dX, \quad \text{with} \quad X \equiv \chi^2. \tag{5}$$

The estimated value of the Chi-square for a fitting is given by the minimum at $\alpha = \alpha^*$ of the function $\chi^2(\alpha) = TSD(\alpha)^2$, as shown in Eq. (1) (least squares). Considering that the Chi-square minimum, $\chi^2(\alpha^*)$, is equivalently referred to all the $M = N$-1 degrees of freedom (for N number of data), then each of them contributes to this minimum by a factor of $\frac{1}{M}\chi^2(\alpha^*)$. This is the estimated value of the reduced Chi-square. For multi-parametrical fitting (Livadiotis 2007) of n free parameters, the degrees of freedom are $M = N$-n. In general, the Chi-square distribution in Eq. (5) is referred to M degrees of freedom.

For testing the goodness of fitting of measurements $\{x_i \pm \sigma_{xi}\}_{i=1}^{N}$ that follow the General Gaussian distribution of shape p, $x_i \sim GG(\mu_{xi}, \sigma_{xi}, p)$, we need to construct the Chi-p distribution connected with L^p fitting methods, where the minimization of $\chi^p(\alpha)$ is given by Eq. (3). The General Gaussian distribution of shape p, $f_i \sim GG\left(\mu_{f_i}, \sigma_{f_i}, p\right)$ (Appendix A). This distribution is parameterized by the mean μ, the variance σ, and the shape parameter p,

$$P(x;\mu,\sigma,p)dx = C_p \cdot e^{-\eta_p \cdot \left|\frac{x-\mu}{\sigma}\right|^p} d\left(\frac{x-\mu}{\sigma}\right), \tag{6}$$

where the involved coefficients are

$$C_p = \sqrt{\frac{p\sin\left(\frac{\pi}{p}\right)}{4\pi(p-1)}}, \eta_p = \left[\frac{\sin\left(\frac{\pi}{p}\right)\Gamma\left(\frac{1}{p}\right)^2}{\pi p(p-1)}\right]^{\frac{p}{2}}. \tag{7}$$

Figure 1 depicts the distribution $P\left(z = \frac{x-\mu}{\sigma}; p\right) \equiv P(x;\mu,\sigma,p)$ for various shape parameters p. Note that the normalized coefficient C_p is derived by setting $\int_{-\infty}^{\infty} P(x;\mu,\sigma,p)\,dx = 1$, while the exponential coefficient η_p is derived so that the L^p-normed variance to equal σ^2. The theory of L^p-normed mean and variance was developed by Livadiotis (2012), which for the case of the General Gaussian distribution (6) leads to the following Propositions:

- *Proposition* 1: The L^p-normed mean of the distribution (6) is $<x>_p = \mu$, $\forall\, p \geq 1$.
- *Proposition* 2: The L^p-normed variance of the distribution (6) is $\sigma_p^2 = \sigma^2$, $\forall\, p \geq 1$.

The proofs of the two Propositions are shown in Appendix A.

We continue with the development of the Chi-p distribution. We start with the following Lemma:

- *Lemma* 1: The surface of the N-dimensional sphere of unit radius in L^p space is given by

$$B_{p,N} = p\left[\left(\frac{2}{p}\right)\Gamma\left(\frac{1}{p}\right)\right]^N / \Gamma\left(\frac{N}{p}\right). \tag{8}$$

The proof is shown in Appendix B.

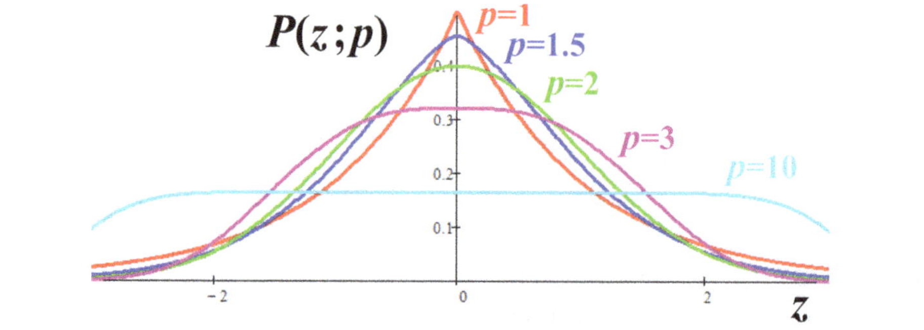

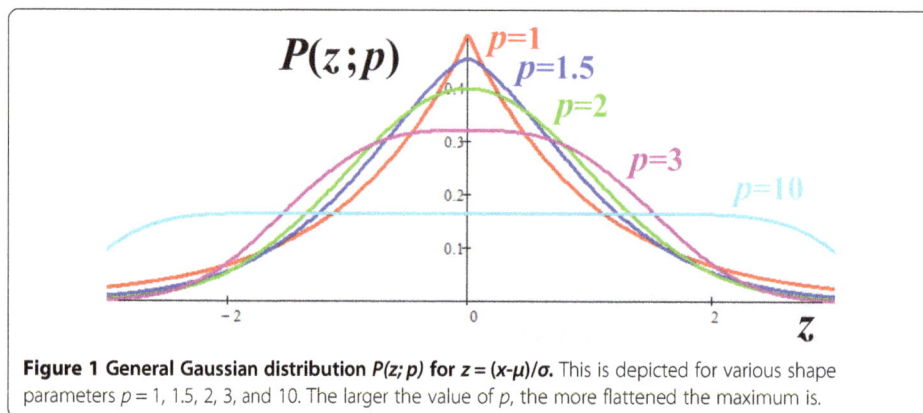

Figure 1 General Gaussian distribution *P(z; p)* for *z* = (*x*-*μ*)/*σ*. This is depicted for various shape parameters $p = 1, 1.5, 2, 3$, and 10. The larger the value of p, the more flattened the maximum is.

– *Theorem* 1:

The Chi-p is given by the sum of absolute values to the exponent p of N independent random variables,

$$\chi^p = \sum_{i=1}^{N} \left| \frac{x_i - \mu_{x_i}}{\sigma_{x_i}} \right|^p . \tag{9}$$

For M degrees of freedom ($M = N\text{-}n$, N number of data, n number of independent variables), the Chi-p distribution is given by

$$P(X; M; p) = \frac{\eta_p^{\frac{M}{p}}}{\Gamma\left(\frac{M}{p}\right)} e^{-\eta_p X} X^{\frac{M}{p}-1}, \tag{10}$$

where the estimated Chi-p value X is given by the minimum at $\alpha = \alpha^*$ of the function $\chi^p(\alpha) = TD(\alpha)^p$, as shown in Eq. (3) (least L^p deviations). Figure 2 plots the Chi-p distribution for various values of the shape parameter p (that correspond to various L^p norms).

– *Proof of Theorem* 1. The distribution of Chi-p can be derived as follows. The normalization of the joint distribution function of all the data is

$$1 = \int_{-\infty}^{+\infty} \prod_{i=1}^{N} \frac{C_p}{\sigma_{x_i}} e^{-\eta_p \left| \frac{x_i - \mu_{x_i}}{\sigma_{x_i}} \right|^p} dx_1 \ldots dx_N, \tag{11}$$

where the coefficients (Livadiotis 2012) are given by Eq. (7).

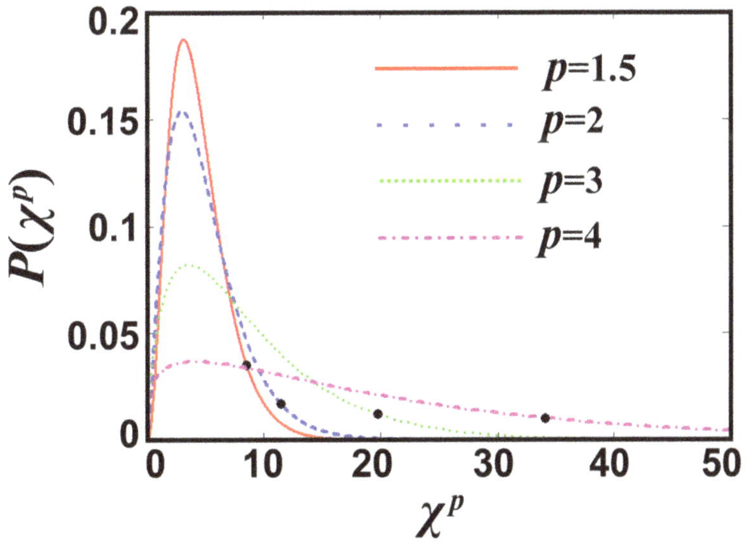

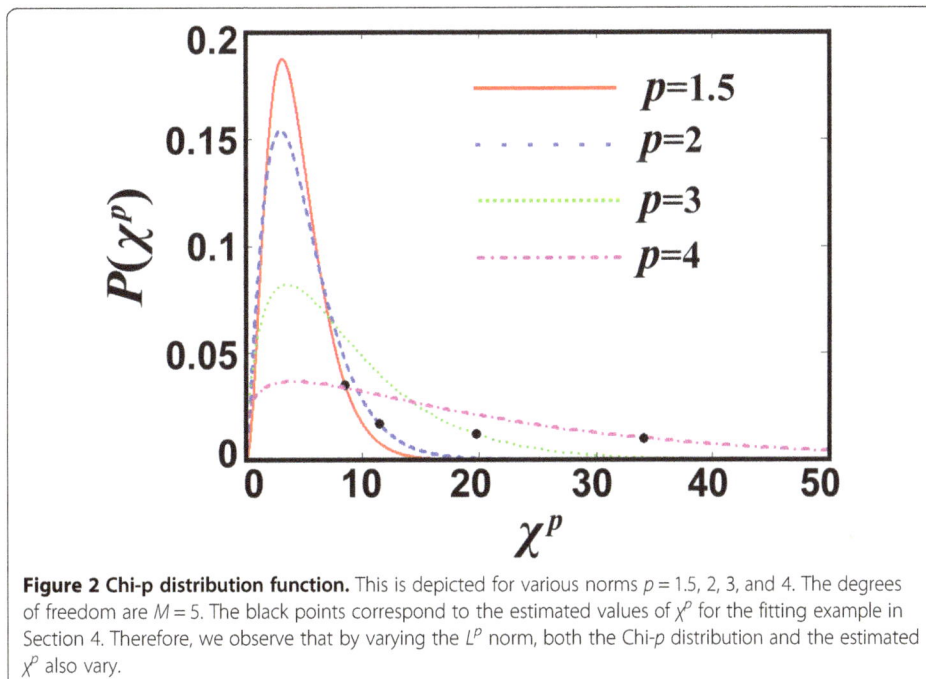

Figure 2 Chi-p distribution function. This is depicted for various norms $p = 1.5$, 2, 3, and 4. The degrees of freedom are $M = 5$. The black points correspond to the estimated values of χ^p for the fitting example in Section 4. Therefore, we observe that by varying the L^p norm, both the Chi-p distribution and the estimated χ^p also vary.

By setting $z_i \equiv \frac{x_i - \mu_x}{\sigma_{x_i}}$, we derive

$$1 = \int_{-\infty}^{+\infty} \prod_{i=1}^{N} C_p e^{-\eta_p |z_i|^p} dz_1 ... dz_N = \int_{-\infty}^{+\infty} C_p^N e^{-\eta_p \sum_{i=1}^{N} |z_i|^p} dz_1 ... dz_N, \tag{12}$$

that is

$$1 = \int_{\vec{z} \in B_{p,N}} d^{N-1} \Omega_N \cdot \int_0^{+\infty} C_p^N e^{-\eta_p Z^p} Z^{N-1} dZ, \tag{13}$$

where we denote $Z^p \equiv \sum_{i=1}^{N} |z_i|^p$, and $B_{p,N} \equiv \int_{\vec{z} \in B_{p,N}} d^{N-1} \Omega_N$ is the surface of the N-dimensional sphere of unit radius in L^p space (*Lemma* 1), so that

$$1 = \int_0^{+\infty} C_p^N B_{p,N} e^{-\eta_p Z^p} Z^{N-1} dZ = \int_0^{+\infty} C_p^N \frac{1}{p} B_{p,N} e^{-\eta_p X} X^{\frac{N}{p}-1} dX \equiv \int_0^{+\infty} P(X; N; p) dX,$$

where we have used the identity $C_p^N \frac{1}{p} B_{p,N} = \eta_p^{\frac{N}{p}} / \Gamma\left(\frac{N}{p}\right)$. Hence, we find

$$P(X; N; p) dX = \frac{\eta_p^{\frac{N}{p}}}{\Gamma\left(\frac{N}{p}\right)} e^{-\eta_p X} X^{\frac{N}{p}-1} dX, \quad \text{with} \quad X \equiv \chi^p. \tag{14}$$

In general, for M degrees of freedom, the Chi-p distribution is given by Eq. (10).

3. Statistical test of a fitting

In order to estimate the goodness of the fitting, we minimize the Chi-p, χ^p,

$$\chi^p = \sum_{i=1}^{N} \sigma_{f_i}^{-p} [f_i - V(x_i; \alpha)]^p, \tag{15}$$

similar to the minimization of the Chi-square, χ^2, for the case of the Euclidean norm,

$$\chi^2 = \sum_{i=1}^{N} \sigma_{f_i}^{-2} [f_i - V(x_i; \alpha)]^2. \tag{16}$$

We begin with the established method of Chi-square, and then we will proceed to the generalized method of Chi-p.

The goodness of a fitting can be estimated by the reduced Chi-square value, $\chi_{\text{red}}^2 = \frac{1}{M} \chi_{\text{test}}^2$, where $M = N$-1 indicates again the degrees of freedom. The meaning of χ_{red}^2 is the portion of χ^2 that corresponds to each of the degrees of freedom, and this has to be ~1 for a good fitting. We can easily understand this, for example, when the given data have equal error σ_f, with $\{f_i \pm \sigma_f\}_{i=1}^{N}$, i.e., $\sigma_{f_i} = \sigma_f$ for all i = 1,...., N. Then, the optimized model value, $V(x_i; \alpha^*)$, gives the expected value of the data point f_i, so that the variance can be approached by $\sigma_f^2 = \frac{1}{M} \sum_{i=1}^{N} [f_i - V(x_i; \alpha^*]^2$ (sample variance). Hence, the derived Chi-square becomes $\chi_{\text{est}}^2 = \sigma_f^{-2} \sum_{i=1}^{N} [f_i - V(x_i; \alpha^*)]^2 = M$, and its reduced value $\chi_{\text{red}}^2 = \frac{1}{M} \chi_{\text{est}}^2 = 1$. Therefore, a fitting can be characterized as "good" when $\chi_{\text{red}}^2 \sim 1$, otherwise there is an overestimation, $\chi_{\text{red}}^2 < 1$, or underestimation, $\chi_{\text{red}}^2 > 1$, of the errors. When the deviations of the data $\{f_i\}_{i=1}^{N}$ from the model values $\{V(x_i; \alpha)\}_{i=1}^{N}$ are small, the fitting is expected to be good. However, this characterization is meaningless if the errors of the data $\{\sigma_{f\,i}\}_{i=1}^{N}$ are either (i) quite larger than their deviations from the model values, i.e., if $\sigma_{fi} >> |f_i - V(x_i; \alpha)|$, or (ii) quite smaller, i.e., if $\sigma_{fi} << |f_i - V(x_i; \alpha)|$ (e.g., see Figure 3). Then, a perfect matching between data and model is useless when the errors of the data are comparably large or small.

Furthermore, a better estimation of the goodness is derived from comparing the calculated χ^2 value and the Chi-square distribution, that is the distribution of all the possible χ^2 values for data with normally distributed errors (parameterized by the degrees of freedom M),

$$P(\chi^2; M)\, d\chi^2 = \frac{2^{-\frac{M}{2}}}{\Gamma(\frac{M}{2})} e^{-\frac{1}{2}\chi^2} (\chi^2)^{\frac{M}{2}-1}\, d\chi^2, \tag{17}$$

(e.g., see Melissinos 1966). The likelihood of having an χ^2 value equal to or smaller than the estimated value χ^2_{est}, is given by the cumulative distribution

$$P(0 \le \chi^2 \le \chi^2_{\text{est}}) = \int_0^{\chi^2_{\text{est}}} P(\chi^2; M)\, d\chi^2 = 1 - \frac{\Gamma(\frac{1}{2}M; \frac{1}{2}\chi^2_{\text{est}})}{\Gamma(\frac{1}{2}M)}, \tag{18}$$

where $\Gamma(x; b) = \int_x^{\infty} e^{-X} X^{b-1} dX$ is the incomplete Gamma function. In addition, the likelihood of having an χ^2 value equal to or larger than the estimated value χ^2_{est}, is given by the complementary cumulative distribution

$$P(\chi^2_{\text{est}} \le \chi^2 < \infty) = \int_{\chi^2_{\text{est}}}^{\infty} P(\chi^2; M)\, d\chi^2 = \frac{\Gamma(\frac{1}{2}M; \frac{1}{2}\chi^2_{\text{est}})}{\Gamma(\frac{1}{2}M)}. \tag{19}$$

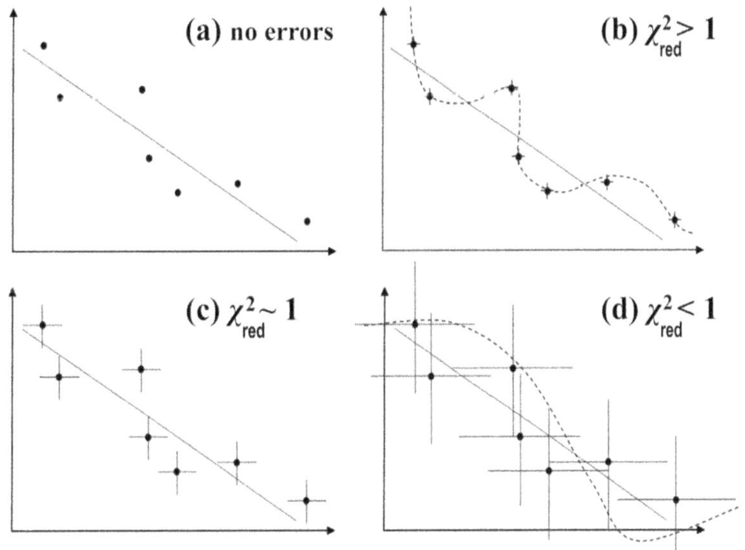

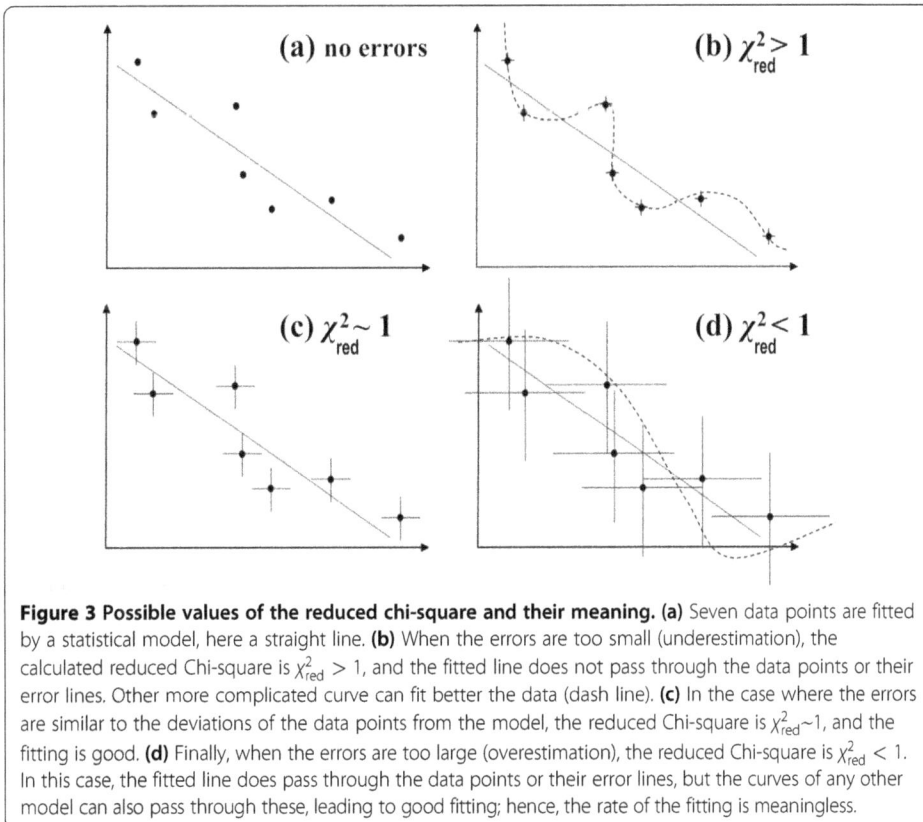

Figure 3 Possible values of the reduced chi-square and their meaning. (a) Seven data points are fitted by a statistical model, here a straight line. **(b)** When the errors are too small (underestimation), the calculated reduced Chi-square is $\chi^2_{red} > 1$, and the fitted line does not pass through the data points or their error lines. Other more complicated curve can fit better the data (dash line). **(c)** In the case where the errors are similar to the deviations of the data points from the model, the reduced Chi-square is $\chi^2_{red} \sim 1$, and the fitting is good. **(d)** Finally, when the errors are too large (overestimation), the reduced Chi-square is $\chi^2_{red} < 1$. In this case, the fitted line does pass through the data points or their error lines, but the curves of any other model can also pass through these, leading to good fitting; hence, the rate of the fitting is meaningless.

The probability of having a result χ^2 larger than the estimated value χ^2_{est}, defines the p-value that equals $P(\chi^2_{\text{est}} \leq \chi^2 < \infty)$. The larger the p-value, the better the fitting is (e.g., Melissinos 1966). However, the p-value test fails when p > 0.5. Indeed, p-values larger than 0.5 correspond to $\chi^2_{\text{est}} < M$ or $\chi^2_{\text{red}} < 1$. Even larger p-values, up to p = 1, correspond to even smaller Chi-squares, down to $\chi^2_{\text{red}} \sim 0$. Thus, an increasing p-value above the threshold of 0.5 cannot lead to a better fitting but to a worse, similar to the indication $\chi^2_{\text{red}} < 1$. For this reason, we use the "p-value of the extremes". According to this, the probability of taking a result χ^2, more extreme than the observed value is given by the p-value that equals the minimum between $P(0 \leq \chi^2 \leq \chi^2_{\text{est}})$ and $P(\chi^2_{\text{est}} \leq \chi^2 < \infty)$, i.e.,

$$\text{p-value} = \min\left[\frac{\Gamma\left(\frac{1}{2}M; \frac{1}{2}\chi^2_{\text{est}}\right)}{\Gamma\left(\frac{1}{2}M\right)}, \quad 1 - \frac{\Gamma\left(\frac{1}{2}M; \frac{1}{2}\chi^2_{\text{est}}\right)}{\Gamma\left(\frac{1}{2}M\right)}\right], \tag{20}$$

(see some applications in Livadiotis and McComas 2013b; Frisch et al. 2013; Funsten et al. 2013). Note that the maximum p-value is 0.5, and this corresponds to the estimated Chi-square $\chi^2_{\text{est},1/2} \cong M - \frac{2}{3}$. This is larger than the Chi-square that maximizes the distribution, $\chi^2_{\text{est,max}} = M - 2$. Hence, $\chi^2_{\text{est,max}} < \chi^2_{\text{est},1/2}$, i.e., the Chi-square that corresponds to p-value = 0.5, is located always at the right of the maximum.

The statistical test of the fitting for the evaluation of its goodness comes from the null hypothesis that the given data are described by the fitted statistical model. If the derived p-value is smaller than the significance level of ~0.05, then the hypothesis is typically rejected, and the hypothesis that the data are described by the examined statistical model is characterized as unlikely.

A convenient rate for a statistical test is to give more detailed characterization than "likely" when p-value > 0.05, or "unlikely" when p-value < 0.05. For this reason, it is necessary to ascribe an 1–1 relation between the domain of p-values $\{p \in [0, 0.5]\}$ and the range of a rating values $\{T \in [-1, 1]\}$, with the correspondence: 1) Impossible $p = 0 \leftrightarrow T = -1$; 2) indefinite $p = 0.05 \leftrightarrow T = 0$; 3) certain $p = 0.5 \leftrightarrow T = 1$. Choosing a power-law function, $(T+1)/2 = (p/p_0)^{\gamma}$, we find $p_0 = 0.5$ and $\gamma =$ log 2, i.e.,

$$(T+1)/2 = (2p)^{\log 2}. \tag{21}$$

We can easily now characterize the testing rates by a linear separation of the values of T, as shown in Table 1.

Table 1 Testing rates and characterizations

p-value	Rate T	Characterization
p ~ 0	T ~ −1	Impossible
0 < p < 0.005	−1 < T < −0.5	Highly unlikely
0.005 ≤ p < 0.05	−0.5 ≤ T < 0	Unlikely
0.05 ≤ p < 0.19	0 ≤ T < 0.5	Likely
0.19 ≤ p < 0.5	0.5 ≤ T < 1	Highly likely
p ~ 0.5	T ~ 1	Certain

In the case of data that follow the General Gaussian distribution of shape p, the derived p-value is dependent on the shape p. Indeed, we have

$$P(\chi^p;M;p)\,d\chi^p = \frac{\eta_p^{\frac{M}{p}}}{\Gamma\left(\frac{M}{p}\right)}\cdot(\chi^p)^{\frac{M}{p}-1}\cdot e^{-\eta_p\chi^p}\,d\chi^p, \tag{22}$$

and

$$P(0\leq\chi^p\leq\chi^p_{\rm est}) = \int_0^{\chi^p_{\rm est}} P(\chi^p;M;p)\,d\chi^p = 1-\frac{\Gamma\left(\frac{1}{p}M;\eta_p\chi^p_{\rm est}\right)}{\Gamma\left(\frac{1}{p}M\right)}, \tag{23}$$

$$P(\chi^p_{\rm est}\leq\chi^p<\infty) = \int_{\chi^p_{\rm est}}^{\infty} P(\chi^p;M;p)\,d\chi^p = \frac{\Gamma\left(\frac{1}{p}M;\eta_p\chi^p_{\rm est}\right)}{\Gamma\left(\frac{1}{p}M\right)}, \tag{24}$$

and the p-value that equals the minimum between $P(0\leq\chi^p\leq\chi^p_{\rm est})$ and $P(\chi^p_{\rm est}\leq\chi^p<\infty)$, i.e.,

$$\text{p-value} = \min\left[\frac{\Gamma\left(\frac{1}{p}M;\eta_p\chi^p_{\rm est}\right)}{\Gamma\left(\frac{1}{p}M\right)}\;,\;1-\frac{\Gamma\left(\frac{1}{p}M;\eta_p\chi^p_{\rm est}\right)}{\Gamma\left(\frac{1}{p}M\right)}\right]. \tag{25}$$

Note that the maximum p-value = 0.5 corresponds to the estimated Chi-square $\chi^p_{\rm est,1/2}\cong\frac{1}{p\eta_p}M-\frac{1}{3\eta_p}$. This is larger than the Chi-square that maximizes the distribution, $\chi^p_{\rm est,1/2}\cong\frac{1}{p\eta_p}M-\frac{1}{\eta_p}$. Hence, again we find $\chi^p_{\rm est,\,max}<\chi^p_{\rm est,1/2}$.

The statistical test has double role in the case of L^p norms. If the shape parameter p is known, then the test can be rated by deriving the p-value and according to Table 1. If the shape parameter is unknown and the fitting is expected to be good, then all the shape values p that correspond to unlikely p-values can be rejected. In fact, the largest p-value corresponds to the most-likely shape parameter p of the examined data. These are shown in the following applications.

4. Applications

Table 2 contains a dataset of observations of the ratio of the umbral area to the whole sunspot area, $\{f_i\}_{i=1}^N$, $N = 6$ (Edwards 1957). Assuming that each of them follows a General Gaussian distribution about their mean, $f_i \sim GG(\mu_i, \sigma_i, p)$, what is the likelihood of these measurements to represent a constant physical quantity? Let this constant be indicated by μ_p, which can be derived from the fitting of $\{f_i \pm \sigma_{f_i}\}_{i=1}^N$, and thus, it is typically depended on the p-norm. However, different values of the p-norm lead to

Table 2 Testing rates and characterizations

Heliographic latitude (degrees)	Ratio of umbral area to whole sunspot area f_i (%)	Standard deviation σ_{f_i} (%)
0-5	0.1708	0.0053
5-10	0.1677	0.0019
10-15	0.1624	0.0016
15-20	0.1610	0.0019
20-25	0.1594	0.0026
>25	0.1627	0.0040

different estimated values of the Chi-p, χ^p_{est}. Thus, the p-value of the null hypothesis (H_o) depends also on the p-norm.

We apply a statistical test to examine whether the data of the sunspot area ratios are dependent with heliolatitude on not. Therefore, the null hypothesis is that the dataset is described by the statistical model of constant value, i.e., $\{V(x_i;\alpha)=\alpha\}_{i=1}^{N}$. We construct and minimize the Chi-p, given by

$$\chi^p(\alpha)=\sum_{i=1}^{N}\left|\frac{f_i-\alpha}{\sigma_{f_i}}\right|^p, \tag{26}$$

so that the L^p-mean value $\alpha_p = \alpha_p(p)$ is implicitly given by

$$\sum_{i=1}^{N}\left|\frac{f_i-\alpha_p}{\sigma_{f_i}}\right|^p sign\left(f_i-\alpha_p\right)=0, \tag{27}$$

and the estimated Chi-p is

$$\chi^p(p)=\sum_{i=1}^{N}\left|\frac{f_i-\alpha_p}{\sigma_{f_i}}\right|^p. \tag{28}$$

Figure 4(a) shows the six data points co-plotted with four values of α_p, that correspond to $p\to 1$, $p\to\infty$, and the two shape parameter values p_1, p_2 for which the p-value is equal to 0.05. The whole diagram of $\alpha_p = \alpha_p(p)$ is shown in Figure 4(b) and the p-value as a function of p is shown in Figure 4(c).

We observe that the function α_p is monotonically increasing converging to some constant value for $p\to\infty$. The corresponding mean value, α_∞, is given by

$$\alpha_\infty=\frac{\frac{x_{\min}}{\sigma_{x_{\min}}}+\frac{x_{\max}}{\sigma_{x_{\max}}}}{\frac{1}{\sigma_{x_{\min}}}+\frac{1}{\sigma_{x_{\max}}}}\cong 0.166. \tag{29}$$

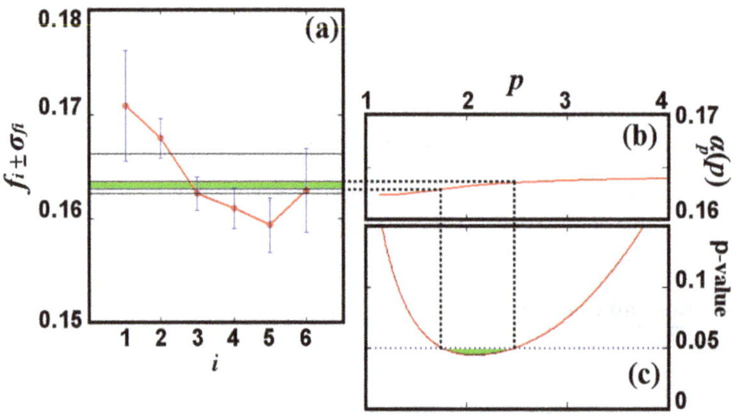

Figure 4 Statistical test for the rate of fitting based on L^p norms. (a) The dataset of Table 2 is co-plotted with four values of α_p, that correspond to $p\to 1$, $p\to\infty$, and the two shape parameter values $p_1 \sim 1.7$ and $p_2 \sim 2.5$ for which the p-value is equal to 0.05. **(b)** The diagram of L^p mean values, $\alpha_p = \alpha_p(p)$. **(c)** The p-value as a function of p. We observe that for the Euclidean norm $p = 2$, the null hypothesis is rejected, i.e., the sunspot area ratio data are not invariant with the heliolatitude. However, if the examined data are expected to be invariant, and thus the null hypothesis to be accepted, then the norms between p_1 and p_2 (green) are rejected because lead to p-value < 0.05.

The p-value has a minimum value at $p \sim 2.08$ and increases for larger shape values p until it reaches $p \sim 5.77$ where becomes p-value ~ 0.5 (not shown in the figure). If the shape p of the dataset is known, e.g., $p = 2$, then the null hypothesis is rejected, i.e., the sunspot area ratio data are dependent on the heliolatitude. On the other hand, if the data are expected to be invariant with the heliolatitude, and thus the null hypothesis to be accepted, then all the norms between $p_1 \sim 1.7$ and $p_2 \sim 2.5$ are rejected, and the norm L^p with $p \sim 5.77$ characterizes better these data points; the respective mean value is given by $\alpha_p(5.77) \sim 0.164$. Therefore, if we know the shape/norm p that characterizes the data, we can proceed and rate the goodness of the fitting. However, if p is unknown, at least we could detect those values of p for which the null hypothesis is accepted or rejected.

One of the most intriguing questions regarding the L^p-normed fitting is how can we determine the characteristic p-norm of the data. This is the suitable norm that should be used for the fitting of those data (Livadiotis 2007). The maximization of the p-value is one promising method. We demonstrate this as follows. We construct $N = 10^4$ data, $\{f_i\}_{i=1}^{N}$, of a random variable that follows the General Gaussian distribution of shape p, $f_i \sim GG(\mu = 0, \sigma = 1, p = 3)$. Figure 5(a) shows that the normalized histogram of these values matches this General Gaussian distribution. The p-value is approximated using the asymptotic behavior of (complete and incomplete) Gamma functions for large degrees of freedom, $M = 9999$. Hence, in order to derive the maximum p-value, it is sufficient to maximize

$$\text{p-value} \sim \left(\frac{e}{M} p \eta_p X_{\text{est}}^p\right)^{\frac{M}{p}} e^{-\eta_p X_{\text{est}}^p}. \tag{30}$$

This is shown in Figure 5(b), where the peak is at $p \cong 2.95 \pm 0.08$. Therefore, the p-value is maximized at the same value of p-norm as the shape of the General Gaussian distribution.

5. Conclusions

This paper (1) presented the derivation of the Chi-p distribution, the analog of Chi-square distribution but for datasets that follow the General Gaussian distribution of

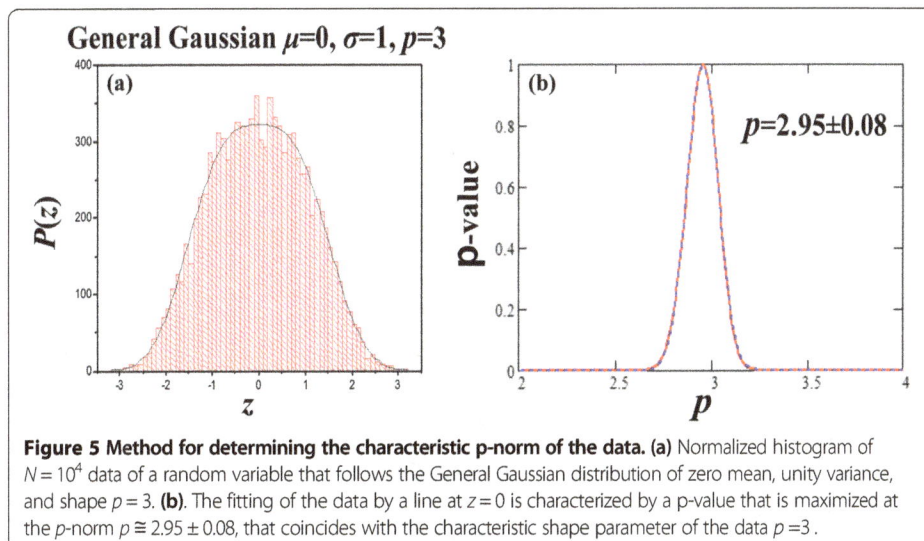

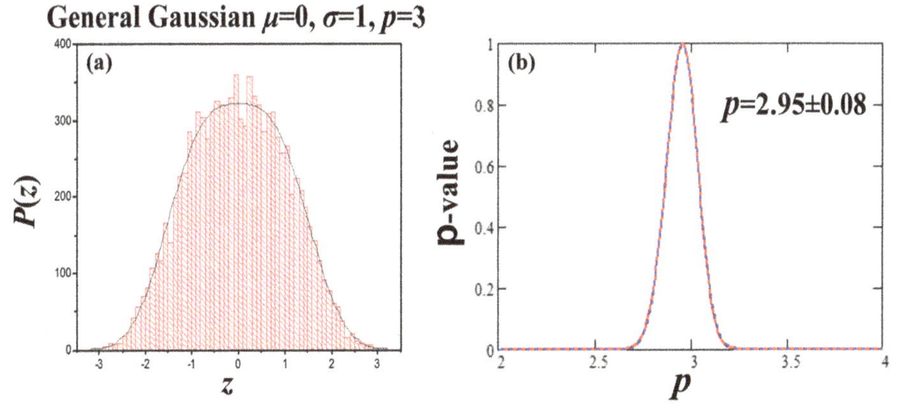

Figure 5 Method for determining the characteristic p-norm of the data. (a) Normalized histogram of $N = 10^4$ data of a random variable that follows the General Gaussian distribution of zero mean, unity variance, and shape $p = 3$. **(b)**. The fitting of the data by a line at $z = 0$ is characterized by a p-value that is maximized at the p-norm $p \cong 2.95 \pm 0.08$, that coincides with the characteristic shape parameter of the data $p = 3$.

shape p, and (2) developed the statistical test for characterizing the goodness of the fitting with L^p norms, which corresponds to datasets that follow the General Gaussian distribution of shape p.

It was shown that the statistical test has double role in the case of L^p norms: (1) If the shape parameter p is fixed and known, then the test can be rated by deriving the p-value. A convenient characterization of the fitting rate was developed. (2) If the shape parameter is unknown and the fitting is expected to be good for some shape parameter value p, a method for estimating p was given by fitting a General Gaussian distribution of shape p to the data, and then use this estimated shape parameter p to the Chi-p distribution to characterize the goodness of fitting. In particular, all the shape values p that correspond to unlikely p-values can be rejected, while the largest p-value corresponds to the most-likely shape parameter p of the examined data. This was verified by an illuminating example where the method of the fitting based on L^p norms was applied.

Appendix A: General Gaussian distribution

According to the theory of L^p-normed mean and variance, developed by Livadiotis (2012), the L^p-normed mean $< x >_p$ of the random variable X with probability distribution $P(x)$, is implicitly defined by

$$\int_{-\infty}^{\infty} P(x)\,|x- < x>_p|^{p-1}\text{sign}(x- < x>_p)\,dx = 0, \tag{A1}$$

where sign(u) returns the sign of u. The L^p-normed variance σ_p^2 is given by

$$\sigma_p^2 = \frac{\int_{-\infty}^{\infty} P(x)\,|x- < x>_p|^{p}\,dx}{(p-1)\int_{-\infty}^{\infty} P(x)\,|x- < x>_p|^{p-2}\,dx}. \tag{A2}$$

Next, we derive the L^p-normed mean and variance of the General Gaussian distribution (6), which are *Propositions* 1 and 2, stated in Section 2.

- *Proposition 1*: Given the distribution (6), we have that the L^p-normed mean is $< x >_p = \mu$, $\forall\, p \geq 1$.

- *Proof.* We have

$$\int_{-\infty}^{\infty} e^{-\eta_p\cdot|z|^p}\,|z- < z>_p|^{p-1}\text{sign}(z- < z>_p)\,dz = 0, \tag{A3}$$

for $z \equiv (x-\mu)/\sigma$, $< z >_p \equiv (< x >_p - \mu)/\sigma$. Let's assume that $< z >_p = 0$. Then, the left-hand side of Eq.(A3) is

$$\int_{-\infty}^{\infty} e^{-\eta_p\cdot|z|^p}\,|z|^{p-1}\text{sign}(z)\,dz = 0, \tag{A4}$$

because the integrant is a product of symmetric and antisymmetric function. Then, (A3) is true for $< z >_p = 0$, and given the uniqueness of the L^p-normed mean for each p, we end up with proposition 1. (Note that it is not surprising that the mean, $< x >_p = \mu$, is independent of p. Livadiotis (2012) showed that symmetric probability distributions lead to L^p-normed means that are independent of p.)

- *Proposition 2*: Given the distribution (6), we have that the L^p-normed variance is $\sigma_p^2 = \sigma^2$, $\forall\, p \geq 1$.

- *Proof.* We have $< |z|^q >= \int_{-\infty}^{\infty} P(z)\,|z|^q\, dz = 0$, i.e.,

$$\int_{-\infty}^{\infty} e^{-\eta_p \cdot |z|^p} |z|^q dz = 2\int_0^{\infty} e^{-\eta_p \cdot z^p} z^q dz = 2\eta_p{}^{-\frac{q+1}{p}} \int_0^{\infty} e^{-w} w^{\frac{q+1}{p}-1} dw$$

$$= 2\eta_p{}^{-\frac{q+1}{p}} \Gamma\left(\frac{q+1}{p}\right), \tag{A5a}$$

or,

$$< z^q >= C_p \frac{2}{p} \eta_p{}^{-\frac{q+1}{p}} \Gamma\left(\frac{q+1}{p}\right). \tag{A5b}$$

Hence, from (A2) we obtain

$$\sigma_p^2 = \frac{\int_{-\infty}^{\infty} P(z)\,|z|^p\, dz}{(p-1)\int_{-\infty}^{\infty} P(z)\,|z|^{p-2} dz} \cdot \sigma^2 = \frac{\eta_p{}^{-\frac{2}{p}} \Gamma\left(1+\frac{1}{p}\right)}{(p-1)\Gamma\left(1-\frac{1}{p}\right)} \cdot \sigma^2 = \sigma^2. \tag{A6}$$

Appendix B: Surface of the *N*-dimensional sphere in L^p space, $B_{p,N}$

This appendix shows the proof of *Lemma* 1, stated in Section 2.

- *Lemma* 1: The surface of the N-dimensional sphere of unit radius in L^p space, $B_{p,N}$, is given by Eq.(8). This is involved in the proof of Chi-p distribution (10), as shown below.

- *Proof of Lemma* 1.

Let the integral

$$1 = \int_{-\infty}^{+\infty} \cdots \int_{-\infty}^{+\infty} F\left(\vec{z}\right) dz_1 \ldots\, dz_N, \tag{B1}$$

where $\vec{z} = (z_1, \ldots, z_N)$, $Z^p \equiv \sum_{i=1}^{N} |z_i|^p$. The magnitude Z is the only quantity with dimensions the same as each of the components z_i. Indeed, if we define $c_i \equiv z_i/\zeta$, where $\zeta \equiv \sqrt{\sum_{i=1}^{N} z_i{}^2}$ is the Euclidean magnitude of $\vec{z}$, then, $Z = \left(\sum_{i=1}^{N} |z_i|^p\right)^{\frac{1}{p}} = \zeta \cdot \left(\sum_{i=1}^{N} |c_i|^p\right)^{\frac{1}{p}}$, i.e., Z and ζ have the same dimensions. (In the previous sections the components z_i were dimensionless by definition, i.e., $z_i \equiv \frac{x_i - \mu_x}{\sigma_{x_i}}$. However, we can still use this dimension analysis, since the components z_i may have dimensions in the generic case). Hence, we write Eq.(B1) as $dz_1 \ldots dz_N = Z^{N-1} dZ\, d^{N-1}\Omega_N$, i.e.,

$$1 = \int_0^{+\infty} \int_{\vec{z} \in B_{p,N}} F(Z; \Omega_N)\, Z^{N-1} dZ\, d^{N-1}\Omega_N, \tag{B2}$$

where $F(\vec{z}) = F(Z; \Omega_N)$; Ω_N symbolizes all the angular dependence, and $d^{N-1}\Omega_N$ denotes the angular infinitesimal. Since $F(Z; \Omega_N) = F(Z)$, we have $B_{p,N} \equiv \int_{\vec{z} \in B_{p,N}} d^{N-1}\Omega_N$, or

$$1 = \int_{\vec{z} \in B_{p,N}} d^{N-1}\Omega_N \cdot \int_0^{+\infty} F(Z) Z^{N-1} dZ = B_{p,N} \cdot \int_0^{+\infty} F(Z) Z^{N-1} dZ$$
$$= C_p^N \frac{1}{p} B_{p,N} \cdot \int_0^{+\infty} F(X^{\frac{1}{p}}) X^{\frac{N}{p}-1} dX,$$

where $F(Z) = C_p^N e^{-\eta_p Z^p}$, $F(X^{\frac{1}{p}}) = C_p^N e^{-\eta_p X}$. Therefore,

$$1 = \int_0^{+\infty} C_p^N \frac{1}{p} B_{p,N} e^{-\eta_p X} X^{\frac{N}{p}-1} dX \equiv \int_0^{+\infty} P(X; N; p) dX,$$

or,

$$P(X; N; p) = C_p^N \frac{1}{p} B_{p,N} e^{-\eta_p X} X^{\frac{N}{p}-1}. \tag{B3}$$

The normalization $\int_0^{+\infty} P(X; N; p) dX = 1$ gives $C_p^N \frac{1}{p} B_{p,N} = \eta_p^{\frac{N}{p}} / \Gamma\left(\frac{N}{p}\right)$, or

$$B_{p,N} = p\eta_p^{\frac{N}{p}} / \left[C_p^N \Gamma\left(\frac{N}{p}\right)\right] = p\left[\left(\frac{2}{p}\right)\Gamma\left(\frac{1}{p}\right)\right]^N / \Gamma\left(\frac{N}{p}\right). \tag{B4}$$

Another way to show Eq.(B4) is through the integration of all the components,

$$\int_{-\infty}^{+\infty} \cdots \int_{-\infty}^{+\infty} F(\vec{z}) dz_1 \ldots dz_N = 2^N \cdot \int_0^{+\infty} \cdots \int_0^{+\infty} F(\vec{z}) dz_1 \ldots dz_N$$
$$= 2^N \cdot \int_0^{+\infty} F(Z) \int_{\substack{\vec{z} \in B_{p,N} \\ z_i \geq 0}} (Z^p - z_2^p - z_3^p \ldots - z_N^p)^{\frac{1}{p}-1} Z^{p-1} dZ dz_2 \ldots dz_N,$$

by substituting $F(\vec{z}) = F(Z)$ and $z_1 = (Z^p - z_2^p - z_3^p \ldots - z_N^p)^{\frac{1}{p}}$ (for $z_i \geq 0$). The integration range $\vec{z} \in B_{p,N}$, $z_i \geq 0$, means $0 \leq z_i \leq \left(Z^p - \sum_{i+1}^N z_j^p \ldots - z_N^p\right)^{\frac{1}{p}}$ for $i = 1, \ldots, N-1$, and $0 \leq z_N \leq Z$. Similar, we have

$$\int_{\substack{\vec{z} \in B_{p,N} \\ z_i \geq 0}} (Z^p - z_2^p - z_3^p \ldots - z_N^p)^{\frac{1}{p}-1} dz_2 \ldots dz_N = a_{1,p} \int_{\substack{\vec{z} \in B_{p,N} \\ z_i \geq 0}} (Z^p - z_3^p \ldots - z_N^p)^{\frac{2}{p}-1} dz_3 \ldots dz_N$$

$$= a_{1,p} a_{2,p} \int_{\substack{\vec{z} \in B_{p,N} \\ z_i \geq 0}} (Z^p - z_4^p \ldots - z_N^p)^{\frac{3}{p}-1} dz_4 \ldots dz_N = \prod_{i=1}^{N-2} a_{i,p} \cdot \int_{\substack{\vec{z} \in B_{p,N} \\ z_i \geq 0}} (Z^p - z_N^p)^{\frac{N-1}{p}-1} dz_N$$

$$= \prod_{i=1}^{N-1} a_{i,p} \cdot Z^{N-p},$$

where

$$a_{i,p} \equiv \int_0^1 (1 - t^p)^{\frac{i}{p}-1} dt. \tag{B5}$$

Hence, we derive

$$\int_{-\infty}^{+\infty} \cdots \int_{-\infty}^{+\infty} F\left(\vec{z}\right) dz_1 \ldots\, dz_N = 2^N \cdot \prod_{i=1}^{N} a_{i,p} \cdot \int_0^{+\infty} F(Z)\, Z^{N-1}\, dZ, \tag{B6}$$

while, on the other hand, we have

$$\begin{aligned} \int_{-\infty}^{+\infty} \cdots \int_{-\infty}^{+\infty} F\left(\vec{z}\right) dz_1 \ldots\, dz_N &= \int_{\vec{z} \in B_{p,N}} d^{N-1}\Omega_N \cdot \int_0^{+\infty} F(Z)\, Z^{N-1} dZ \\ &= B_{p,N} \cdot \int_0^{+\infty} F(Z)\, Z^{N-1} dZ, \end{aligned} \tag{B7}$$

thus,

$$B_{p,N} = 2^N \cdot \prod_{i=1}^{N-1} a_{i,p}. \tag{B8}$$

We easily find that

$$a_{i,p} = \frac{1}{p} \int_0^1 y^{\frac{1}{p}-1} (1-y)^{\frac{i}{p}-1} dy = \frac{1}{p} B\left(\frac{1}{p}, \frac{i}{p}\right), \tag{B9}$$

where $B(x,y) \equiv \Gamma(x)\Gamma(y)/\Gamma(x+y)$ is the Beta function. Hence, we have

$$B_{p,N} = p \left(\frac{2}{p}\right)^N \Gamma\left(\frac{1}{p}\right)^{N-1} \cdot \prod_{i=1}^{N-1} \Gamma\left(\frac{i}{p}\right) / \Gamma\left(\frac{i+1}{p}\right). \tag{B10}$$

Since, $\prod_{i=1}^{N-1} \Gamma\left(\frac{i}{p}\right) / \Gamma\left(\frac{i+1}{p}\right) = \Gamma\left(\frac{1}{p}\right) / \Gamma\left(\frac{N}{p}\right)$, finally, we end up with Eq.(B4).

Competing interests

The authors declare that they have no competing interests.

References

Adèr, HJ: Modelling (Chapter 12). In: Adèr, HJ, Mellenbergh, GJ, with contributions by D.J. Hand (eds.) Advising on Research Methods: A consultant's companion, pp. 271–304. Johannes van Kessel Publishing, Huizen, The Netherlands (2008)

Burden, RL, Faires, JD: Numerical Analysis, pp. 437–438. PWS Publishing Company, Boston, MA (1993)

Edwards, AWF: The proportion of umbra in large sunspots, 1878–1954. The Observatory **77**, 69–70 (1957)

Frisch, PC, Bzowski, M, Livadiotis, G, McComas, DJ, Möbius, E, Mueller, HR, Pryor, WR, Schwadron, NA, Sokól, JM, Vallerga, JV, Ajello, JM: Decades-long changes of the interstellar wind through our solar system. Science **341**, 1080 (2013)

Funsten, HO, Frisch, PC, Heerikhuisen, J, Higdon, DM, Janzen, P, Larsen, BA, Livadiotis, G, McComas, DJ, Möbius, E, Reese, CS, Reisenfeld, DB, Schwadron, NA, Zirnstein, E: The circularity of the IBEX Ribbon of enhanced energetic neutral atom flux. Astrophys. J. **776**, 30 (2013)

Livadiotis, G: Approach to general methods for fitting and their sensitivity. Physica A **375**, 518–536 (2007)

Livadiotis, G: Approach to the block entropy modeling and optimization. Physica A **387**, 2471–2494 (2008)

Livadiotis, G: Expectation values and Variance based on L^p norms. Entropy **14**, 2375–2396 (2012)

Livadiotis, G, McComas, DJ: Fitting method based on correlation maximization: Applications in Astrophysics. J. Geophys. Res. **118**, 2863–2875 (2013a)

Livadiotis, G, McComas, DJ: Evidence of large scale phase space quantization in plasmas". Entropy **15**, 1116–1132 (2013b)

Livadiotis, G, Moussas, X: The sunspot as an autonomous dynamical system: A model for the growth and decay phases of sunspots. Physica A **379**, 436–458 (2007)

McCullagh, P: What is statistical model? Ann. Stat. **30**, 1225–1310 (2002)

Melissinos, AC: Experiments in Modern Physics, pp. 464–467. Academic Press Inc, London, UK (1966)

Sengupta, A: A rational function approximation of the singular eigenfunction of the monoenergetic neutron transport equation. J. Phys. A **17**, 2743–2758 (1984)

12

Univariate and multivariate Pareto models

Barry C Arnold

Correspondence:
barry.arnold@ucr.edu
Department of Statistics, University of California, Riverside, USA

Abstract

The Pareto distribution has long been recognized as a suitable model for many non-negative socio-economic variables. Univariate and multivariate variations abound. Some unification is possible by representing the Pareto variables in terms of independent gamma distributed components. Further unification is sometimes possible since some of the frequently used multivariate Pareto models share the same copula. In some cases, inference strategies can be developed to take advantage of the stochastic representations in terms of gamma components.

Keywords: Inequality; Heavy tails; Generalized Pareto; Feller-Pareto; Kumaraswamy distribution; Hidden truncation; Conditional specification

1 Introduction

Discussion of Pareto and Pareto-like distributions can be traced back to Vilfredo Pareto's Economics textbook published in Rome in 1897. His observation that the number of persons in a population whose incomes exceed x is often well approximated by $Cx^{-\alpha}$ for some positive C and some positive α, led inexorably to consideration of the following well known model used for fitting univariate income data.

$$P(X > x) = \overline{F}_X(x) = (x/\sigma)^{-\alpha}, \quad x > \sigma. \tag{1}$$

Here σ, the scale parameter, is positive and α (Pareto's index of inequality) is also positive.

This model is typically referred to as the classical Pareto model. Improved fitting of data is encountered when more general Pareto-like distributions are considered. In this survey, the classical model will be embedded in a hierarchy of more complicated Pareto models. In this hierarchy of generalized Pareto distributions, the classical model will be called the Pareto (I) distribution. Multivariate income distributions are also of interest and, in that arena, a hierarchy of multivariate Pareto distributions is available, paralleling and closely related to the univariate hierarchy.

Even more flexible models have been proposed using these univariate and multivariate Pareto models as building blocks. Several of these will be described in this paper.

The end result is an impressively flexible array of income models from which the researcher can select a parsimonious model for the particular data set at hand. The emphasis in this survey will be on distributional properties of the models but some attention will be paid to estimation and inference strategies.

2 A hierarchy of generalized Pareto models

As the basic distribution in our hierarchy of generalized Pareto models we use the classical Pareto distribution, called here the Pareto (I) distribution. Its survival function is of the form (1). In practice, α is frequently assumed to be larger than 1, so that the distribution has a finite mean. If a random variable X has (1) as its survival function, then we write $X \sim P(I)(\sigma, \alpha)$. In this basic model, the parameter σ has a dual role. It is indeed a scale parameter, but also it determines the lower bound of the support of the distribution and, in a sense, plays to some extent the role of a location parameter. A slightly more general model, which separates the roles of location and scale parameters has then been frequently used.

This distribution will be called the Pareto (II) distribution. Its survival function is of the form

$$\overline{F}(x) = \left[1 + \left(\frac{x - \mu}{\sigma}\right)\right]^{-\alpha}, \quad x > \mu \tag{2}$$

where μ, the location parameter, is real valued, σ is positive and α is positive. In most applications μ will be non-negative, but negative values for μ pose no problems. If X has (2) as its survival function, we will write $X \sim P(\text{II})(\mu, \sigma, \alpha)$.

There is an intimate relation between the Pareto (II) distribution and the Pickands generalized Pareto model, which is much used in the study of extreme values and peaks over thresholds. A good general reference for discussion of the role of the Pickands generalized Pareto distribution is the book by Falk et al. (2011). The density of the Pickands generalized Pareto model is

$$f(x; \sigma, k) = \frac{1}{\sigma}\left(1 - \frac{kx}{\sigma}\right)^{(1-k)/k} I(x > 0, (kx)/\sigma < 1). \tag{3}$$

where $\sigma > 0$ and $-\infty < k < \infty$. The density corresponding to $k = 0$ is obtained by taking the limit as $k \uparrow 0$ in (3). This model, (3), includes three sub-models. When $k < 0$, it yields a Pareto (II) density (with $\mu = 0$), when $k = 0$ it yields an exponential density, while for $k > 0$, it corresponds to a scaled Beta distribution (of the first or standard kind). In fact, several results for the Pareto (II) distribution can be proved to remain valid in the more general context of the Pickands generalized Pareto model. In an income modeling setting, as Pareto observed, heavy tailed distributions are typically encountered and, for this reason, we will concentrate on the Pareto (II) sub-model. We remark, in passing, that despite Pareto's insistence on the ubiquity of heavy tails, several authors have utilized scaled Beta distributions as income models, and some have even argued in favor of the exponential distribution as a model. The most general model in our hierarchy, the Feller-Pareto model will be seen to actually include some light-tailed distributions corresponding to scaled Beta models with an additional location parameter. Applications of such light tailed models are more likely to be encountered outside of the income distribution context.

Truncated versions of the Pareto (II) distribution, which clearly are not heavy tailed, are sometimes appropriate models for data sets which, for some reason, exclude large values. Such distributions are of the form

$$\begin{aligned} F(x) &= 0, & x \leq \mu, \\ &= \frac{1 - \left(1 + \frac{x-\mu}{\sigma}\right)^{-\alpha}}{1 - \left(1 + \frac{\tau-\mu}{\sigma}\right)^{-\alpha}}, & \mu < x < \tau, \\ &= 1, & x \geq \tau, \end{aligned} \tag{4}$$

where $-\infty < \mu < \tau < \infty,\ \sigma > 0$, and $\alpha > 0$. Some discussion of such models may be found in Aban et al. (2006) and Arnold and Austin (1987).

The Pareto (III) distribution is a variant model with tail behavior comparable to that of the Pareto (II) distribution. Its survival function is of the form

$$\overline{F}(x) = \left[1 + \left(\frac{x-\mu}{\sigma}\right)^{1/\gamma}\right]^{-1}, \quad x > \mu \tag{5}$$

where μ is real, σ is positive and γ is positive. We will call γ the inequality parameter. If $\mu = 0$ and $\gamma \leq 1$, then γ turns out to be precisely the Gini index of inequality for this distribution. If X has (5) as its survival function, we will write $X \sim P(\mathrm{III})(\mu, \sigma, \gamma)$.

If we introduce both a shape and an inequality parameter, we arrive at the Pareto (IV) family:

$$\overline{F}(x) = \left[1 + \left(\frac{x-\mu}{\sigma}\right)^{1/\gamma}\right]^{-\alpha}, \quad x > \mu \tag{6}$$

where μ (location) is real, σ (scale) is positive, γ (inequality) is positive and α (shape) is positive. Although we continue to call γ the inequality parameter it will only be identifiable with the Gini index when $\alpha = 1$ and $\mu = 0$. One might argue instead that in the P(IV) model both γ and α would be best described as shape parameters, since neither of them has a direct inequality interpretation. An anonymous referee points out that the two parameters γ and α govern the behavior of the P(IV) density as x approaches μ from above and as x approaches infinity. Thus, $f(x; \mu, \sigma, \gamma, \alpha) \sim x^{-\alpha/\gamma - 1}$ as $x \to \infty$ and $f(x-\mu; \mu, \sigma, \gamma, \alpha) \sim (x-\mu)^{1/\gamma - 1}$ as $x \to \mu$. He suggests that an argument might be advanced in favor of a reparameterization in which we define $\beta = \alpha/\gamma$, to highlight the roles of α and β in determining the limiting behavior of the density. However, in this paper, to be consistent with the notation in Arnold (1983), we will continue with the $\mu, \sigma, \gamma, \alpha$ parameterization and continue to call γ the inequality parameter. If a random variable X has (6) as its survival function, we will write $X \sim P(\mathrm{IV})(\mu, \sigma, \gamma, \alpha)$. Note that the Pareto (IV) distribution, with $\mu = 0$, is also known as a Burr Type XII distribution.

The three more specialized families. P(I)-P(III), may be identified as special cases of the Pareto (IV) family as follows:

$$\begin{aligned} P(\mathrm{I})(\sigma, \alpha) &= P(\mathrm{IV})(\sigma, \sigma, 1, \alpha), \\ P(\mathrm{II})(\mu, \sigma, \alpha) &= P(\mathrm{IV})(\mu, \sigma, 1, \alpha), \\ P(\mathrm{III})(\mu, \sigma, \gamma) &= P(\mathrm{IV})(\mu, \sigma, \gamma, 1). \end{aligned} \tag{7}$$

Feller (1971), p. 49, suggested a different definition of a Pareto distribution. It can be recognized as the distribution of a ratio of two independent gamma variables (a distribution also known as Beta distribution of the second kind). By considering a linear function of a power of such a random variable, we arrive at a very general family, called the Feller–Pareto family. Thus if $X_i \sim \Gamma(\delta_i, 1)\ i = 1, 2$, are independent random variables, and if for μ real, $\sigma > 0$ and $\gamma > 0$ we define

$$W = \mu + \sigma(X_2/X_1)^{\gamma}, \tag{8}$$

then W has a Feller–Pareto distribution, and we write $W \sim FP(\mu, \sigma, \gamma, \delta_1, \delta_2)$.

It may be verified that the Pareto (IV) distributions are identifiable with the Feller–Pareto distributions with $\delta_2 = 1$, i.e.,

$$P(\text{IV})(\mu, \sigma, \gamma, \alpha) = FP(\mu, \sigma, \gamma, \alpha, 1). \tag{9}$$

The density of the general Feller–Pareto distribution defined by (8) is of the form

$$f_W(w) = \left(\frac{w-\mu}{\sigma}\right)^{(\delta_2/\gamma)-1} \left[1 + \left(\frac{w-\mu}{\sigma}\right)^{1/\gamma}\right]^{-\delta_1-\delta_2} \Big/ \left[\gamma\sigma B(\delta_1, \delta_2)\right], \quad w > \mu. \tag{10}$$

The corresponding survival function is obtainable from tables of the incomplete beta function. For many computations it is simpler to work directly with the representation (8). The Pareto (IV) distributions correspond to the case in which X_2 has a gamma distribution while X_1 has an exponential distribution. The Pareto (III) distributions are encountered when both X_1 and X_2 are exponential variables.

Kalbfleisch and Prentice (1980) call the Feller-Pareto density (with $\mu = 0$) a generalized F density. Instead we might describe a Feller Pareto variable as being a location and scale transform of a generalized beta variable of the second kind. Recall that a beta variable of the second kind is just a ratio of independent gamma variables.

An even more general model might be built using independent variables $X_i \sim \Gamma(\delta_i, 1)$, $i = 1, 2$. One could define

$$W = \mu + \sigma \left(\frac{X_2^{\gamma_2}}{X_1^{\gamma_1}}\right). \tag{11}$$

The additional flexibility provided by the introduction of such a sixth parameter in the model has not been investigated.

The full array of generalized univariate Pareto distributions to be considered in this paper are subsumed in the Feller–Pareto family and a unified derivation of many distributional results is possible. However, in the case of Pareto (I)-(IV) distributions, some alternative representations are also useful.

A random variable X has a $P(\text{I})(\sigma, \alpha)$ distribution if it is of the form

$$X = \sigma e^{V/\alpha} \tag{12}$$

where V is a standard exponential random variable. An analogous representation of a Pareto (II) variable in terms of an exponential random variable is possible, i.e.,

$$X = \mu + \sigma\left(e^{V/\alpha} - 1\right). \tag{13}$$

Likewise Pareto (III) and (IV) variables can be represented as $X = \mu + \sigma\left(e^{V} - 1\right)^{\gamma}$ and $X = \mu + \sigma\left(e^{V/\alpha} - 1\right)^{\gamma}$ respectively. The representation (12) for a classical Pareto variable (i.e., Pareto (I)) highlights the useful observation that the logarithm of such a variable has a shifted exponential distribution. This will permit the recognition of many distributional properties of Pareto (I) variables as reflections of parallel properties of exponential variables.

A second important representation of the Pareto (II) distribution, known to Maguire et al. (1952), is as a mixture of exponentials. We may describe it in terms of the conditional survival function, given an auxiliary gamma distributed random variable Z. Thus, if

$$P(X > x \mid Z = z) = e^{-z(x-\mu)/\sigma}, \quad x > \mu,$$

i.e., a (translated) exponential distribution, and if $Z \sim \Gamma(\alpha, 1)$, then it follows that unconditionally $X \sim P(\text{II})(\mu, \sigma, \alpha)$. Alternatively, this can be viewed as being equivalent to the representation in (8) after setting $\gamma = \delta_2 = 1$.

This representation of the Pareto (II) distribution as a gamma mixture of exponential distributions is often encountered in reliability and survival contexts, see e.g., Keiding et al. (2002). It is also familiar in Bayesian analysis of exponential data, where the gamma density enters as a convenient prior. In this context the Pareto (II) distribution is sometimes called the Lomax distribution.

The Pareto (III) distribution was apparently first considered by (Fisk 1961a; 1961b) who called it a sech^2 distribution. It is closely related to the logistic distribution. We say that a random variable X has a logistic (μ, σ) distribution, if its distribution function assumes the form

$$F_X(x) = \left[1 + e^{-(x-\mu)/\sigma}\right]^{-1}, \quad -\infty < x < \infty$$

and we write $X \sim L(\mu, \sigma)$. It is not difficult to verify that

$$X \sim L(\mu, \sigma) \Leftrightarrow e^X \sim P(\text{III})(0, e^{\mu}, \sigma). \tag{14}$$

It is as a consequence of the relation (14) that the Pareto (III) distribution, with $\mu = 0$, is sometimes called the log-logistic distribution.

Remark 1. *Johnson et al. (1994) refer to a Pareto distribution of the third kind that is not to be confused with the Pareto (III) distribution discussed in this paper. The survival function of this "third kind" distribution is of the form*

$$\overline{F}(x) = \left(1 + \frac{x}{\sigma}\right)^{-\alpha} e^{-\beta x}, \quad x > 0. \tag{15}$$

This distribution, which was suggested by Pareto (1897), was proposed to accommodate cases in which the basic Pareto model (1) was inadequate for fitting certain data configurations. This model is closely related to the Pareto (II) distribution, but with an additional exponential factor. Note that it could be viewed as the distribution of the minimum of a Pareto (II) variable (with $\mu = 0$) and an independent exponential variable. This model has been used infrequently, but recently it has reappeared, this time called a tapered Pareto distribution (Kagan and Schoenberg 2001).

2.1 Distributional properties

The Feller–Pareto distributions are unimodal. The mode is at μ if $\gamma > \delta_2$, while if $\gamma \leq \delta_2$, we find (here $W \sim FP(\mu, \sigma, \gamma, \delta_1, \delta_2)$)

$$\text{mode}(W) = \mu + \sigma\left[(\delta_2 - \gamma)/(\delta_1 + \gamma)\right]^{\gamma} \tag{16}$$

In order to compute moments of the Pareto distributions, it is convenient to work with the representation (8). With $W \sim FP(\mu, \sigma, \gamma, \delta_1, \delta_2)$, if we define $W^* = (W - \mu)/\sigma$, then $W^* \sim FP(0, 1, \gamma, \delta_1, \delta_2)$, i.e., $W^* =^d (X_2/X_1)^{\gamma}$ where $X_i \sim \Gamma(\delta_i, 1)$ $i = 1, 2$, are independent random variables. It then can be readily verified that for a real number τ, the τ'th moment of W^* when it exists is of the form

$$E\left(W^{*\tau}\right) = \Gamma(\delta_1 - \gamma\tau)\Gamma(\delta_2 + \gamma\tau)/\Gamma(\delta_1)\Gamma(\delta_2), \quad -(\delta_2/\gamma) < \tau < (\delta_1/\gamma). \tag{17}$$

From this expression moments for the Feller-Pareto and the P(II)-P(IV) distributions are readily obtained.

Moments of the Pareto (I) distribution cannot be obtained in this way since, for it, $\mu = \sigma \neq 0$. They are obtainable by direct integration:

$$(\text{Pareto (I)}) \quad E(X^\tau) = \sigma^\tau \left(1 - \frac{\tau}{\alpha}\right)^{-1}, \quad \tau < \alpha. \tag{18}$$

Sums of independent Pareto variables typically do not have analytically tractable distributions. If we multiply independent Pareto variables rather than adding them, it is sometimes possible to get simple expressions for the density of the resulting product. In the case of the Pareto I distribution the key lies in utilization of representation (12). Thus, if $X_1, X_2, \ldots, X_n$ are independent Pareto I variables with $X_i \sim P(\mathrm{I})(\sigma_i, \alpha_i)$, then their product W has the representation

$$W = \left(\prod_{i=1}^{n} \sigma_i\right) \exp\left(\sum_{i=1}^{n} (V_i/\alpha_i)\right) \tag{19}$$

where the V_i's are independent standard exponential variables. In some cases expressions are available for the distribution of $\sum_{i=1}^{n} V_i/\alpha_i$. In particular, if $\alpha_i = \alpha, (i = 1, 2, \ldots, n)$, then $\sum_{i=1}^{n} V_i/\alpha \sim \Gamma(n, 1/\alpha)$, and we may readily obtain the density of W.

A second case in which simple closed form expressions are available is one in which all the α_i's are distinct. In this situation we can use a result for weighted sums of exponentials given in, for example, Feller (1971), p. 40, and write the survival function of the product in the form:

$$P(W > w) = \sum_{i=1}^{n} \left(\frac{w}{\sigma}\right)^{-\alpha_i} \prod_{\substack{k=1 \\ k \neq i}}^{n} \left(\frac{\alpha_k}{\alpha_i - \alpha_k}\right), \quad w > \sigma \tag{20}$$

where $\sigma = \prod_{i=1}^{n} \sigma_i$ and $\alpha_i \neq \alpha_j$ if $i \neq j$. The distribution of products of independent Pareto (IV) variables with μ_i's equal to 0 can, via the representation (8), be reduced to a problem involving the distribution of products of powers of independent gamma random variables. Unlike the Pareto (I) case, closed form expressions for the resulting density are apparently not obtainable, although moments of such products are readily available.

The Pareto (IV) family is closed under minimization when certain parameters are common to the minimands. Thus, if X_1 and X_2 are independent random variables with $X_i \sim P(\mathrm{IV})(\mu, \sigma, \gamma, \alpha_i), i = 1, 2$, then

$$\min(X_1, X_2) \sim P(\mathrm{IV})(\mu, \sigma, \gamma, \alpha_1 + \alpha_2). \tag{21}$$

Note that in this situation the X_i's share common values for the parameters μ, σ and γ.

Pareto (III) variables exhibit an interesting closure property with respect to geometric minimization and maximization. Indeed, this was used as a justification for use of the Pareto (III) distribution as a suitable model for income distributions based on a scenario involving competitive bidding for employment (Arnold and Laguna 1976). For this, consider a sequence $X_1, X_2, \ldots$ of i.i.d. Pareto $(\mathrm{III})(\mu, \sigma, \gamma)$ random variables. Suppose that for some $p \in (0, 1)$, N_p is independent of the X_i's and has a geometric(p) distribution, i.e., $P(N = n) = p(1 - p)^{n-1}, n = 1, 2, \ldots$. Define the corresponding random extrema by

$$U_p = \min\{X_1, X_2, \ldots, X_{N_p}\}, \tag{22}$$

and

$$V_p = \max\{X_1, X_2, \ldots, X_{N_p}\}. \tag{23}$$

It is readily verified, by conditioning on N_p, that U_p and V_p each have Pareto (III) distributions. Thus

$$U_p \sim P(III)(\mu, \sigma p^{\gamma}, \gamma), \tag{24}$$

and

$$V_p \sim P(III)(\mu, \sigma p^{-\gamma}, \gamma). \tag{25}$$

Observe that, if $\mu = 0$, then

$$p^{-\gamma} U_p \stackrel{d}{=} p^{\gamma} V_p \stackrel{d}{=} X_1.$$

Some characterization results based on this observation were discussed in Arnold et al. (1986).

It is possible to write down expressions for the densities of order statistics from a Pareto (IV) sample. The corresponding distribution functions will involve incomplete beta functions. Simulation of such order statistics may be accomplished by utilizing the relatively simple form of the Pareto (IV) quantile function, i.e.,

$$F^{-1}(u) = \mu + \sigma \left[(1-u)^{-1/\alpha} - 1 \right]^{\gamma}. \tag{26}$$

From this we have that if $X_{i:n}$ is the ith order statistic from a sample of size n from a Pareto (IV) distribution, then

$$X_{i:n} \stackrel{d}{=} F^{-1}(U_{i:n}) \tag{27}$$

where $\stackrel{d}{=}$ means that the two random variables are identically distributed, where F^{-1} is as given in (26), and where $U_{i:n}$ is the ith order statistic of a sample of size n from a uniform (0,1) distribution. It is well known (see e.g. David and Nagaraja 2003) that $U_{i:n} \sim \text{Beta}(i, n-i+1)$.

In some special cases the density of the ith order statistic (27) assumes a known form. For example:

$$X_i\text{'s} \sim P(\text{III})(\mu, \sigma, \gamma) \Rightarrow X_{i:n} \sim FP(\mu, \sigma, \gamma, n-i+1, i). \tag{28}$$

Another case involves minima:

$$X_i\text{'s} \sim P(\text{IV})(\mu, \sigma, \gamma, \alpha) \Rightarrow X_{1:n} \sim P(\text{IV})(\mu, \sigma, \gamma, n\alpha). \tag{29}$$

3 Some related extensions

A variety of models have been proposed to add more flexibility to the generalized Pareto models discussed in Section 2. Most of them include Pareto models as special cases. In this Section we will make note of a selection of these models.

Many early researchers modeled the logarithm of income (called income power by Champernowne (1937)). Thus, instead of postulating a simple distribution for income, a relatively simple distribution was assumed for some function of income. More flexibility may be introduced by considering a parametric family of monotonic transformations of the income data whose parameters must be estimated from the data. For example, we might begin with a parametric family of increasing functions $\psi(x; \underline{\tau})$ with corresponding inverse functions $\psi^{-1}(x; \underline{\tau})$ and assume that $\psi(X; \underline{\tau})$ has a Pareto (IV)$(0, \sigma, \gamma, \alpha)$

distribution. If we denote the corresponding $P(IV)(0, \sigma, \gamma, \alpha)$ distribution by $F_{\sigma,\gamma,\alpha}(x)$ then the distribution of X will be

$$F_X(x; \sigma, \gamma, \alpha, \underline{\tau}) = F_{\sigma,\gamma,\alpha}\left(\psi^{-1}(x; \underline{\tau})\right). \tag{30}$$

A parallel extension involves quantile functions instead of distribution functions. For this the quantile function of X is assumed be of the form

$$F_X^{-1}(u; \sigma, \gamma, \alpha, \underline{\tau}) = F_{\sigma,\gamma,\alpha}^{-1}\left(\widetilde{\psi}(u; \underline{\tau})\right). \tag{31}$$

where $\widetilde{\psi}(u; \underline{\tau})$ is a parametric family of monotone functions mapping $(0, 1)$ onto $(0, 1)$. A popular model of this genre, introduced by Jones (2004), makes use of the family of quantile functions of Beta distributions. The density function of this Beta-generalized Pareto distribution is given by

$$f_X(x; \sigma, \gamma, \alpha, \lambda_1, \lambda_2) = \frac{\alpha\left[1 - \left(1 + \left(\frac{x}{\sigma}\right)^{1/\gamma}\right)^{-\alpha}\right]^{\lambda_1 - 1}\left(1 + \left(\frac{x}{\sigma}\right)^{1/\gamma}\right)^{-\alpha\lambda_2 - 1}\left(\frac{x}{\sigma}\right)^{(1/\gamma)-1}}{\sigma\gamma B(\lambda_1, \lambda_2)}, \tag{32}$$

where $x \in (0, \infty)$. Of course, if $\lambda_1 = \lambda_2 = 1$, the Beta-generalized distribution simplifies to become a Pareto distribution.

Another popular model of the form (31) involves the simple choice $\widetilde{\psi}(u) = u^{1/\theta}$ where $\theta > 0$. In such a case we have

$$F_X(x; \sigma, \gamma, \alpha, \theta) = \left[F_{\sigma,\gamma,\alpha}(x)\right]^{\theta}, \quad x > 0, \tag{33}$$

and the distribution is usually called the exponentiated generalized Pareto distribution. It can be recognized as a special case of the Beta-generalized Pareto model with the parameters chosen to be $\lambda_1 = \theta$ and $\lambda_2 = 1$.

Instead of the Beta distribution, one might use the Kumaraswamy distribution to obtain an alternative generalized Pareto distribution. First,we must recall the definition of the Kumaraswamy distribution. We say that X has a Kumaraswamy (λ_1, λ_2) distribution if its density and distribution functions are :

$$f_K(x) = \lambda_1\lambda_2 x^{\lambda_1 - 1}\left(1 - x^{\lambda_1}\right)^{\lambda_2 - 1}, \quad 0 < x < 1, \tag{34}$$

and

$$F_K(x) = 1 - \left(1 - x^{\lambda_1}\right)^{\lambda_2}, \quad 0 < x < 1. \tag{35}$$

See Jones (2009) for a comprehensive introduction to the Kumaraswamy distribution. Let $F_P(x)$ denote the Pareto (IV) distribution function and suppose that K has a Kumaraswamy (λ_1, λ_2) distribution. Define $Y = F_P^{-1}(K)$, then Y has a Kumaraswamy-Pareto (IV) distribution with corresponding density

$$f_Y(y) = f_K(F_P(y))f_P(y).$$

Akinsete et al. (2008) consider some special subcases of the Beta-generalized Pareto distribution, while Paranaiba et al. (2013) discuss the Kumaraswamy-generalized Pareto distribution. Submodels of the Kumaraswamy-generalized Pareto model are often of interest. For example the exponentiated generalized Pareto distribution (33) is such a submodel.

And, of course, one can concatenate these constructions and consider a Beta-Kumaraswamy-Pareto distribution. Going one step further we would arrive at a generalized-Beta-Kumaraswamy-Pareto model. Each generalization adds flexibility at the cost of introducing more parameters. Some degree of parsimony is evidently called for here.

Pillai (1991) suggested an extension of the Pareto (III) distribution, motivated by its closure under geometric minimization. A random variable is said to have a semi-Pareto (III) distribution if its survival function is of the form

$$\overline{F}(x;\mu,\sigma,\gamma,p)=\left[1+\left(\frac{x-\mu}{\sigma}\right)^{1/\gamma}h\left(\frac{x-\mu}{\sigma}\right)\right]^{-1},\quad x>\mu, \tag{36}$$

where $\mu\in(-\infty,\infty)$, $\sigma,\gamma\in(0,\infty)$, $p\in(0,1)$ and $h(x)$ is a periodic function of $\ln x$ with period $-2\pi/[\gamma\ln p]$, and with $h(0)=1$. The case in which $h(x)\equiv 1$ for every x corresponds to the usual Pareto (III) model. More generality can be arrived at if $h(x)$ is replaced by a suitable parametric family of periodic functions. Note that, in order for (36) to be a valid survival function, it must be the case that $x^{1/\gamma}h(x)$ is a non-decreasing function of x.

Hidden truncation or selection models may sometimes provide alternative models that are more suitable than basic Pareto models. The corresponding scenario is one in which the variable X is observed only if a covariable Y takes on a value less than some threshold value. Thus the distribution of the observed X's is of the form $P(X\le x|Y\le y_0)$. With this in mind, consider the case in which (X,Y) has a bivariate Pareto (IV) distribution with the following joint survival function.

$$P(X>x,Y>y)=\left[1+\left(\frac{x-\mu}{\sigma}\right)^{1/\gamma}+\left(\frac{y-\nu}{\tau}\right)^{1/\delta}\right]^{-\alpha},\quad x>\mu, y>\nu. \tag{37}$$

(Such distributions will be discussed in more detail in Section 5). This distribution has Pareto (IV) marginals and has Pareto (IV) conditionals. After suitable reparameterization the corresponding hidden truncation density for X, given that it can only be observed if Y is not too large, is

$$\begin{aligned} f_{HT}(x;\mu,\sigma,\gamma,\alpha,\theta) &= \frac{\alpha\left(\frac{x-\mu}{\sigma}\right)^{(1/\gamma)-1}}{\gamma\sigma[1-(1+\theta)^{-\alpha}]} \\ &\quad\times\left[\left(1+\left(\frac{x-\mu}{\sigma}\right)^{1/\gamma}\right)^{-(\alpha+1)}\right. \\ &\quad\left.-\left(1+\left(\frac{x-\mu}{\sigma}\right)^{1/\gamma}+\theta\right)^{-(\alpha+1)}\right],\quad x>\mu, \end{aligned} \tag{38}$$

where a new parameter $\theta=[(y_0-\nu)/\tau]^{1/\delta}$ has been introduced. In this model, μ is a real valued parameter, often positive, while all of the other parameters, σ,γ,α

and θ are positive valued. An alternative representation of this density is possible as follows.

$$f_{HT}(x;\mu,\sigma,\gamma,\alpha,\theta)$$

$$= \frac{1}{1-(1+\theta)^{-\alpha}}\left[\frac{\alpha}{\gamma\sigma}\left(\frac{x-\mu}{\sigma}\right)^{(1/\gamma)-1}\left[1+\left(\frac{x-\mu}{\sigma}\right)^{1/\gamma}\right]^{-(\alpha+1)}\right] \tag{39}$$

$$-\frac{(1+\theta)^{-\alpha}}{1-(1+\theta)^{-\alpha}}\left[\frac{\alpha}{\gamma\sigma_1}\left(\frac{x-\mu}{\sigma_1}\right)^{(1/\gamma)-1}\left[1+\left(\frac{x-\mu}{\sigma_1}\right)^{1/\gamma}\right]^{-(\alpha+1)}\right]$$

where $\sigma_1 = \sigma(1+\theta)^{\gamma}$. This is recognizable as a linear combination of two Pareto (IV) densities. Note that the density is a linear combination of two Pareto (IV) densities, but it is not a convex combination since, although the coefficients add up to 1, the second coefficient is negative. Motivated by this example, one might also consider k-component linear combinations of Pareto (IV) densities as income models, allowing k to be greater than 2. Such models with positive coefficients are natural candidates for fitting multi-modal income data sets which may well have a mixture genesis. Note that, by testing the hypothesis that $\theta = 0$ one can decide whether or not the data set at hand has been subject to hidden truncation. More detailed discussion of these hidden truncation Pareto models, in the Pareto (II) case, may be found in Arnold and Ghosh (2011).

4 Inference, briefly

Suppose that $X_1, X_2, \ldots, X_n$ are independent identically distributed random variables with a common Pareto (IV) distribution. The sample size should be reasonably large, since we have four parameters to estimate. The sample minimum, or some minor corrected version of it, will be a suitable estimate of the location parameter μ. After subtracting it from each of the observations, the remaining three parameters may be estimated using maximum likelihood. The corresponding Fisher information matrix is available (as indeed is the Fisher information matrix for the Feller-Pareto model with $\mu = 0$).

Either a global search or numerical solution of the likelihood equations will be required to identify the location of the maximum of the likelihood function. In the Pareto (I) case, a variety of alternative estimates are available including best unbiased estimates. Alternatively, in the Pareto (I) case one can take logarithms of the data and arrive at a shifted exponential model, for which many estimation strategies have been developed.

A diffuse prior Bayesian analysis can be used for Pareto (IV) data. It will, predictably, yield results similar to those obtained via maximum likelihood. In the Pareto (I) case, Lwin (1972) introduced a conjugate family of priors for (σ, α) which can be used to incorporate some degree of prior knowledge of the parameters. Arnold et al. (1998) suggest use of a more flexible family of what they call conditionally conjugate priors in this setting. These priors are tailor-made for subsequent use of Gibbs sampling algorithms to generate realizations from the corresponding posterior distribution.

More details on parametric inference for Pareto models may be found in Arnold (1983) and Arnold (2008).

5 Multivariate Pareto models

The first author to systematically study k-dimensional Pareto distributions was Mardia (1962). Mardia's type I multivariate Pareto distribution has the attractive feature that both marginals and conditional distributions are Paretian in nature. We will say that a k-dimensional random vector $\underline{X}$ has a type I multivariate Pareto distribution, if the joint survival function is of the form

$$\bar{F}_{\underline{X}}(\underline{x}) = \left[\sum_{i=1}^{k}(x_i/\sigma_i) - k + 1\right]^{-\alpha}, \quad x_i > \sigma_i \tag{40}$$

and we write $\underline{X} \sim MP^{(k)}(\mathrm{I})(\underline{\sigma}, \alpha)$. The σ_i's are non-negative marginal scale parameters. The non-negative parameter α is an inequality parameter (common to all marginals). It follows from (40) that the one-dimensional marginals are classical Pareto distributions. Thus $X_i \sim P(\mathrm{I})(\sigma_i, \alpha)$, $i = 1, 2, \ldots, k$. By setting selected x_i's equal to σ_i in (40), it is apparent that, for any $k_1 < k$, all k_1 dimensional marginals are again multivariate Pareto. If we use the notational device $\underline{X} = (\dot{\underline{X}}, \ddot{\underline{X}})$ where $\dot{\underline{X}}$ is k_1 dimensional, with an analogous partition of the vector $\underline{\sigma} = (\dot{\underline{\sigma}}, \ddot{\underline{\sigma}})$, we may write

$$\dot{\underline{X}} \sim MP^{(k_1)}(\mathrm{I})(\dot{\underline{\sigma}}, \alpha). \tag{41}$$

Conditional distributions are also of the form (40), but with a change of location.

It is natural to extend this basic multivariate Pareto model by the introduction of location, scale, inequality and shape parameters in a manner parallel to that used to develop the univariate Pareto (II)-(IV) distributions, as follows:

($MP^{(k)}(II)$) We will say that $\underline{X}$ has a k-dimensional Pareto distribution of type II, if its joint survival function is of the form

$$\bar{F}_{\underline{X}}(\underline{x}) = \left[1 + \sum_{i=1}^{k}\left(\frac{x_i - \mu_i}{\sigma_i}\right)\right]^{-\alpha}, \quad x_i > \mu_i, \quad i = 1, 2, \ldots, k \tag{42}$$

and we write $\underline{X} \sim MP^{(k)}(\mathrm{II})(\underline{\mu}, \underline{\sigma}, \alpha)$.

($MP^{(k)}(III)$) $\underline{X}$ has a k-dimensional Pareto distribution of type III, if its joint survival function is of the form

$$\bar{F}_{\underline{X}}(\underline{x}) = \left[1 + \sum_{i=1}^{k}\left(\frac{x_i - \mu_i}{\sigma_i}\right)^{1/\gamma_i}\right]^{-1}, \quad x_i > \mu_i, \quad i = 1, 2, \ldots, k \tag{43}$$

and we write $\underline{X} \sim MP^{(k)}(\mathrm{III})(\underline{\mu}, \underline{\sigma}, \underline{\gamma})$.

($MP^{(k)}(IV)$) $\underline{X}$ has a k-dimensional Pareto distribution of type IV, if its joint survival function is of the form

$$\bar{F}_{\underline{X}}(\underline{x}) = \left[1 + \sum_{i=1}^{k}\left(\frac{x_i - \mu_i}{\sigma_i}\right)^{1/\gamma_i}\right]^{-\alpha}, \quad x_i > \mu_i, \quad i = 1, 2, \ldots, k \tag{44}$$

and we write $\underline{X} \sim MP^{(k)}(\mathrm{IV})(\underline{\mu}, \underline{\sigma}, \underline{\gamma}, \alpha)$.

The marginals and conditionals of an $MP^{(k)}(II)$ distribution are again of the $MP^{(k)}(II)$ form. An $MP^{(k)}(III)$ distribution has $MP^{(k)}(III)$ marginals, but not conditionals. However an $MP^{(k)}(IV)$ distribution does have both its marginals and conditionals of the $MP^{(k)}(IV)$ form. Specifically, in the $MP^{(k)}(IV)$ case, using the dot – double dot notation, we have

$$\dot{\underline{X}} \sim MP^{(k_1)}(\text{IV})(\dot{\underline{\mu}}, \dot{\underline{\sigma}}, \dot{\underline{\gamma}}, \alpha) \tag{45}$$

and

$$\dot{\underline{X}}|\ddot{\underline{X}} = \ddot{\underline{x}} \sim MP^{(k_1)}(\text{IV})(\dot{\underline{\mu}}, \underline{\dot{\tau}}, \dot{\underline{\gamma}}, \alpha + k - k_1), \tag{46}$$

where

$$\tau_i = \sigma_i \left[1 + \sum_{j=k_1+1}^{k} \left(\frac{x_j - \mu_j}{\sigma_j} \right)^{1/\gamma_j} \right]^{\gamma_i}, \quad i = 1, 2, \ldots, k_1. \tag{47}$$

Takahasi (1965) discussed the $MP^{(k)}(IV)$ distribution with $\underline{\mu} = \underline{0}$ and $\underline{\sigma} = \underline{1}$. He called it a multivariate Burr's distribution, and noted that the marginal and conditional distributions were of the same form.

As in the univariate case, distributional properties of these multivariate Pareto distributions and possible further extensions are more transparent if one uses a representation of the variables as functions of certain independent gamma variables. Thus if $\underline{X}$ has a k-dimensional Pareto distribution of type IV, we may act as if the X_i's have the representation

$$X_i = \mu_i + \sigma_i (W_i/Z)^{\gamma_i}, \quad i = 1, 2, \ldots, k \tag{48}$$

where the W_i's are independent identically distributed $\Gamma(1,1)$ variables (i.e., standard exponential variables) and Z, independent of the W_i's, has a $\Gamma(\alpha, 1)$ distribution. This representation, for example, makes it easy to compute the means, variances and covariances of $\underline{X}$.

A generalization of the representation (48) is one in which the W_i's are gamma rather than exponential variables. The resulting distribution will be called k-dimensional Feller-Pareto, since its marginals are of the Feller-Pareto form. Thus $\underline{X} \sim FP^{(k)}(\underline{\mu}, \underline{\sigma}, \underline{\gamma}, \alpha, \underline{\beta})$ if

$$X_i = \mu_i + \sigma_i (W_i/Z)^{\gamma_i}, \quad i = 1, 2, \ldots, k \tag{49}$$

where the W_i's and Z are independent random variables with $W_i \sim \Gamma(\beta_i, 1)$, ($i = 1, 2, \ldots, k$), and $Z \sim \Gamma(\alpha, 1)$. The marginal and conditional distributions of this multivariate Feller-Pareto distribution are again multivariate Feller-Pareto. The covariance structure can be readily obtained from the representation (49). Parallel to the situation in one dimension, there exist alternative names that could be applied to multivariate Feller-Pareto variables. They could be called multivariate generalized F variables or multivariate generalized beta of the second kind variables. An evident drawback of the multivariate Feller-Pareto model (and its various submodels) is the presence of a common value of α which appears in each marginal density. The consequences of this homogeneity are not easy to pin down. Certainly a model with Feller Pareto marginals with different α's for each of the marginals would be desirable, if one can be developed with attractive distributional properties (e,g., "nice" conditional distributions).

5.1 Other multivariate Pareto distributions

Although the title of this section promises discussion of multivariate models, only the bivariate case will be treated. It will be left to the reader to visualize the, usually straightforward, extension to the multivariate case. Notational complexity is avoided to a great extent by focusing on the case $k = 2$.

It is not difficult to verify that a $P(\mathrm{IV})(\mu, \sigma, \gamma, \alpha)$ distribution can be represented as a scale mixture of Weibull distributions. Equivalently, as remarked earlier, that a *P(IV)* random variable admits a representation as

$$X = \mu + \sigma(U/Z)^{\gamma}$$

where $U \sim exp(1)$ and $Z \sim \Gamma(\alpha, 1)$ are independent variables. A natural bivariate version of this construction begins with (U_1, U_2) having a bivariate exponential distribution with standard exponential marginals, perhaps one of the Marshall-Olkin type with parameters 1, 1 and λ. Then, with $Z \sim \Gamma(\alpha, 1)$ independent of (U_1, U_2), we define (X_1, X_2) by

$$X_i = \mu_i + \sigma_i(U_i/Z)^{\gamma_i}, \quad i = 1, 2. \tag{50}$$

Observe that in any bivariate Pareto (IV) distribution generated by this method, the marginals share a common value of α.

A second approach to generating bivariate *P(IV)* distributions makes use of the following representation of a *P(IV)* variable. Suppose that $U \sim exp(1)$, then

$$\mu + \sigma\left(e^{U/\alpha} - 1\right)^{\gamma} \sim P(\mathrm{IV})(\mu, \sigma, \gamma, \alpha) \tag{51}$$

Here too then, we can begin with (U_1, U_2) having an arbitrary bivariate distribution with standard exponential marginals and construct a variable (X_1, X_2) with a bivariate Pareto (IV) distribution by defining

$$X_i = \mu_i + \sigma_i\left(e^{U_i/\alpha_i} - 1\right)^{\gamma_i}, \quad i = 1, 2. \tag{52}$$

A third approach makes use of the fact that minima of independent Pareto (IV) random variables themselves have Pareto (IV) distributions. Thus if X_i, $i = 1, 2$, are independent with $X_i \sim P(\mathrm{IV})(\mu, \sigma, \gamma, \alpha_i)$, then $\min(X_1, X_2) \sim P(\mathrm{IV})(\mu, \sigma, \gamma, \alpha_1 + \alpha_2)$. We then begin with three independent random variables Y_1, Y_2, Y_3 with $Y_i \sim P(\mathrm{IV})(\mu, \sigma, \gamma, \alpha_i)$ and define

$$\begin{aligned} X_1 &= \min(Y_1, Y_3), \\ X_2 &= \min(Y_2, Y_3) \end{aligned} \tag{53}$$

(this approach is often called the method of trivariate reduction). In addition to having Pareto IV marginals, it is clear that the distribution described by (54) has the property that $\min(X_1, X_2) \sim P(\mathrm{IV})(\mu, \sigma, \gamma, \alpha_1 + \alpha_2 + \alpha_3)$. This distribution has the perhaps undesirable property that $P(X_1 = X_2) > 0$, and has another unfortunate property in that the marginals share common values of μ, σ and γ. This latter problem can be avoided to some extent by assuming that the Y_i's have $P(\mathrm{IV})(0, 1, \gamma, \alpha_i)$ distributions and then defining

$$\begin{aligned} X_1 &= \mu_1 + \sigma_1 \min(Y_1, Y_3), \\ X_2 &= \mu_2 + \sigma_2 \min(Y_2, Y_3). \end{aligned} \tag{54}$$

In this case the X_i's share only a common value of γ.

Finally we mention the popular Copula based approach to constructing bivariate distributions with given marginals. For this, we begin with an analytically tractable bivariate distribution for (Z_1, Z_2) and apply marginal transformations to produce a bivariate distribution with Pareto (IV) marginals. A popular choice for the distribution of (Z_1, Z_2) is a bivariate normal with standard normal marginals and correlation ρ, but of course any other bivariate distribution can be used in its place. Now using $F_{\mu,\sigma,\gamma,\alpha}$ to denote the distribution function of a $P(\text{IV})(\mu, \sigma, \gamma, \alpha)$ random variable and Φ to denote a standard normal distribution function, we define

$$\begin{aligned} X_1 &= F^{-1}_{\mu_1,\sigma_1,\gamma_1,\alpha_1}(\Phi(Z_1)), \\ X_2 &= F^{-1}_{\mu_2,\sigma_2,\gamma_2,\alpha_2}(\Phi(Z_2)). \end{aligned} \tag{55}$$

The correlation structure of the X_i's is inherited from the correlation structure of the Z_i's. In this case the extension to k dimensions is particularly transparent. Note also that the model has one dependence parameter which, if set equal to 0, yields a model with independent $P(IV)$ marginals. It will be noted that this feature of having a single dependence parameter is shared by the other bivariate models introduced in this Section.

6 Multivariate extensions

Several of the univariate extensions, discussed in Section 3, can be readily modified to yield k-dimensional versions. For example a random variable with the univariate Beta-generalized-Pareto (IV) distribution can be viewed as being defined by

$$X = F^{-1}_{\mu,\sigma,\gamma,\alpha}(V), \tag{56}$$

where $V \sim Beta(\lambda_1, \lambda_2)$. For a bivariate version of this construction, we begin with (V_1, V_2) having a bivariate Beta distribution, perhaps of the the type introduced by Arnold and Ng (2011), and make suitable marginal transformations. Thus we define

$$\begin{aligned} X_1 &= F^{-1}_{\mu_1,\sigma_1,\gamma_1,\alpha_1}(V_1), \\ X_2 &= F^{-1}_{\mu_2,\sigma_2,\gamma_2,\alpha_2}(V_2). \end{aligned} \tag{57}$$

Higher dimensional versions of this construction require only the identification of a suitable k-dimensional Beta distribution. A Dirichlet distribution might be used here. Some other alternatives are described in Arnold and Ng (2011).

To identify a suitable bivariate analog of the Kumaraswamy-Pareto (IV) distribution, all that is required is a bivariate-Kumaraswamy distribution. One possible such distribution was suggested by Nadarajah et al. (2011).

Hidden truncation models, likewise, can be considered in higher dimensions. For example we may begin with $\underline{X} \sim MP^{(k)}(\text{IV})(\underline{\mu}, \underline{\sigma}, \underline{\gamma}, \alpha)$. Then, using our dot – double dot notation, we have

$$f_{HT}(\dot{\underline{x}}) = f_{\dot{\underline{X}}|\ddot{\underline{X}} \leq \ddot{\underline{x}}}(\dot{\underline{x}}) = f_{\dot{\underline{X}}}(\dot{\underline{x}}) \frac{P(\ddot{\underline{X}} \leq \ddot{\underline{x}} | \dot{\underline{X}} = \dot{\underline{x}})}{P(\ddot{\underline{X}} \leq \ddot{\underline{x}})} \tag{58}$$

which is not difficult to evaluate, since the conditional distribution of $\ddot{\underline{X}}$ given that $\dot{\underline{X}} = \dot{\underline{x}}$ is of the $MP^{(k-k_1)}(IV)$ form (refer to equation (46), being careful to switch the roles of $\dot{\underline{X}}$ and $\ddot{\underline{X}}$).

We conclude this section by noting the availability of multivariate distributions with Pareto conditionals rather than Pareto marginals. Detailed discussion of such models may be found in Arnold et al. (1999), Chapter 5.

7 Envoi

The survey presented in this paper is far from complete. A more detailed and extensive survey (though somewhat out of date) can be found in Arnold (1983). A revision of that book is, however, currently in preparation. In the interim, see Arnold (2008) for a more up-to-date presentation and, as mentioned in Section 4, for more details on inferential strategies. More work is still needed on the development of estimation and hypothesis testing strategies, especially for multivariate Pareto data. Creative Bayesian analyses involving informative priors, in multivariate settings and in cases involving covariates, are also notable for their absence. Finally, I apologize to those readers whose important contributions have been overlooked in this survey. I excuse myself by repeating that the survey is necessarily incomplete. However, please do advise me of any glaring omissions that you might note.

Competing interests
The author declares that he has no competing interests.

Acknowledgement
The constructive suggestions supplied by anonymous referees have resulted in a much improved manuscript.

References

Aban, IB, Meerschaert, MM, Panorska, AK: Parameter estimation for the truncated Pareto distribution. J. Amer. Statist. Assoc. **101**, 270–277 (2006)

Akinsete, A, Famoye, F, Lee, C: The beta-Pareto distribution. Statistics. **42**, 547–563 (2008)

Arnold, BC: Pareto Distributions. International Cooperative Publishing House, Burtonsville, MD (1983)

Arnold, BC: Pareto and generalized Pareto distributions. In: Chotikapanich, D (ed.) Modeling Income Distributions and Lorenz Curves, pp. 119–145. Springer, New York, (2008)

Arnold, BC, Austin, K: Truncated Pareto distributions: flexible, tractable and familiar. Technical Report #150, Department of Statistics, University of California, Riverside, CA (1987)

Arnold, BC, Castillo, E, Sarabia, JM: Bayesian analysis for classical distributions using conditionally specified priors. Sankhya: Ind. J. Stat. Series B. **60**, 228–245 (1998)

Arnold, BC, Castillo, E, Sarabia, JM: Conditional Specification of Statistical Models. Springer, New York (1999)

Arnold, BC, Ghosh, I: Inference for Pareto data subject to hidden truncation. J. Ind. Soc. Probability Stat. **13**, 1–16 (2011)

Arnold, BC, Laguna, L: A stochastic mechanism leading to asymptotically Paretian distributions. In: Business and Economic Statistics Section, Proceedings of the American Statistical Association, pp. 208–210, (1976)

Arnold, BC, Ng, HKT: Flexible bivariate beta distributions. J. Multivariate Anal. **102**, 1194–1202 (2011)

Arnold, BC, Robertson, CA, Yeh, HC: Some properties of a Pareto-type distribution. Sankhya: Ind. J. Stat. Series A. **48**, 404–408 (1986)

Champernowne, DG: The theory of income distribution. Econometrica. **5**, 379–381 (1937)

David, HA, Nagaraja, HN: Order Statistics. Third edition. Wiley, Hoboken, NJ (2003)

Falk, M, Hüsler, J, Reiss, R: Laws of Small Numbers, Extremes and Rare Events. Third edition. Birkhäuser/Springer, Basel (2011)

Feller, W: An Introduction to Probability Theory and its Applications, Vol. 2. Second edition. Wiley, New York (1971)

Fisk, PR: The graduation of income distributions. Econometrica. **29**, 171–185 (1961a)

Fisk, PR: Estimation of location and scale parameters in a truncated grouped sech-square distribution. J. Am. Stat. Assoc. **56**, 692–702 (1961b)

Johnson, NL, Kotz, S, Balakrishnan, N: Continuous Univariate Distributions, Vol. 1. Second edition. Wiley, New York (1994)

Jones, MC: Families of distributions arising from distributions of order statistics. TEST. **13**, 1–43 (2004)

Jones, MC, Kumaraswamy's distribution: A beta-type distribution with some tractability advantages. Stat. Methodol. **6**, 70–81 (2009)

Kagan, YY, Schoenberg, F: Estimation of the upper cutoff parameter for the tapered Pareto distribution. J. Appl. Probability. **38A**, 158–175 (2001)

Kalbfleisch, JD, Prentice, RL: The Statistical Analysis of Failure Time Data. Wiley, New York (1980)

Keiding, N, Kvist, K, Hartvig, H, Tvede, M, Juul, S: Estimating time to pregnancy from current durations in a cross-sectional sample. Biostatistics. **3**, 565–578 (2002)

Lwin, T: Estimation of the tail of the Paretian law. Skand. Aktuarietidskr. **55**, 170–178 (1972)

Maguire, BA, Pearson, ES, Wynn, AHA: The time intervals between industrial accidents. Biometrika. **39**, 168–180 (1952)

Mardia, KV: Multivariate Pareto distributions. Ann. Math. Stat. **33**, 1008–1015 (1962)
Nadarajah, S, Cordeiro, GM, Ortega, EMM: General results for the Kumaraswamy-G distribution. J. Stat. Comput. Simul. **82**, 951–979 (2011)
Pareto, V: Cours d'economie Politique, Vol. II, F. Rouge, Lausanne (1897)
Paranaiba, PF, Ortega, EMM, Cordeiro, GM, de Pascoa, MAR: The Kumaraswamy Burr XII distribution: theory and practice. J. Stat. Comput. Simul. **83**, 2117–2143 (2013)
Pillai, RN: Semi-Pareto processes. J. Appl. Probab. **28**, 461–465 (1991)
Takahasi, K: Note on the multivariate Burr's distribution. Ann. Inst. Statist. Math. **17**, 257–260 (1965)

Permissions

All chapters in this book were first published in JSDA, by Springer; hereby published with permission under the Creative Commons Attribution License or equivalent. Every chapter published in this book has been scrutinized by our experts. Their significance has been extensively debated. The topics covered herein carry significant findings which will fuel the growth of the discipline. They may even be implemented as practical applications or may be referred to as a beginning point for another development.

The contributors of this book come from diverse backgrounds, making this book a truly international effort. This book will bring forth new frontiers with its revolutionizing research information and detailed analysis of the nascent developments around the world.

We would like to thank all the contributing authors for lending their expertise to make the book truly unique. They have played a crucial role in the development of this book. Without their invaluable contributions this book wouldn't have been possible. They have made vital efforts to compile up to date information on the varied aspects of this subject to make this book a valuable addition to the collection of many professionals and students.

This book was conceptualized with the vision of imparting up-to-date information and advanced data in this field. To ensure the same, a matchless editorial board was set up. Every individual on the board went through rigorous rounds of assessment to prove their worth. After which they invested a large part of their time researching and compiling the most relevant data for our readers.

The editorial board has been involved in producing this book since its inception. They have spent rigorous hours researching and exploring the diverse topics which have resulted in the successful publishing of this book. They have passed on their knowledge of decades through this book. To expedite this challenging task, the publisher supported the team at every step. A small team of assistant editors was also appointed to further simplify the editing procedure and attain best results for the readers.

Apart from the editorial board, the designing team has also invested a significant amount of their time in understanding the subject and creating the most relevant covers. They scrutinized every image to scout for the most suitable representation of the subject and create an appropriate cover for the book.

The publishing team has been an ardent support to the editorial, designing and production team. Their endless efforts to recruit the best for this project, has resulted in the accomplishment of this book. They are a veteran in the field of academics and their pool of knowledge is as vast as their experience in printing. Their expertise and guidance has proved useful at every step. Their uncompromising quality standards have made this book an exceptional effort. Their encouragement from time to time has been an inspiration for everyone.

The publisher and the editorial board hope that this book will prove to be a valuable piece of knowledge for researchers, students, practitioners and scholars across the globe.

List of Contributors

Sterling Sawaya
Institute for Behavioral Genetics, University of Colorado, Boulder, Colorado, USA
Formerly: Department of Anatomy, and Allan Wilson Centre for Molecular Ecology and Evolution, University of Otago, Dunedin, New Zealand

Steffen Klaere
Department of Statistics and School of Biological Sciences, University of Auckland, Auckland, New Zealand

Marco AR Ferreira
Department of Statistics, University of Missouri, Columbia, USA

Esther Salazar
Department of Electrical and Computer Engineering, Duke University, Durham, USA

Werner Hürlimann
Feldstrasse 145, CH-8004 Zürich, Switzerland

Alfred Akinsete
Department of Mathematics, Marshall University, Huntington, West Virginia 25755, USA

Felix Famoye
Department of Mathematics, Central Michigan University, Mount Pleasant, Michigan 48859, USA

Carl Lee
Department of Mathematics, Central Michigan University, Mount Pleasant, Michigan 48859, USA

Wan-Chen Lee
Department of Statistics, University of Manitoba, Winnipeg, Canada

Wolf-Dieter Richter
Institute of Mathematics, University of Rostock, Ulmenstraße 69, Haus 3, 18057 Rostock, Germany

Mohammad Z Raqab
Department of Statistics and Operations Research, Kuwait University, P.O. Box 5969 Safat, 13060 Kuwait City, Kuwait

Ayman Alzaatreh
Department of Mathematics and Statistics, Austin Peay State University, Clarksville, TN 37044, USA

Carl Lee
Department of Mathematics, Central Michigan University, Mount Pleasant, MI 48859, USA

Felix Famoye
Department of Mathematics, Central Michigan University, Mount Pleasant, MI 48859, USA

Ayman Alzaatreh
Department of Mathematics and Statistics, Austin Peay State University, Clarksville, TN 37044, USA

Carl Lee
Department of Mathematics, Central Michigan University, Mount Pleasant, MI 48859, USA

Felix Famoye
Department of Mathematics, Central Michigan University, Mount Pleasant, MI 48859, USA

Evdokia Xekalaki
Department of Statistics, Athens University of Economics and Business, 76 Patision St., Athens, Greece

Manoel Santos-Neto
Departamento de Estatística, Universidade Federal de Campina Grande, Bodocongó, 58429-970, Campina Grande, PB, BraziL

Marcelo Bourguignon
Departamento de Estatística, Universidade Federal do Piauí, Ininga, 64049-550, Teresina, PI, Brazil

Luz M Zea
Departamento de Estatística, Universidade Federal do Rio Grande do Norte, Campus Universitário Lagoa Nova, 59078-970, Natal, RN, Brazil

Abraão DC Nascimento
Departamento de Estatística, Universidade Federal de Pernambuco, Cidade Universitária, 50740-540, Recife, PE, Brazil

Gauss M Cordeiro
Departamento de Estatística, Universidade Federal de Pernambuco, Cidade Universitária, 50740-540, Recife, PE, Brazil

George Livadiotis
Southwest Research Institute, San Antonio, TX, USA

Barry C Arnold
Department of Statistics, University of California, Riverside, USA

www.ingramcontent.com/pod-product-compliance
Lightning Source LLC
Chambersburg PA
CBHW080254230326
41458CB00097B/4447
* 9 7 8 1 6 8 2 8 5 1 6 4 7 *